# FUNDAMENTALS OF ELECTRICAL ENGINEERING

# FUNDAMENTALS OF ELECTRICAL ENGINEERING

Ana Oprisan
Linda Nie
Thomas Baker

Kruger Brentt
Publishers

2025

Kruger Brentt Publishers UK. LTD.
Company Number 9728962

Regd. Office: 68 St Margarets Road, Edgware, Middlesex HA8 9UU



ISBN: 978-1-78715-371-4



For information on all our publications visit our website at http://krugerbrentt.com/

# PREFACE

**"Fundamentals of Electrical Engineering"** provides a comprehensive overview of the foundational principles, theories, and applications that form the basis of electrical engineering. This book serves as an essential resource for students, educators, and professionals seeking to build a solid understanding of the core concepts and practices in electrical engineering.

Electrical engineering is a broad and dynamic field that encompasses the study and application of electricity, electronics, and electromagnetism. From power generation and distribution to telecommunications, electronics, and computer systems, electrical engineering plays a critical role in shaping modern society and advancing technological innovation.

As authors of this book, we have endeavored to compile a diverse array of perspectives, methodologies, and case studies from leading experts and practitioners in the field of electrical engineering. Through a blend of theoretical foundations, practical insights, and real-world examples, this book covers a wide range of topics, including circuit analysis, electromagnetics, power systems, electronics, digital signal processing, and control systems.

Our overarching goal in presenting this book is to provide readers with a comprehensive understanding of the principles and applications of electrical engineering. By offering insights into the latest advancements, best practices, and emerging trends in the field, we aim to inspire curiosity, critical thinking, and innovation in electrical engineering education and practice.

We extend our sincere gratitude to all the contributors who have generously shared their expertise, experiences, and insights in the field of electrical engineering. It is our hope that this book will serve as a valuable reference and guide for students, educators, and professionals as they navigate the complexities of electrical engineering and contribute to the advancement of technology and society.

**Ana Oprisan**

**Linda Nie**

**Thomas Baker**

# CONTENTS

# 1 Chapter

# INTRODUCTION TO ELECTRICAL ENGINEERING

## 1.1 THEMES

Since its inception in the late 1800s, electrical engineering has expanded its scope from primarily dealing with electrical circuits for electricity, telegraphy, and telephony to include a wide array of specialties. Nevertheless, the fundamental concepts remain pertinent in the present day: For the past 150 years, electrical engineering has revolved around the concepts of power generation, transmission, and information. This course focuses on the latter topic: the depiction, manipulation, transfer, and reception of information using electrical methods. This course provides an explanation of the concept of information, the methods engineers use to measure information, and the way in which electrical impulses are used to convey information.

Information can manifest in diverse formats. During a conversation with a buddy, your brain converts your thoughts into motor commands, which then coordinate the movement of different parts of your vocal tract, such as the jaw, tongue, and lips. Information originates in one's thoughts and is conveyed through speech, which necessitates a clearly defined and widely recognized structure in order for others to comprehend one's message. Utterances transmit information through sound pressure waves that travel to your friend's ear. At that point, the sound energy is transformed into brain activity, and if your statement is logical, she comprehends your words. Your remarks may have been stored on a compact disc (CD), sent to your friend via mail, and played back on her audio system. Information can be represented as a textual document that you input into your word processing software. You can transmit the file to a friend by email, who will then peruse and comprehend its contents. From an information theoretic perspective, all of these scenarios are considered to be equal, despite the fact that the types of information representation, such as sound waves, plastic, and computer files, differ significantly.

Engineers classify information into two distinct forms: analog and digital, without considering its substance. Analog information is characterized by continuous values, such as those seen in audio and video signals. Digital information is characterized by distinct values, such as text (like the one you are currently reading) and DNA sequences.

Energy conversion, also referred to as transduction, is the process of transforming signals containing information from one kind of energy to another. Although conversion systems experience inefficiency because to the dissipation of input energy as heat, this does not necessarily result in the loss of transmitted information. From a conceptual standpoint, any type of energy could be utilized to symbolize information. However, electric signals possess distinct advantages when it comes to representing, transmitting (via antennas or wires), and manipulating information (through noise reduction circuits and computer-based modifications). Thus, we will be concerned with how to

- Represent all forms of information with electrical signals,
- Encode information as00 voltages, currents, and electromagnetic waves,
- Manipulate information-bearing electric signals with circuits and computers, and
- Receive electric signals and convert the information expressed by electric signals back into a useful form.

Telegraphy is the oldest electrical information method, originating in 1837. During that period, electrical research relied heavily on empirical methods, and only those with practical experience and intuition were able to design telegraph systems. The field of electrical research reached maturity when James Clerk Maxwell declared in 1864 a series of equations that he asserted governed all electrical processes. These equations accurately forecasted that light was an electromagnetic wave and that energy could spread. The telephone's development in 1876 was primarily attributed to empirical effort, as Maxwell's presentation was intricate. After Heinrich Hertz verified Maxwell's anticipation of what is now referred to as radio waves around 1882, Oliver Heaviside and other individuals simplified Maxwell's equations, which subsequently gained widespread recognition. This comprehension of basic principles resulted in a rapid sequence of innovations—the wireless telegraph (1899), the vacuum tube (1905), and radio broadcasting—which signified the genuine advent of the communications era.

In the early twentieth century, a thorough understanding of circuit theory and electromagnetic theory was sufficient for an electrical engineer to be considered competent and capable of creating high-quality designs. Therefore, circuit theory played a crucial role as the basis and structure of all electrical engineering education. During the middle of the century, three significant innovations altered the fundamental principles. The initial public display of the inaugural electronic computer (1946), the creation of the transistor (1947), and the release of Claude

Shannon's A Mathematical Theory of Communication (1948) marked significant milestones in technological advancement. Although first developed independently, these innovations were essential in ushering in the information age, characterized by the coexistence and rivalry of digital and analog communication systems in terms of design preferences. Approximately two decades later, the laser was created, thereby expanding the range of design options even further. Consequently, the main emphasis transitioned from the methods of constructing communication systems (during the circuit theory era) to the objectives that communications systems were designed to achieve. An implementation can only be selected once the intended system is specified. Modern electrical engineers must prioritize the overall objective of the system and comprehend the compromises involved in choosing between digital and analog options, as well as between hardware and software setups when developing information systems.

Please be aware that Rice University's Disability Support Services has made this collection accessible in a Braille-printable format. Kindly click on the provided link to initiate the download of a compressed file (.zip) that includes all the essential files in .dxb format as well as image files.

## 1.2 SIGNALS REPRESENT INFORMATION

The signal, which is the fundamental quantity in electrical engineering, represents information whether it be in analog or digital form. Mathematically speaking, a signal can be defined as a function. Analog signals possess a continuous range of values, while digital signals possess a discrete range of values. The signal's independent variable can be categorized as time (such as voice), space (like visuals), or integers (representing the sequential arrangement of letters and numbers in a football score).

### 1.2.1 Analog Signals

Analog signals are usually signals defined over continuous independent variable(s). Speech is produced by your vocal cords exciting acoustic resonances in your vocal tract. The result is pressure waves propagating in the air, and the speech signal thus corresponds to a function having independent variables of space and time and a value corresponding to air pressure: s(x,t) (Hence we use vector notation x to denote spatial coordinates). When you record someone talking, you are evaluating the speech signal at a particular spatial location, x0 say. An example of the resulting waveform s (x0, t) is shown in this figure (Figure 1.1: Speech Example).

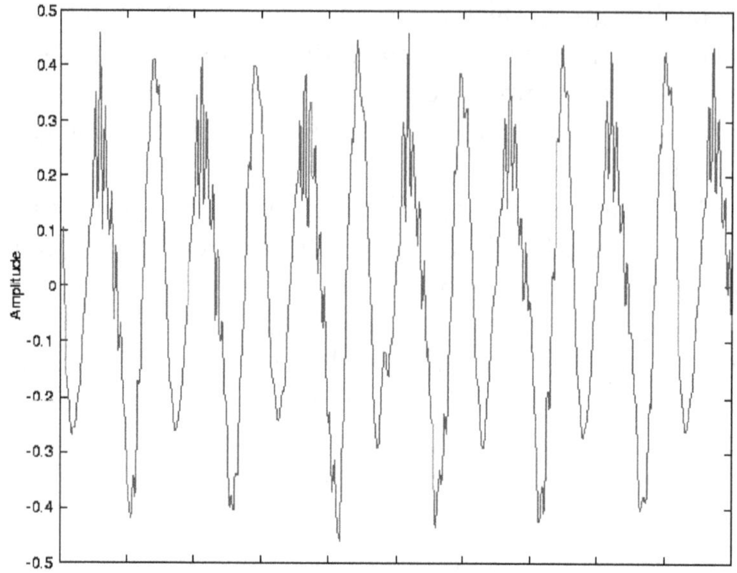


**Figure 1.1 A speech signal's amplitude relates to tiny air pressure variations. Shown is a recording of the vowel "e" (as in "speech").**

Photographs are static, and are continuous-valued signals defined over space. Black-and-white images have only one value at each point in space, which amounts to its optical reflection properties. In the Figure 1.2 an image is shown, demonstrating that it (and all other images as well) are functions of two independent spatial variables.

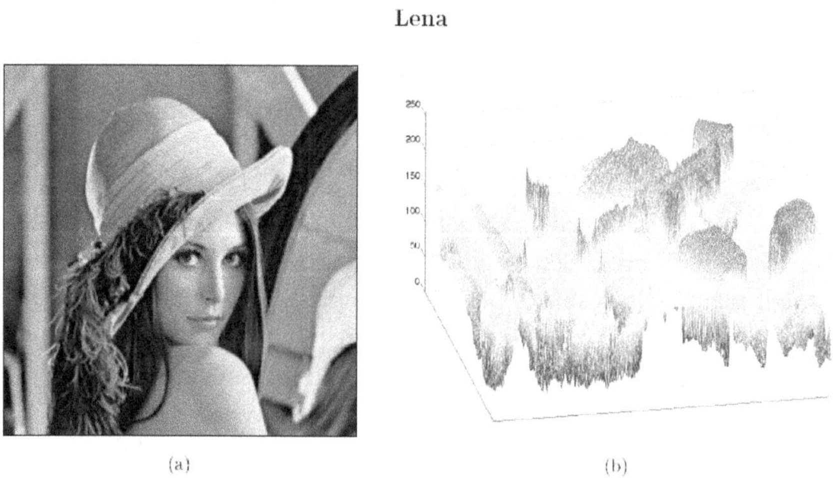


**Figure 1.2: On the left is the classic Lena image, which is used ubiquitously as a test image. It contains straight and curved lines, complicated texture, and a face. On the right is a perspective display of the Lena image as a signal: a function of two spatial variables. The colors merely help show what signal values are about the same size. In this image, signal values range between 0 and 255; why is that?**

Color images have values that express how reflectivity depends on the optical spectrum. Painters long ago found that mixing together combinations of the so-called primary colors-red, yellow and blue--can produce very realistic color images. Thus, images today are usually thought of as having three values at every point in space, but a different set of colors is used: How much of red, green and blue is present. Mathematically, color pictures are multivalued vector-valued-signals: $s(x)=(r(x),g(x),b(x))T$.

There are many intriguing instances where the analog signal relies on a discrete variable rather than a continuous variable, such as time. For instance, temperature readings collected at regular intervals of one hour exhibit continuous analog values, yet the independent variable of the signal is fundamentally represented by integers.

### 1.2.2. Digital Signals

The term "digital" refers to a signal that is discrete-valued and has an independent variable that takes on integer values. Digital information include numerical values and symbolic representations, such as characters inputted via a keyboard. Computers depend on the digital encoding of data to manipulate and alter information. Symbols lack a numerical value and are each denoted by a distinct number. The ASCII character code encompasses both uppercase and lowercase characters, integers, punctuation marks, and several other symbols, all of which are represented by a seven-bit integer. As an illustration, the ASCII code denotes the letter A with the numerical value of 65. The ASCII Table displays the global standard for mapping characters to integers.

**Table 1.1 The ASCII translation table shows how standard keyboard characters are represented by integers. In pairs of columns, this table displays first the so-called 7-bit code (how many characters in a seven-bit code?), then the character the number represents. The numeric codes are represented in hexadecimal (base-16) notation. Mnemonic characters correspond to control characters, some of which may be familiar (like cr for carriage return) and some not (bel means a "bell").**

| 00 | nul | 01 | soh | 02 | stx | 03 | etx | 04 | eot | 05 | enq | 06 | ack | 07 | bel |
|---|---|---|---|---|---|---|---|---|---|---|---|---|---|---|---|
| 08 | bs | 09 | ht | 0A | nl | 0B | vt | 0C | np | 0D | cr | 0E | so | 0F | si |
| 10 | dle | 11 | dc1 | 12 | dc2 | 13 | dc3 | 14 | dc4 | 15 | nak | 16 | syn | 17 | etb |
| 18 | car | 19 | cm | 1A | sub | 1B | esc | 1C | fs | 1D | gs | 1E | rs | 1F | us |
| 20 | sp | 21 | ! | 22 | " | 23 | # | 24 | $ | 25 | % | 26 | & | 27 | ' |
| 28 | ( | 29 | ) | 2A | * | 2B | + | 2C | , | 2D | - | 2E | . | 2F | / |
| 30 | 0 | 31 | 1 | 32 | 2 | 33 | 3 | 34 | 4 | 35 | 5 | 36 | 6 | 37 | 7 |
| 38 | 8 | 39 | 9 | 3A | : | 3B | ; | 3C | < | 3D | = | 3E | > | 3F | ? |

| 40 | @ | 41 | A | 42 | B | 43 | C | 44 | D | 45 | E | 46 | F | 47 | G |
|---|---|---|---|---|---|---|---|---|---|---|---|---|---|---|---|
| 48 | H | 49 | I | 4A | J | 4B | K | 4C | L | 4D | M | 4E | N | 4F | 0 |
| 50 | P | 51 | Q | 52 | R | 53 | S | 54 | T | 55 | U | 56 | V | 57 | W |
| 58 | X | 59 | Y | 5A | Z | 5B | [ | 5C | \ | 5D | ] | 5E | ^ | 5F | _ |
| 60 | ‘ | 61 | a | 62 | b | 63 | c | 64 | d | 65 | e | 66 | f | 67 | g |
| 68 | h | 69 | i | 6A | j | 6B | k | 6C | l | 6D | m | 6E | n | 6F | o |
| 70 | p | 71 | q | 72 | r | 73 | s | 74 | t | 75 | u | 76 | v | 77 | w |
| 78 | x | 79 | y | 7A | z | 7B | { | 7C | \| | 7D | } | 7E | ~ | 7F | del |

## 1.3 STRUCTURES OF COMMUNICATION SYSTEMS

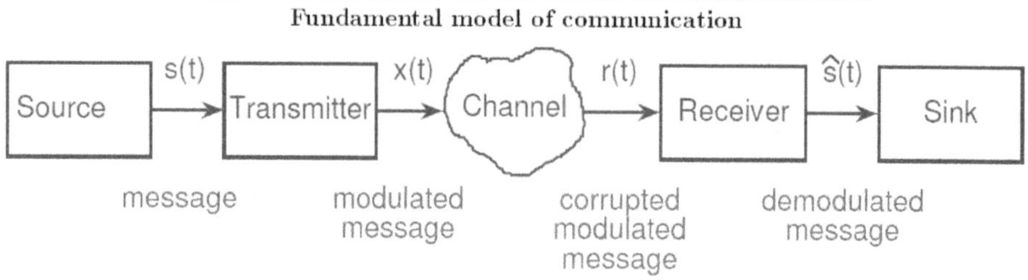


**Figure 1.3: The Fundamental Model of Communication.**

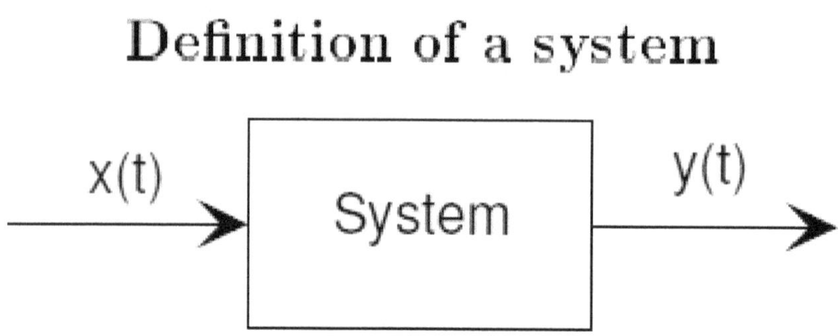


**Figure 1.4: A system operates on its input signal x(t) to produce an output y(t).**

The fundamental model of communications is portrayed in Figure 1.3 (Fundamental model of communication). In this fundamental model, each message-bearing signal, exemplified by s (t), is analog and is a function of time. A system operates on zero, one, or several signals to produce more signals or to simply absorb them (Figure 1.4 (Definition of a system)). In electrical engineering, we represent a system as a box, receiving input signals (usually coming from the left) and producing from them new output signals. This graphical representation is known as a block diagram. We denote input signals by lines having arrows pointing into the box, output signals by arrows pointing away. As typified by the communications model, how information flows, how it is corrupted and manipulated, and how it is ultimately received is summarized by interconnecting block diagrams: The outputs of one or more systems serve as the inputs to others.

In the communications model, the source produces a signal that will be absorbed by the sink. Examples of time-domain signals produced by a source are music, speech, and characters typed on a keyboard. Signals can also be functions of two variables—an image is a signal that depends on two spatial variables—or more—television pictures (video signals) are functions of two spatial variables and time. Thus, information sources produce signals. In physical systems, each signal corresponds to an electrical voltage or current. To be able to design systems, we must understand electrical science and technology. However, we first need to understand the big picture to appreciate the context in which the electrical engineer works.

In communication systems, messages—signals produced by sources—must be recast for transmission. The block diagram has the message s(t) passing through a block labeled transmitter that produces the signal x(t). In the case of a radio transmitter, it accepts an input audio signal and produces a signal that physically is an electromagnetic wave radiated by an antenna and propagating as Maxwell's equations predict. In the case of a computer network, typed characters are encapsulated in packets, attached with a destination address, and launched into the Internet. From the communication systems "big picture" perspective, the same block diagram applies although the systems can be very different. In any case, the transmitter should not operate in such a way that the message s(t) cannot be recovered from x(t). In the mathematical sense, the inverse system must exist, else the communication system cannot be considered reliable. (It is ridiculous to transmit a signal in such a way that no one can recover the original. However, clever systems exist that transmit signals so that only the "in crowd" can recover them. Such cryptographic systems underlie secret communications.)

Transmitted signals next pass through the next stage, the evil channel. Nothing good happens to a signal in a channel: It can become corrupted by noise, distorted, and attenuated among many possibilities. The channel cannot be escaped (the real world is cruel), and transmitter design and receiver design focus on how best to jointly fend off the channel's effects on signals. The channel is another system in our block diagram, and produces r(t), the signal received by the receiver. If the channel were benign (good luck finding such a channel in the real world), the receiver would serve as the inverse system to the transmitter, and yield the message with no distortion. However, because of the channel, the receiver must do its best to produce a received message $\hat{s}(t)$ that resembles s(t) as much as possible. Shannon showed in his 1948 paper that reliable—for the moment, take this word to mean error-free—digital communication was possible over arbitrarily noisy channels. It is this result that modern communications systems exploit, and why many communications systems are going "digital." The module on Information Communication (Section 6.1) details Shannon's theory of information, and there we learn of Shannon's result and how to use it.

Finally, the received message is passed to the information sink that somehow makes use of the message. In the communications model, the source is a system having no input but producing an output; a sink has an input and no output.

Comprehending signal creation and system functioning involves grasping the essence of signals, the characteristics of the information they convey, the conversion of information from analog to digital formats, and the processing of information by systems that operate on signals carrying information. This comprehension necessitates expertise in two distinct domains of knowledge. One field of study is electrical science: What is the method for representing and manipulating signals using electricity? The second aspect is signal science, which focuses on understanding the structure of signals, regardless of their origin. This includes analyzing their information content and determining the specific capabilities that this structure imposes on communication systems.

## 1.4 THE FUNDAMENTAL SIGNAL

### 1.4.1 The Sinusoid

The most ubiquitous and important signal in electrical engineering is the sinusoid.

#### Sine Definition

$$s\left(t\right) = A\cos\left(2\pi ft + \phi\right) \mathrm{or} A\cos\left(\omega t + \phi\right) \tag{1.1}$$

A is known as the sinusoid's amplitude, and determines the sinusoid's size. The amplitude conveys the sinusoid's physical units (volts, lumens, etc). The frequency f has unitsof Hz (Hertz) or s-1, and determines how rapidly the sinusoid oscillates per unit time. The temporal variable t always has units of seconds, and thus the frequency determines how many oscillations/second the sinusoid has. AM radio stations have carrier frequencies of about 1 MHz (one mega-hertz or 106 Hz), while FM stations have carrier frequencies of about 100 MHz. Frequency can also be expressed by the symbol ω, which has units of radians/second. Clearly, ω=2πf. In communications, we most often express frequency in Hertz. Finally, φ is the phase, and determines the sine wave's behavior at the origin (t=0). It has units of radians, but we can express it in degrees, realizing that in computations we must convert from degrees to radians. Note that if $\phi = -\frac{\pi}{2}$, the sinusoid corresponds to a sine function, having a zero value at the origin.

$$A\sin(2\pi ft+\phi)=A\cos\left(2\pi ft+\phi-\frac{\pi}{2}\right) \tag{1.2}$$

Thus, the only difference between a sine and cosine signal is the phase; we term either a sinusoid.

We can also define a discrete-time variant of the sinusoid: Acos(2πfn+φ). Here, the independent variable is n and represents the integers. Frequency now has no dimensions, and takes on values between 0 and 1.

### Exercise 1.4.1

Show that cos(2πfn)=cos(2π(f+1)n) which means that a sinusoid having a frequency larger than one corresponds to a sinusoid having a frequency less than one.

Notice that we shall call either sinusoid an analog signal. Only when the discrete-time signal takes on a finite set of values can it be considered a digital signal.

### Exercise 1.4.2

Can you think of a simple signal that has a finite number of values but is defined in continuous time? Such a signal is also an analog signal.

## 1.4.2 Communicating Information with Signals

The basic idea of communication engineering is to use a signal's parameters to represent either real numbers or other signals. The technical term is to modulate the carrier signal›s parameters to transmit information from one place to another. To explore the notion of modulation, we can send a real number (today›s temperature, for example) by changing a sinusoid›s amplitude accordingly. If we wanted to send the daily temperature, we would keep the frequency constant (so the receiver would know what to expect) and change the amplitude at midnight. We could relate temperature to amplitude by the formula A=A0(1+kT),where A0 and k are constants that the transmitter and receiver must both know.

If we had two numbers we wanted to send at the same time, we could modulate the sinusoid's frequency as well as its amplitude. This modulation scheme assumes we can estimate the sinusoid's amplitude and frequency; we shall learn that this is indeed possible.

Now suppose we have a sequence of parameters to send. We have exploited all of the sinusoid's two parameters. What we can do is modulate them for a limited time (say T seconds), and send two parameters every T. This simple notion corresponds to how a modem works. Here, typed characters are encoded into eight bits, and the individual bits are encoded into a sinusoid's amplitude and frequency. We'll learn how this is done in subsequent modules, and more importantly, we'll learn what the limits are on such digital communication schemes.

## 1.5 INTRODUCTION PROBLEMS

### Problem 1.1: RMS Values

The rms (root-mean-square) value of a periodic signal is defined to be

$$s = \sqrt{\frac{1}{T}\int_0^T s^2(t)\,dt}$$

where T is defined to be the signal›s period: the smallest positive number such that s(t)=s(t+T).

- What is the period of s(t)=Asin(2πf0t+φ)?
- What is the rms value of this signal? How is it related to the peak value?
- What is the period and rms value of the depicted square wave, generically denoted by sq(t)?
- By inspecting any device you plug into a wall socket, you'll see that it is labeled "110 volts AC". What is the expression for the voltage provided by a wall socket? What is its rms value?

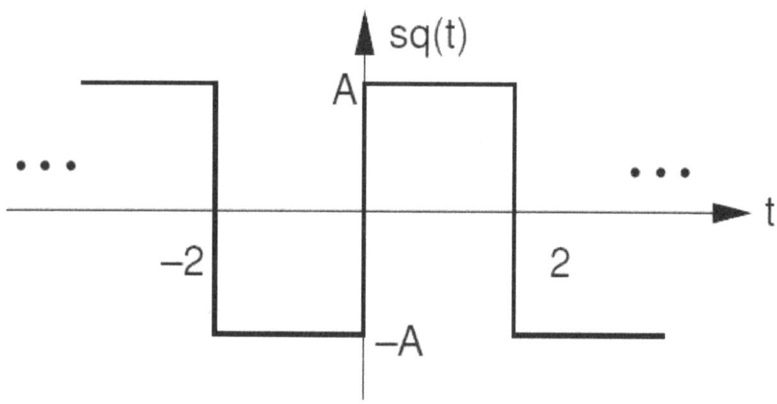


Figure 1.5

### Problem 1.2: Modems

The word "modem" is short for "modulator-demodulator." Modems are used not only for connecting computers to telephone lines, but also for connecting digital (discrete-valued) sources to generic channels. In this problem, we explore a simple kind of modem, in which binary information is represented by the presence or absence of a sinusoid (presence representing a "1" and absence a "0").

Consequently, the modem's transmitted signal that represents a single bit has the form

$$x(t) = A\sin(2\pi f_0 t)\ ,0 \le t \le T$$

Within each bit interval T, the amplitude is either A or zero.

- What is the smallest transmission interval that makes sense with the frequency f0?
- Assuming that ten cycles of the sinusoid comprise a single bit's transmission interval, what is the datarate of this transmission scheme?
- Now suppose instead of using "on-off" signaling, we allow one of several different values for the amplitude during any transmission interval. If N amplitude values are used, what is the resulting data rate?
- The classic communications block diagram applies to the modem. Discuss how the transmitter must interface with the message source since the source is producing letters of the alphabet, not bits.

## Problem 1.3: Advanced Modems

To transmit symbols, such as letters of the alphabet, RU computer modems use two frequencies (1600 and 1800 Hz) and several amplitude levels. A transmission is sent for a period of time T (known as the transmission or baud interval) and equals the sum of two amplitude-weighted carriers.

$$x(t) = A_1\sin(2\pi f_1 t) + A_2\sin(2\pi f_2 t)\ ,0 \le t \le T$$

We send successive symbols by choosing an appropriate frequency and amplitude combination, and sending them one after another.

- What is the smallest transmission interval that makes sense to use with the frequencies given above? In other words, what should T be so that an integer number of cycles of the carrier occurs?
- Sketch (using Matlab) the signal that modem produces over several transmission intervals. Make sure you axes are labeled.
- Using your signal transmission interval, how many amplitude levels are needed to transmit ASCII characters at a datarate of 3,200 bits/s? Assume use of the extended (8-bit) ASCII code.

**Note:** We use a discrete set of values for A1 and A2. If we have N1 values for amplitude A1, and N2 values for A2, we have N1N2 possible symbols that can be sent during each T second interval. To convert this number into bits (the fundamental unit of information engineers use to qualify things), compute: log2(N1N2)

## Solutions to Exercises in Chapter 1

### *Solution to Exercise 1.4.1 (p. 7)*

As $\cos(\alpha + \beta) = \cos(\alpha)\cos(\beta) - \sin(\alpha)\sin(\beta)$, $\cos(2\pi(f+1)n) = \cos(2\pi fn)\cos(2\pi n) - \sin(2\pi fn)\sin(2\pi n) = \cos(2\pi fn)$.

### *Solution to Exercise 1.4.2 (p. 7)*

A square wave takes on the values 1 and −1 alternately. See the plot in the module Elemental Signals (Section 2.2.6: Square Wave).

# 2 Chapter

# FUNDAMENTALS SIGNALS AND SYSTEMS

## 2.1 COMPLEX NUMBERS

The primary signal utilized in electrical engineering is the sinusoid, which may be formally represented in terms of an even more fundamental signal: the complex exponential. Expressing sinusoids in terms of complex exponentials is not a mathematical anomaly. Mastery of complex numbers and rational functions of complex variables is an essential talent that all engineers must acquire. Comprehending information and designing power systems, as well as creating novel systems, all rely on the utilization of complex numbers. Essentially, they are indispensable to contemporary electrical engineering, a realization that was made more than a century ago.

### 2.1.1 Definitions

The notion of the square root of -1 originated with the quadratic formula: the solution of certain quadratic equations mathematically exists only if the so-called imaginary quantity $\sqrt{-1}$ could be defined. Euler first used i for the imaginary unit but that notation did not take hold until roughly Ampère›s time. Ampère used the symbol i to denote current (intensité de current). It wasn't until the twentieth century that the importance of complex numbers to circuit theory became evident. By then, using i for current was entrenched and electrical engineers chose i for writing complex numbers.

An imaginary number has the form $jb = \sqrt{-b^2}$. A complex number, z, consists of the ordered pair (a,b), a is the real component and b is the imaginary component (the j is suppressed because the imaginary component of the pair is always in the second position). The imaginary number jb equals (0,b). Note that a and b are real-valued numbers.

Figure 2.1 (The Complex Plane) shows that we can locate a complex number in what we call the complex plane. Here, a, the real part, is the x-coordinate and b, the imaginary part, is the y-coordinate.

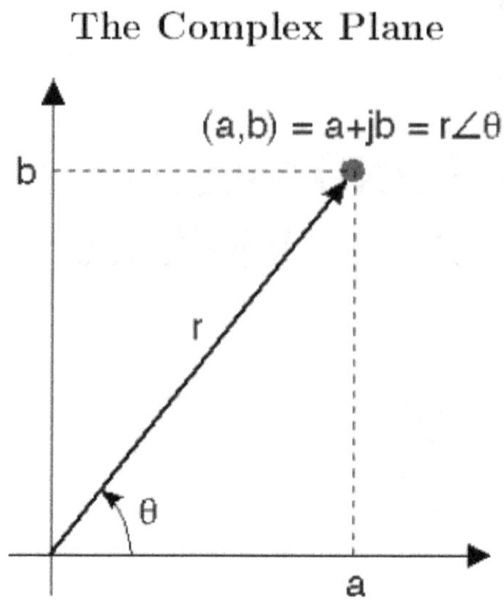


**Figure 2.1: A complex number is an ordered pair (a,b) that can be regarded as coordinates in the plane. They can also be expressed in polar coordinates as r∠θ.**

From analytic geometry, we know that locations in the plane can be expressed as the sum of vectors, with the vectors corresponding to the x and y directions. Consequently, a complex number z can be expressed as the (vector) sum z=a+ jb where i indicates the y-coordinate. This representation is known as the Cartesian form of z. An imaginary number can't be numerically added to a real number; rather, this notation for a complex number represents vector addition, but it provides a convenient notation when we perform arithmetic manipulations.

Some obvious terminology. The real part of the complex number z=a+jb, written as Re(z), equals a. We consider the real part as a function that works by selecting that component of a complex number not multiplied by i. The imaginary part of z, Im (z), equals b: that part of a complex number that is multiplied by i. Again, both the real and imaginary parts of a complex number are real-valued.

The complex conjugate of z, written as z*, has the same real part as z, but an imaginary part of the opposite sign.

$$\begin{aligned} z &= Re(z) + j\mathrm{Im}(z) \\ z^* &= Re(z) - j\mathrm{Im}(z) \end{aligned} \tag{2.1}$$

Using Cartesian notation, the following properties easily follow.

- If we add two complex numbers, the real part of the result equals the sum of the real parts and the imaginary part equals the sum of the imaginary parts. This property follows from the laws of vector addition.

$$a_1 + jb_1 + a_2 + jb_2 = a_1 + a_2 + j\left(b_1 + b_2\right)$$

In this way, the real and imaginary parts remain separate.

- The product of i and a real number is an imaginary number: ia. The product of i and an imaginary number is a real number: i(ib) = –b because j2 = –1. Consequently, multiplying a complex number by i rotates the number›s position by 90 degrees.

### Exercise 2.1.1

Use the definition of addition to show that the real and imaginary parts can be expressed as a sum/difference of a complex number and its conjugate. $Re\left(z\right) = \frac{z+z^*}{2}$ and $\mathrm{Im}\left(z\right) = \frac{z-z^*}{2j}$.

Complex numbers can also be expressed in an alternate form, polar form, which we will find quite useful. Polar form arises arises from the geometric interpretation of complex numbers. The Cartesian form of a complex number can be re-written as

$$a + jb = \sqrt{a^2 + b^2}\left(\frac{a}{\sqrt{a^2 + b^2}} + j\frac{b}{\sqrt{a^2 + b^2}}\right)$$

By forming a right triangle having sides a and b, we see that the real and imaginary parts correspond to the cosine and sine of the triangle's base angle. We thus obtain the polar form for complex numbers.

$$z = a + jb = r\angle\theta$$

$$r = |z| = \sqrt{a^2 + b^2}$$

$$a = r\cos\left(\theta\right)$$

$$b = r\sin\left(\theta\right)$$

$$\theta = \arctan\left(\frac{b}{a}\right)$$

The quantity r is known as the magnitude of the complex number z, and is frequently written as |z| . The quantity θ is the complex number's angle. In using the arc-tangent formula to find the angle, we must take into account the quadrant in which the complex number lies.

### Exercise 2.1.2

Convert 3-2j to polar form.

## 2.1.2 Euler's Formula

Surprisingly, the polar form of a complex number z can be expressed mathematically as

$$z = re^{j\theta} \tag{2.2}$$

To show this result, we use Euler's relations that express exponentials with imaginary arguments in terms of trigonometric functions.

$$e^{j\theta} = \cos(\theta) + j\sin(\theta) \tag{2.3}$$

$$\cos(\theta) = \frac{e^{j\theta} + e^{-(j\theta)}}{2} \tag{2.4}$$

$$\sin(\theta) = \frac{e^{j\theta} - e^{-(j\theta)}}{2j}$$

The first of these is easily derived from the Taylor's series for the exponential.

$$e^{x} = 1 + \frac{x}{1!} + \frac{x^2}{2!} + \frac{x^3}{3!} + \dots$$

Substituting jθ for x, we find that

$$e^{j\theta} = 1 + j\frac{\theta}{1!} - \frac{\theta^2}{2!} - j\frac{\theta^3}{3!} + \dots$$

Because j2=-1,j3=-j, and j4=1. Grouping separately the real-valued terms and the imaginary-valued ones,

$$e^{j\theta} = 1 - \frac{\theta^2}{2!} + \dots + j\left(\frac{\theta}{1!} - \frac{\theta^3}{3!} + \dots\right)$$

The real-valued terms correspond to the Taylor's series for cos(θ), the imaginary ones to sin(θ), and Euler's first relation results. The remaining relations are easily derived from the first. We see that multiplying the exponential in above equation by a real constant corresponds to setting the radius of the complex number to the constant.

### 2.1.3 Calculating with Complex Numbers

Adding and subtracting complex numbers expressed in Cartesian form is quite easy: You add (subtract) the real parts and imaginary parts separately.

$$z_1 \pm z_2 = (a_1 \pm a_2) + j(b_1 \pm b_2) \tag{2.5}$$

To multiply two complex numbers in Cartesian form is not quite as easy, but follows directly from following the usual rules of arithmetic.

$$\begin{aligned} z_1 z_2 &= (a_1 + jb_1)(a_2 + jb_2) \\ &= a_1 a_2 - b_1 b_2 + j(a_1 b_2 + a_2 b_1) \end{aligned} \tag{2.6}$$

Note that we are, in a sense, multiplying two vectors to obtain another vector. Complex arithmetic provides a unique way of defining vector multiplication.

### Exercise 2.1.3

What is the product of a complex number and its conjugate?

Division requires mathematical manipulation. We convert the division problem into a multiplication problem by multiplying both the numerator and denominator by the conjugate of the denominator.

$$\begin{aligned} \frac{z_1}{z_2} &= \frac{a_1+jb_1}{a_2+jb_2} \\ &= \frac{a_1+jb_1}{a_2+jb_2}\frac{a_2-jb_2}{a_2-jb_2} \\ &= \frac{(a_1+jb_1)(a_2-jb_2)}{{a_2}^2+{b_2}^2} \\ &= \frac{a_1a_2+b_1b_2+j(a_2b_1-a_1b_2)}{{a_2}^2+{b_2}^2} \end{aligned} \tag{2.7}$$

Because the final result is so complicated, it's best to remember how to perform division—multiplying numerator and denominator by the complex conjugate of the denominator—than trying to remember the final result.

The properties of the exponential make calculating the product and ratio of two complex numbers much simpler when the numbers are expressed in polar form.

$$\begin{aligned} z_1 z_2 &= r_1 e^{j\theta_1} r_2 e^{j\theta_2} \\ &= r_1 r_2 e^{j(\theta_1+\theta_2)} \end{aligned} \tag{2.8}$$

$$\frac{z_1}{z_2} = \frac{r_1 e^{j\theta_1}}{r_2 e^{j\theta_2}} = \frac{r_1}{r_2} e^{j(\theta_1-\theta_2)}$$

In order to perform multiplication, the radius is equivalent to the multiplication of the radii, and the angle is equal to the addition of the angles. In order to divide, the radius is equal to the ratio of the radii, and the angle is equal to the difference between the angles. When the original complex numbers are expressed in Cartesian form, it is often beneficial to convert them into polar form before carrying out multiplication or division, particularly in the case of division. The process of adding and subtracting polar forms involves transforming them to Cartesian form, performing the arithmetic operation, and then converting the result back to polar form.

### Exercise 2.1

When we solve circuit problems, the crucial quantity, known as a transfer function, will always be expressed as the ratio of polynomials in the variable s=j2πf. What we'll need to understand the circuit's effect is the transfer function in polar form. For instance, suppose the transfer function equals

$$\frac{s+2}{s^2+s+1} \tag{2.9}$$

$$s = j2\pi f \tag{2.10}$$

Performing the required division is most easily accomplished by first expressing the numerator and denominator each in polar form, then calculating the ratio. Thus,

$$\frac{s+2}{s^2+s+1} = \frac{j2\pi f+2}{-4\pi^2 f^2 + j2\pi f + 1} \tag{2.11}$$

$$= \frac{\sqrt{4+4\pi^2 f^2}e^{j\arctan(\pi f)}}{\sqrt{\left(1-4\pi^2 f^2\right)^2 + 4\pi^2 f^2}e^{j\arctan\left(\frac{2\pi f}{1-4\pi^2 f^2}\right)}} \tag{2.12}$$

$$= \sqrt{\frac{4+4\pi^2 f^2}{1-4\pi^2 f^2+16\pi^4 f^4}}e^{j\left(\arctan(\pi f)-\arctan\left(\frac{2\pi f}{1-4\pi^2 f^2}\right)\right)} \tag{2.13}$$

## 2.2 ELEMENTAL SIGNALS

Elemental signals are the building blocks with which we build complicated signals. By definition, elemental signals have a simple structure. Exactly what we mean by the «structure of a signal» will unfold in this section of the course. Signals are nothing more than functions defined with respect to some independent variable, which we take to be time for the most part. Very interesting signals are not functions solely of time; one great example of which is an image. For it, the independent variables are x and y (two-dimensional space). Video signals are functions of three variables: two spatial dimensions and time. Fortunately, most of the ideas underlying modern signal theory can be exemplified with one-dimensional signals.

### 2.2.1 Sinusoids

Perhaps the most common real-valued signal is the sinusoid.

$$s(t) = A\cos\left(2\pi f_0 t + \phi\right) \tag{2.14}$$

For this signal, A is its amplitude, f0 its frequency, and φ its phase.

### 2.2.2 Complex Exponentials

The most important signal is complex-valued, the complex exponential.

$$\begin{aligned} s(t) &= Ae^{j(2\pi f_0 t+\phi)} \\ &= Ae^{j\phi}e^{j2\pi f_0 t} \end{aligned} \quad (2.15)$$

Here, j denotes $\sqrt{-1}$. Aejφ is known as the signal's complex amplitude. Considering the complex amplitude as a complex number in polar form, its magnitude is the amplitude A and its angle the signal phase. The complex amplitude is also known as a phasor. The complex exponential cannot be further decomposed into more elemental signals, and is the most important signal in electrical engineering! Mathematical manipulations at first appear to be more difficult because complex-valued numbers are introduced. In fact, early in the twentieth century, mathematicians thought engineers would not be sufficiently sophisticated to handle complex exponentials even though they greatly simplified solving circuit problems. Steinmetz introduced complex exponentials to electrical engineering, and demonstrated that "mere" engineers could use them to good effect and even obtain right answers! See Complex Numbers (Section 2.1) for a review of complex numbers and complex arithmetic.

The complex exponential defines the notion of frequency: it is the only signal that contains only one frequency component. The sinusoid consists of two frequency components: one at the frequency f0 and the other at –f0.

**Euler Relation:** This decomposition of the sinusoid can be traced to Euler's relation.

$$\cos(2\pi ft) = \frac{e^{j2\pi ft} + e^{-(j2\pi ft)}}{2} \quad (2.16)$$

$$\sin(2\pi ft) = \frac{e^{j2\pi ft} - e^{-(j2\pi ft)}}{2j} \quad (2.17)$$

$$e^{j2\pi ft} = \cos(2\pi ft) + j\sin(2\pi ft) \quad (2.18)$$

Decomposition: The complex exponential signal can thus be written in terms of its real and imaginary parts using Euler's relation. Thus, sinusoidal signals can be expressed as either the real or the imaginary part of a complex exponential signal, the choice depending on whether cosine or sine phase is needed, or as the sum of two complex exponentials. These two decompositions are mathematically equivalent to each other.

$$A\cos(2\pi ft + \phi) = Re\left(Ae^{j\phi}e^{j2\pi ft}\right) \quad (2.19)$$

$$A\sin\left(2\pi ft+\phi\right)=\mathrm{Im}\left(Ae^{j\phi}e^{j2\pi ft}\right) \tag{2.20}$$

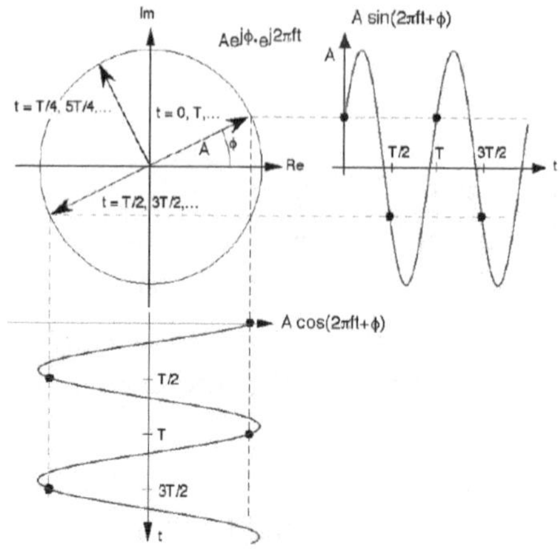


**Figure 2.2 Graphically, the complex exponential scribes a circle in the complex plane as time evolves. Its real and imaginary parts are sinusoids. The rate at which the signal goes around the circle is the frequency f and the time taken to go around is the period T. A fundamental relationship is $T=\frac{1}{f}$.**

Using the complex plane, we can envision the complex exponential's temporal variations as seen in the above figure (Figure 2.2). The magnitude of the complex exponential is A, and the initial value of the complex exponential at t = 0 has an angle of φ. As time increases, the locus of points traced by the complex exponential is a circle (it has constant magnitude of A). The number of times per second we go around the circle equals the frequency f. The time taken for the complex exponential to go around the circle once is known as its period T, and equals $\frac{1}{f}$. The projections onto the real and imaginary axes of the rotating vector representing the complex exponential signal are the cosine and sine signal of Euler's relation ((2.16)).

### 2.2.3 Real Exponentials

As opposed to complex exponentials which oscillate, real exponentials (Figure 2.3) decay.

$$s\left(t\right)=e^{-\frac{t}{\tau}} \tag{2.21}$$

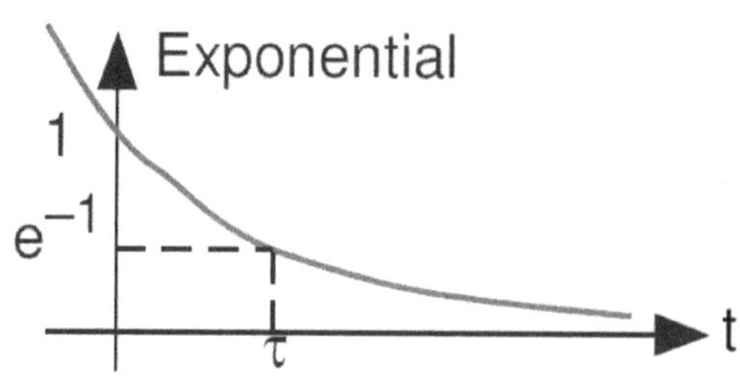


**Figure 2.3: The real exponential**

The quantity τ is known as the exponential›s time constant, and corresponds to the time required for the exponential to decrease by a factor of $\frac{1}{e}$, which approximately equals 0.368. A decaying complex exponential is the product of a real and a complex exponential.

$$\begin{aligned} s(t) &= Ae^{j\phi}e^{-\frac{t}{\tau}}e^{j2\pi ft} \\ &= Ae^{j\phi}e^{\left(-\frac{1}{\tau}+j2\pi f\right)t} \end{aligned} \quad (2.22)$$

In the complex plane, this signal corresponds to an exponential spiral. For such signals, we can define complex frequency as the quantity multiplying t.

## 2.2.4 Unit Step

The unit step function is denoted by u(t), and is defined to be

$$u(t) = \begin{cases} 0 \text{ if } t < 0 \\ 1 \text{ if } t > 0 \end{cases} \quad (2.23)$$

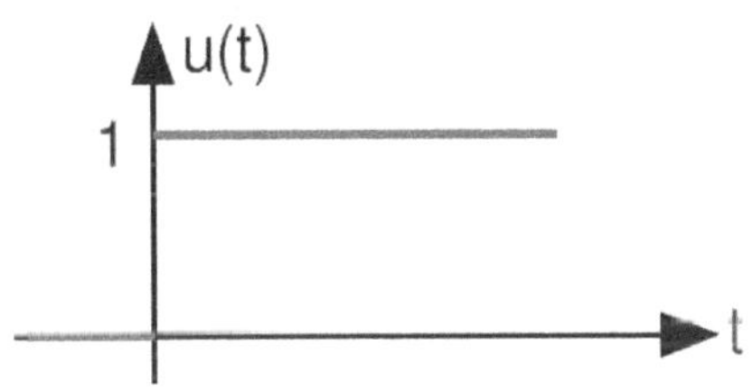


**Figure 2.4: Unit Step**

Origin Warning: This signal is discontinuous at the origin. Its value at the origin need not be defined, and doesn't matter in signal theory.

This kind of signal is used to describe signals that "turn on" suddenly. For example, to mathematically represent turning on an oscillator, we can write it as the product of a sinusoid and a step: s(t)=Asin(2πft)u(t)

## 2.2.5 Pulse

The unit pulse describes turning a unit-amplitude signal on for a duration of Δ seconds, then turning it off.

$$p_{\Delta}(t) = \begin{cases} 0 \text{ if } t < 0 \\ 1 \text{ if } 0 < t < \Delta \\ 0 \text{ if } t > \Delta \end{cases} \quad (2.24)$$

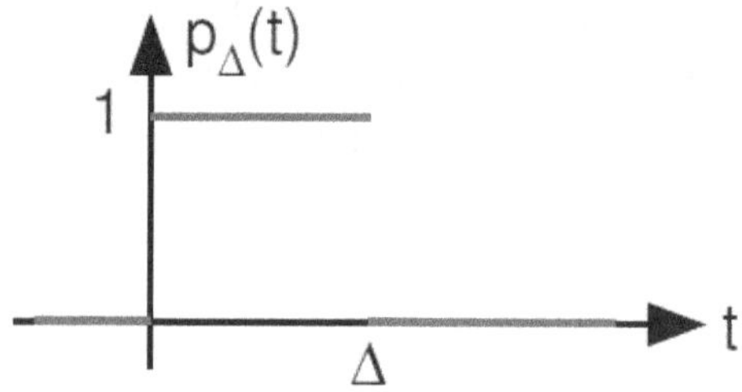


**Figure 2.5: The Pulse**

We will find that this is the second most important signal in communications.

### 2.2.6 Square Wave

The square wave, represented by Figure 2.6, is a periodic signal similar to the sinusoid. The signal also possesses an amplitude and a time, both of which need to be given in order to describe the signal accurately. It is later discovered that the sine wave is a more straightforward signal compared to the square wave.

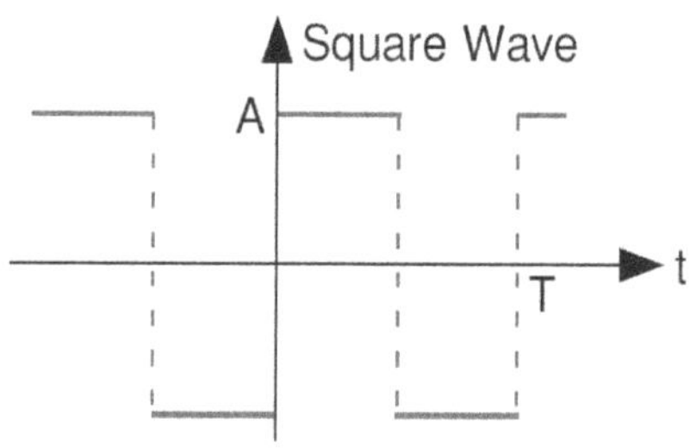


**Figure 2.6: The Square wave**

## 2.3 SIGNAL DECOMPOSITION

The complexity of a signal is not correlated with the level of waviness. Instead, a signal expert seeks methods to break down a particular signal into a combination of simpler signals, which we refer to as signal decomposition. The complexity of a signal cannot be precisely computed, but it is primarily determined by the number of terms in its decomposition. When expressing a signal as the sum of component signals, we have the ability to modify the gain of each component signal by multiplying it with a constant and adjusting its timing. Complex decompositions may include derivatives or integrals of elementary signals.

### Example 2.2

As an example of signal complexity, we can express the pulse $p_\Delta(t)$ as a sum of delayed unit steps.

$$p_\Delta(t) = u(t) - u(t - \Delta) \tag{2.25}$$

Thus, the pulse is a more complex signal than the step. Be that as it may, the pulse is very useful to us.

### Example 2.3.1

Express a square wave having period T and amplitude A as a superposition of delayed and amplitude-scaled pulses.

Because the sinusoid is a superposition of two complex exponentials, the sinusoid is more complex. We could not prevent ourselves from the pun in this statement. Clearly, the word "complex" is used in two different ways here. The complex exponential can also be written (using Euler's relation (2.16)) as a sum of a sine and a cosine. We will discover that virtually every signal can be decomposed into a sum of complex exponentials, and that this decomposition is very useful. Thus, the complex exponential is more fundamental, and Euler›s relation docs not adequately reveal its complexity.

## 2.4 DISCRETE-TIME SIGNALS

So far, we have treated what are known as analog signals and systems. Mathematically, analog signals are functions having continuous quantities as their independent variables, such as space and time. Discrete-time signals are functions defined on the integers; they are sequences. One of the fundamental results of signal theory will detail conditions under which an analog signal can be converted into a discrete-time one and retrieved without error. This result is important because discrete-time signals can be manipulated by systems instantiated as computer programs. Subsequent modules describe how virtually all analog signal processing can be performed with software.

As important as such results are, discrete-time signals are more general, encompassing signals derived from analog ones and signals that aren›t. For example, the characters forming a text file form a sequence, which is also a discrete-time signal. We must deal with such symbolic valued signals and systems as well.

As with analog signals, we seek ways of decomposing real-valued discrete-time signals into simpler components. With this approach leading to a better understanding of signal structure, we can exploit that structure to represent information (create ways of representing information with signals) and to extract information (retrieve the information thus represented). For symbolic-valued signals, the approach is different: We develop a common representation of all symbolic-valued signals so that we can embody the information they contain in a unified way. From an information representation perspective, the most important issue becomes, for both real-valued and symbolic-valued signals, efficiency; What is the most parsimonious and compact way to represent information so that it can be extracted later.

### 2.4.1 Real- and Complex-valued Signals

A discrete-time signal is represented symbolically as s(n), where n = {...,-1,0,1...}. We usually draw discrete-time signals as stem plots to emphasize the fact they are functions defined only on the integers. We can delay a discrete-time signal by an integer just as with analog ones. A delayed unit sample has the expression δ(n−m), and equals one when n = m.

Discrete-Time Cosine Signal

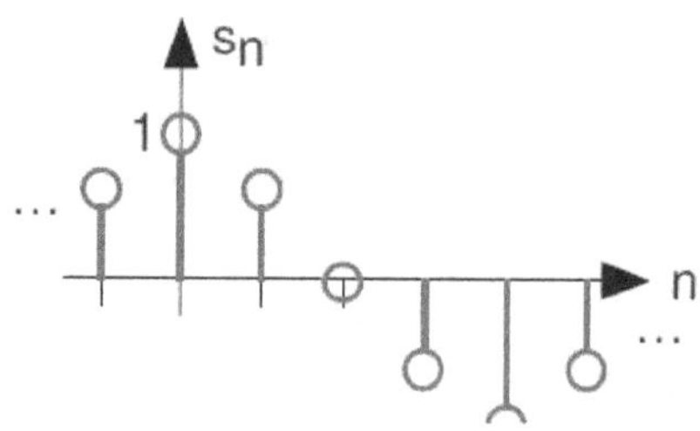


**Figure 2.7: The discrete-time cosine signal is plotted as a stem plot. Can you find the formula for this signal?**

### 2.4.2 Complex Exponentials

The most important signal is, of course, the complex exponential sequence.

$$s(n) = e^{j2\pi fn} \tag{2.26}$$

### 2.4.3 Sinusoids

Discrete-time sinusoids have the obvious form s(n)=Acos(2πfn+φ). As opposed to analog complex exponentials and sinusoids that can have their frequencies be any real value, frequencies of their discrete-time counterparts yield unique waveforms only when f lies in the interval $\left(-\frac{1}{2}, \frac{1}{2}\right]$. This property can be easily understood by noting that adding an integer to the frequency of the discrete-time complex exponential has no effect on the signal's value.

$$\begin{aligned} e^{j2\pi(f+m)n} &= e^{j2\pi fn}e^{j2\pi mn} \\ &= e^{j2\pi fn} \end{aligned} \tag{2.27}$$

This derivation follows because the complex exponential evaluated at an integer multiple of 2π equals one.

### 2.4.4 Unit Sample

The second-most important discrete-time signal is the unit sample, which is defined to be

$$\delta(n) = \begin{cases} 1 & \text{if } n = 0 \\ 0 & \text{otherwise} \end{cases} \quad (2.28)$$

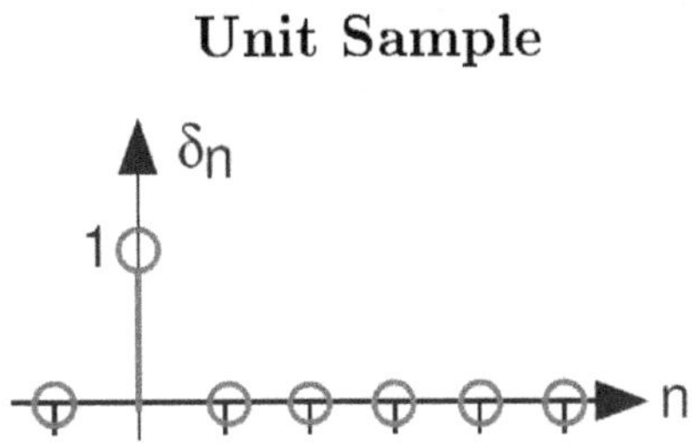


**Figure 2.8: The unit sample**

Examination of a discrete-time signal's plot, like that of the cosine signal shown in Figure 2.7, reveals that all signals consist of a sequence of delayed and scaled unit samples. Because the value of a sequence at each integer m is denoted by s(m) and the unit sample delayed to occur at m is written δ(n-m), we can decompose any signal as a sum of unit samples delayed to the appropriate location and scaled by the signal value.

$$s(n) = \sum_{m=-\infty}^{\infty} s(m)\,\delta(n-m) \quad (2.29)$$

This kind of decomposition is unique to discrete-time signals, and will prove useful subsequently.

Discrete-time systems can act on discrete-time signals in ways similar to those found in analog signals and systems. Because of the role of software in discrete-time systems, many more different systems can be envisioned and "constructed" with programs than can be with analog signals. In fact, a special class of analog signals can be converted into discrete-time signals, processed with software, and converted back into an analog signal, all without the incursion of error. For such signals, systems can be easily produced in software, with equivalent analog realizations difficult, if not impossible, to design.

## 2.4.5 Symbolic-valued Signals

Another interesting aspect of discrete-time signals is that their values do not need to be real numbers. We do have real-valued discrete-time signals like the sinusoid, but we also have signals that denote the sequence of characters typed on the keyboard. Such characters certainly aren't real numbers, and as a collection of possible signal values, they have little mathematical structure other than that they are members of a set. More formally, each element of the symbolic-valued signal s(n) takes on one of the values {a1,....aK} which comprise the alphabet A. This technical terminology does not mean we restrict symbols to being members of the English or Greek alphabet. They could represent keyboard characters, bytes (8-bit quantities), integers that convey daily temperature. Whether controlled by software or not, discrete-time systems are ultimately constructed from digital circuits, which consist entirely of analog circuit elements. Furthermore, the transmission and reception of discrete-time signals, like e-mail, is accomplished with analog signals and systems. Understanding how discrete-time and analog signals and systems intertwine is perhaps the main goal of this course.

## 2.5 INTRODUCTION TO SYSTEMS

Signals are manipulated by systems. Mathematically, we represent what a system does by the notation y(t)=S(x(t)),with x representing the input signal and y the output signal.

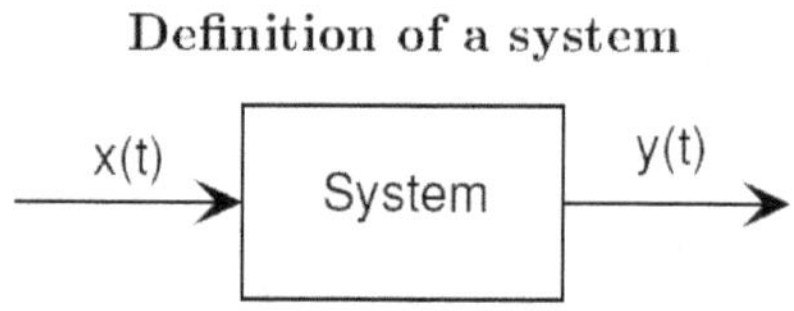


**Figure 2.9 The system depicted has input x(t) and output y(t). Mathematically, systems operate on function(s) to produce other function(s). In many ways, systems are like functions, rules that yield a value for the dependent variable (our output signal) for each value of its independent variable (its input signal). The notation y(t)=S(x(t)) corresponds to this block diagram. We term S(•) the input-output relation for the system.**

This notation mimics the mathematical symbology of a function: A system's input is analogous to an independent variable and its output the dependent variable. For the mathematically inclined, a system is a functional: a function of a function (signals are functions).

Simple systems can be connected together--one system's output becomes another's input--to accomplish some overall design. Interconnection topologies can be quite complicated, but usually consist of weaves of three basic interconnection forms.

### 2.5.1 Cascade Interconnection

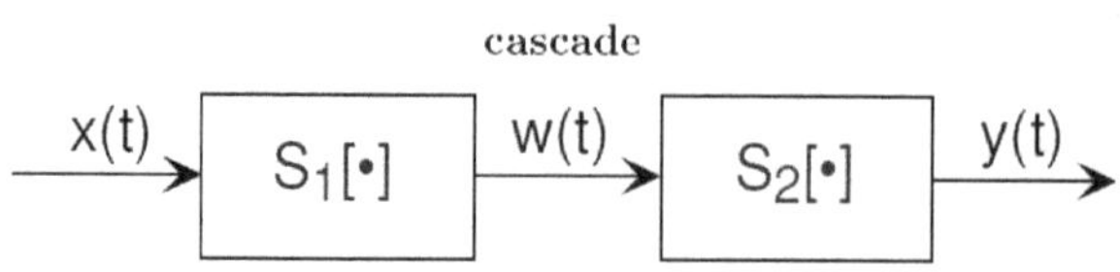


**Figure 2.10: The most rudimentary ways of interconnecting systems are shown in the figures in this section. This is the cascade configuration.**

The simplest form is when one system's output is connected only to another's input. Mathematically, w(t)=S1(x(t)), and y(t)=S2(x(t)), with the information contained in x(t) processed by the first, then the second system. In some cases, the ordering of the systems matter, in others it does not. For example, in the fundamental model of communication the ordering most certainly matters.

### 2.5.2 Parallel Interconnection

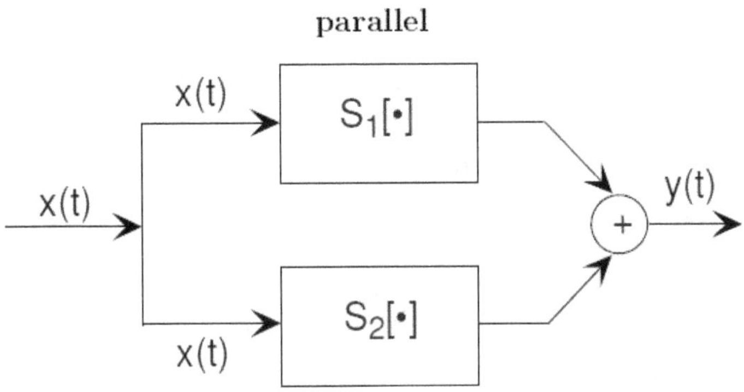


**Figure 2.11: The parallel configuration**

A signal x(t)is routed to two (or more) systems, with this signal appearing as the input to all systems simultaneously and with equal strength. Block diagrams have the convention that signals going to more than one system are not split into pieces along the way. Two or more systems operate on x(t) and their outputs are added together to create the output y(t). Thus, y(t)=S1(x(t))+S2(x(t)), and the information in x(t) is processed separately by both systems.

### 2.5.3 Feedback Interconnection

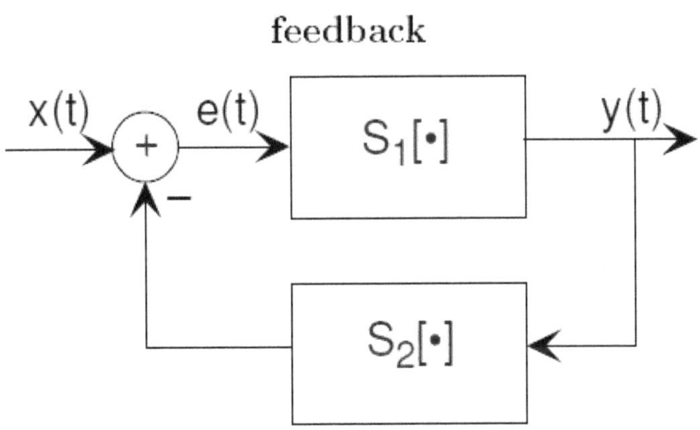


**Figure 2.12: The feedback configuration**

The subtlest interconnection configuration has a system's output also contributing to its input. Engineers would say the output is "fed back" to the input through system 2, hence the terminology. The mathematical statement of the feedback interconnection is that the feed-forward system produces the output: y(t)=S1(e(t)). The input e(t) equals the input signal minus the output of some other system's output to: y(t):e(t)=x(t)-S2(y(t)).Feedback systems are omnipresent in control problems, with the error signal used to adjust the output to achieve some condition defined by the input (controlling) signal.

For example, in a car's cruise control system, x(t) is a constant representing what speed you want, and y(t) is the car's speed as measured by a speedometer. In this application, system 2 is the identity system (output equals input).

## 2.6 SIMPLE SYSTEMS

Systems modify signals to generate output signals based on their input signals. The reasons for categorizing the following as "simple" will only become clear near the end of the course.

### 2.6.1 Sources

Sources produce signals without having input. We like to think of these as having controllable parameters, like amplitude and frequency. Examples would be oscillators that produce periodic signals like sinusoids and square waves and noise generators that yield signals with erratic waveforms (more about noise subsequently). Simply writing an expression for the signals they produce specifies sources. A sine wave generator might be specified by: y(t)=Asin(2πf0t)u(t), which says that the source was turned on at t=0 to produce a sinusoid of amplitude A and frequently f0.

### 2.6.2 Amplifiers

An amplifier multiplies its input by a constant known as the amplifier gain.

$$y(t) = Gx(t) \tag{2.30}$$

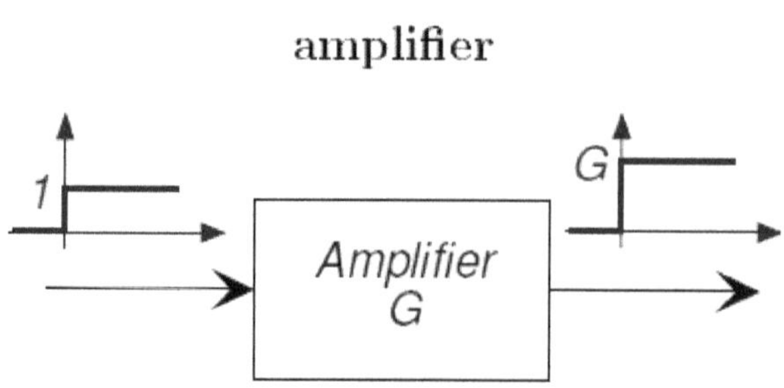


**Figure 2.13: An amplifier**

The gain can be positive or negative (if negative, we would say that the amplifier inverts its input) and its magnitude can be greater than one or less than one. If less than one, the amplifier actually attenuates. A real-world example of an amplifier is your home stereo. You control the gain by turning the volume control.

### 2.6.3 Delay

A system serves as a time delay when the output signal equals the input signal at an earlier time.

$$y(t) = x(t - \tau) \tag{2.31}$$

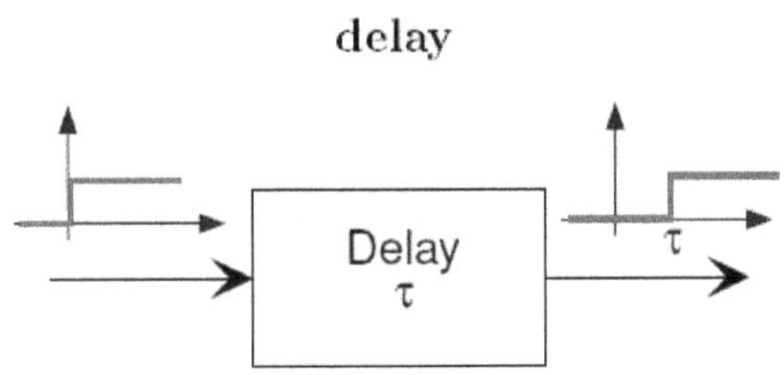


**Figure 2.14: A delay**

Here, τ is the delay. The way to understand this system is to focus on the time origin: The output at time t = τ equals the input at time t = 0 Thus, if the delay is positive, the output emerges later than the input, and plotting the output amounts to shifting the input plot to the right. The delay can be negative, in which case we say the system advances its input. Such systems are difficult to build (they would have to produce signal values derived from what the input will be), but we will have occasion to advance signals in time.

### 2.6.4 Time Reversal

Here, the output signal equals the input signal flipped about the time origin.

$$y(t) = x(-t) \tag{2.32}$$

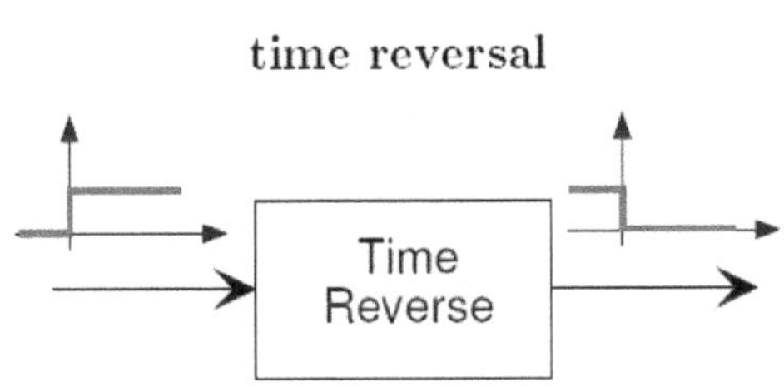


**Figure 2.15: A time reversal system**

Again, such systems are difficult to build, but the notion of time reversal occurs frequently in communications systems.

### Exercise 2.6.1

Mentioned earlier was the issue of whether the ordering of systems mattered. In other words, if we have two systems in cascade, does the output depend on which comes first? Determine if the ordering matters for the cascade of an amplifier and a delay and for the cascade of a time-reversal system and a delay.

## 2.6.5 Derivative Systems and Integrators

Systems that perform calculus-like operations on their inputs can produce waveforms significantly different than present in the input. Derivative systems operate in a straightforward way: A first-derivative system would have the input-output relationship $y(t) = \frac{d}{dt}x(t)$. Integral systems have the complication that the integral's limits must be defined. It is a signal theory convention that the elementary integral operation have a lower limit of -∞ and that the value of all signals at t=-∞ equals zero. A simple integrator would have input-output relation:

$$y(t) = \int_{-\infty}^{t} x(\alpha)\, d\alpha \qquad (2.33)$$

## 2.6.6 Linear Systems

Linear systems are a class of systems rather than having a specific input-output relation. Linear systems form the foundation of system theory, and are the most important class of systems in communications. They have the property that when the input is expressed as a weighted sum of component signals, the output equals the same weighted sum of the outputs produced by each component. When S(•) is linear, for all choices of signals and gains.

$$S(G_1 x_1(t) + G_2 x_2(t)) = G_1 S(x_1(t)) + G_2 S(x_2(t)) \qquad (2.34)$$

This general input-output relation property can be manipulated to indicate specific properties shared by all linear systems.

- S(Gx(t))=GS(x(t))The colloquialism summarizing this property is "Double the input, you double the output." Note that this property is consistent with alternate ways of expressing gain changes: Since 2x(t) also equals x(t)+x(t), the linear system definition provides the same output no matter which of these is used to express a given signal.
- S(0)=0 If the input is identically zero for all time, the output of a linear system must be zero. This property follows from the simple derivation: S(0)=S(x(t)-x(t))=S(x(t))-S(x(t))=0

Just why linear systems are so important is related not only to their properties, which are divulged throughout this course, but also because they lend themselves to relatively simple mathematical analysis. Said another way, "They're the only systems we thoroughly understand!"

We can find the output of any linear system to a complicated input by decomposing the input into simple signals. The equation above says that when a system is linear, its output to a decomposed input is the sum of outputs to each input. for example if,

$$x(t) = e^{-t} + \sin(2\pi f_0 t)$$

The output S(x(t)) of any linear system equals

$$y(t) = S(e^{-t}) + S(\sin(2\pi f_0 t))$$

## 2.6.7 Time-Invariant Systems

Systems that don't change their input-output relation with time are said to be time-invariant. The mathematical way of stating this property is to use the signal delay concept described in Simple Systems (Section 2.6.3: Delay)

$$(y(t) = S(x(t))) \Rightarrow (y(t-\tau) = S(x(t-\tau))) \tag{2.35}$$

If you delay (or advance) the input, the output is similarly delayed (advanced). Thus, a time-invariant system responds to an input you may supply tomorrow the same way it responds to the same input applied today; today's output is merely delayed to occur tomorrow.

The collection of linear, time-invariant systems are the most thoroughly understood systems. Much of the signal processing and system theory discussed here concentrates on such systems. For example, electric circuits are, for the most part, linear and time-invariant. Nonlinear ones abound, but characterizing them so that you can predict their behavior for any input remains an unsolved problem.

**Table 2.1 Linear, Time-Invarient Table**

| | | |
|---|---|---|
| $y(t) = \frac{d}{dt}(x)$ | Yes | Yes |
| $y(t) = \frac{d^2}{dt^2}(x)$ | Yes | Yes |
| $y(t) = \left(\frac{d}{dt}(x)\right)^2$ | no | Yes |
| $y(t) = \frac{dx}{dt} + x$ | Yes | Yes |
| $y(t) = x_1 + x_2$ | Yes | Yes |
| $y(t) = x(t-\tau)$ | Yes | Yes |
| $y(t) = \cos(2\pi f t)\, x(t)$ | Yes | no |
| $y(t) = x(-t)$ | Yes | no |

| | | |
|---|---|---|
| $y(t) = x^2(t)$ | no | Yes |
| $y(t) = \|x(t)\|$ | no | Yes |
| $y(t) = mx(t) + b$ | no | Yes |

## 2.7 SIGNALS AND SYSTEMS PROBLEMS

### Problem 2.1: Complex Number Arithmetic

Find the real part, imaginary part, the magnitude and angle of the complex numbers given by the following expressions.

- -1
- $\frac{1+\sqrt{3}j}{2}$
- $1 + j + e^{j\frac{\pi}{2}}$
- $e^{j\frac{\pi}{3}} + e^{j\pi} + e^{-\left(j\frac{\pi}{3}\right)}$

### Problem 2.2: Discovering Roots

Complex numbers expose all the roots of real (and complex) numbers. For example, there should be two square-roots, three cube-roots, etc. of any number. Find the following roots.

- What are the cube-roots of 27? In other words, what is $27^{\frac{1}{3}}$?
- What are the fifth roots of 3 $\left(3^{\frac{1}{5}}\right)$?
- What are the fourth roots of one?

### Problem 2.3: Cool Exponentials

Simplify the following (cool) expressions.

- jj
- j 2j
- jjj

### Problem 2.4: Complex-valued Signals

Complex numbers and phasors play a very important role in electrical engineering. Solving systems for complex exponentials is much easier than for sinusoids, and linear systems analysis is particularly easy.

- Find the phasor representation for each, and re-express each as the real and imaginary parts of a complex exponential. What is the frequency (in Hz) of each? In general, are your answers unique? If so, prove it; if not, find an alternative answer for the complex exponential representation.

- 3sin(24t)
- $\sqrt{2}\cos\left(2\pi 60t + \frac{\pi}{4}\right)$
- $2\cos\left(t + \frac{\pi}{6}\right) + 4\sin\left(t - \frac{\pi}{3}\right)$

⦿ Show that for linear systems having real-valued outputs for real inputs, that when the input is the real part of a complex exponential, the output is the real part of the system's output to the complex exponential (see Figure 2.16).

$$S\left(Re\left(Ae^{j2\pi ft}\right)\right) = Re\left(S\left(Ae^{j2\pi ft}\right)\right)$$

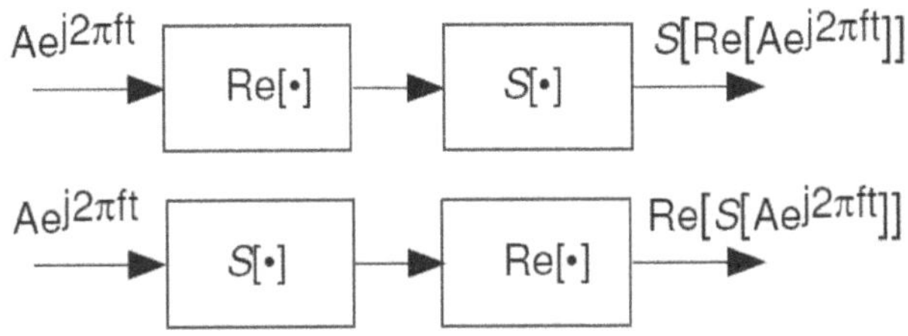


**Figure 2.16 Complex system**

**Problem 2.5:**

For each of the indicated voltages, write it as the real part of a complex exponential v(t)=R(Vest)). Explicitly indicate the value of the complex amplitude V and the complex frequency s. Represent each complex amplitude as a vector in the V-plane, and indicate the location of the frequencies in the complex s-plane.

⦿ v(t)=cos(5t)

⦿ $v(t) = \sin\left(8t + \frac{\pi}{4}\right)$

⦿ v(t)=e-t

⦿ $v(t) = e^{-(3t)}\sin\left(4t + \frac{3\pi}{4}\right)$

⦿ $v(t) = 5e^{(2t)}\sin\left(8t + 2\pi\right)$

⦿ v(t)=-2

⦿ v(t)=4sin(2t)+3cos(2t)

⦿ $v(t) = 2\cos\left(100\pi t + \frac{\pi}{6}\right) - \sqrt{3}\sin\left(100\pi t + \frac{\pi}{2}\right)$

**Problem 2.6:**

Express each of the following signals (Figure 2.17) as a linear combination of delayed and weighted step functions and ramps (the integral of a step).

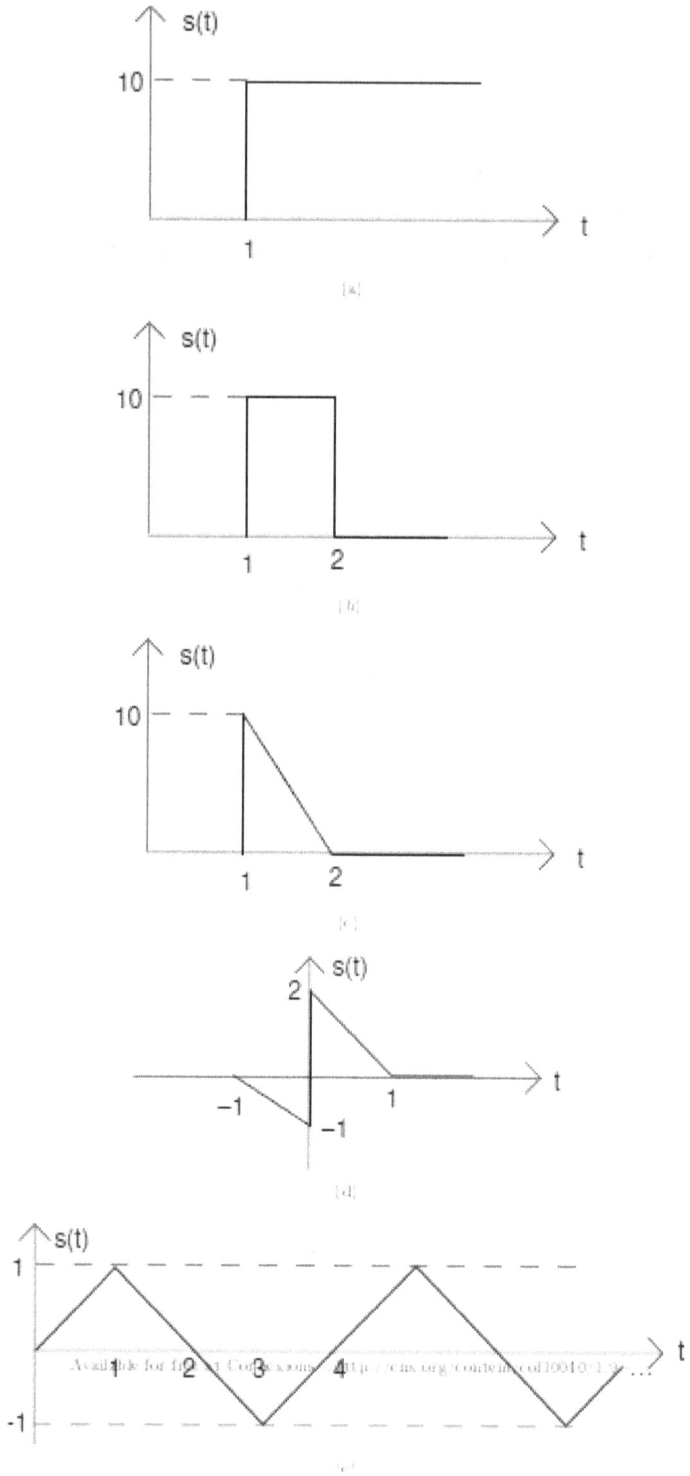


**Figure 2.17**

## Problem 2.7: Linear, Time-Invariant Systems

When the input to a linear, time-invariant system is the signal x(t), the output is the signal y(t), (Figure 2.18).

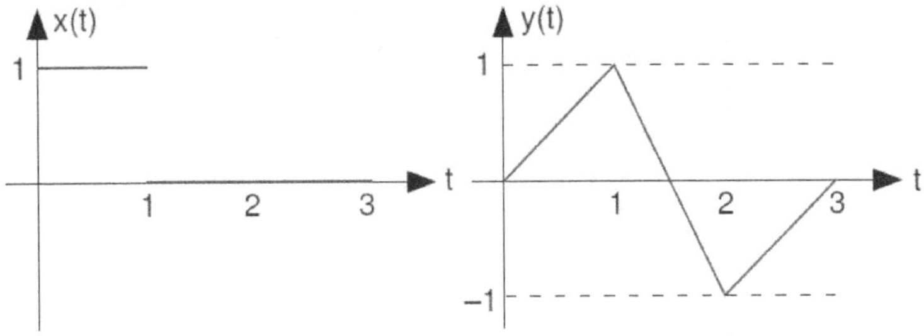


**Figure 2.18: Time -invariant system**

- Find and sketch this system's output when the input is the depicted signal: (Figure 2.19).
- Find and sketch this system's output when the input is a unit step.

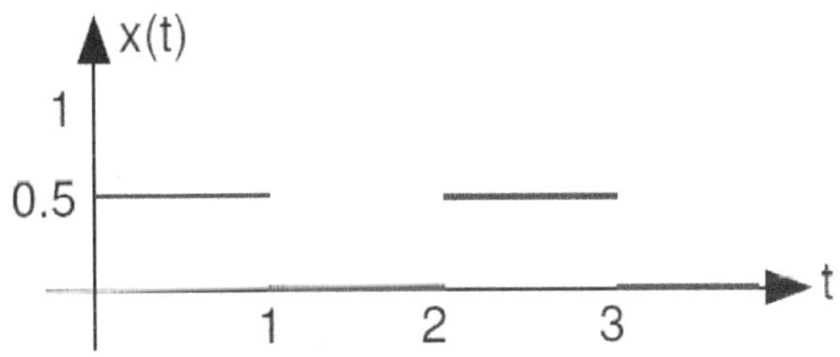


**Figure 2.19**

## Problem 2.8: Linear Systems

The depicted input (Figure 2.20) x(t) to a linear, time-invariant system yields the output y(t).

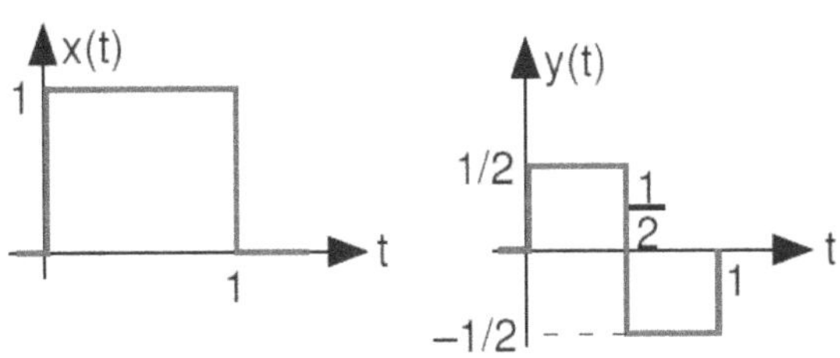


**Figure 2.20: Linear system**

- What is the system's output to a unit step input u(t)?
- What will the output be when the input is the depicted sqaure wave: (Figure 2.21)

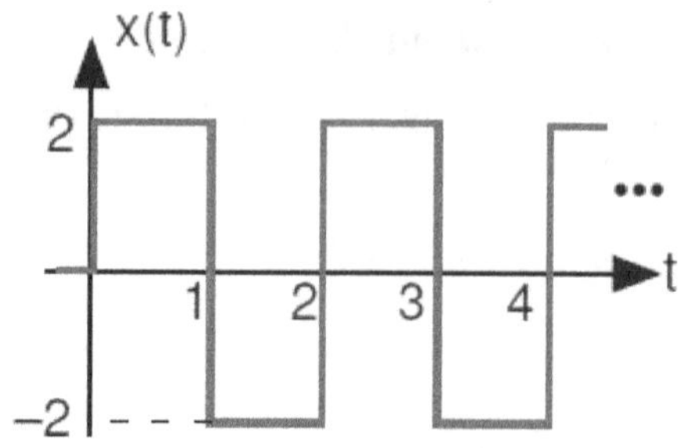


Figure 2.21

## Problem 2.9: Communication Channel

A particularly interesting communication channel can be modeled as a linear, time-invariant system. When the transmitted signal x(t) is a pulse, the received signal r(t) is as shown: (Figure 2.22)

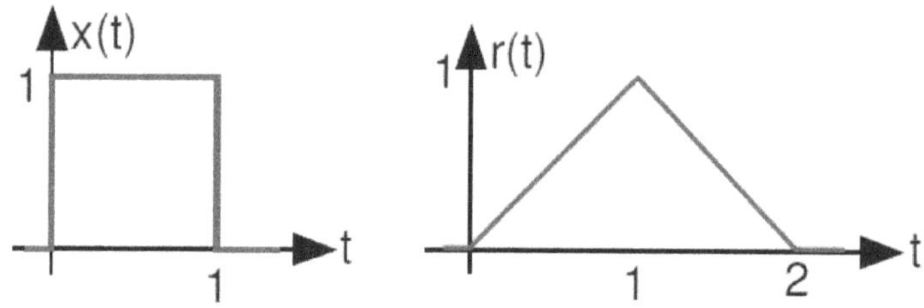


Figure 2.22: Communication channel and pulse sequence

- What will be the received signal when the transmitter sends the pulse sequence (Figure 2.23)x1(t)?
- What will be the received signal when the transmitter sends the pulse signal (Figure 2.23) x2(t) that has half the duration as the original?

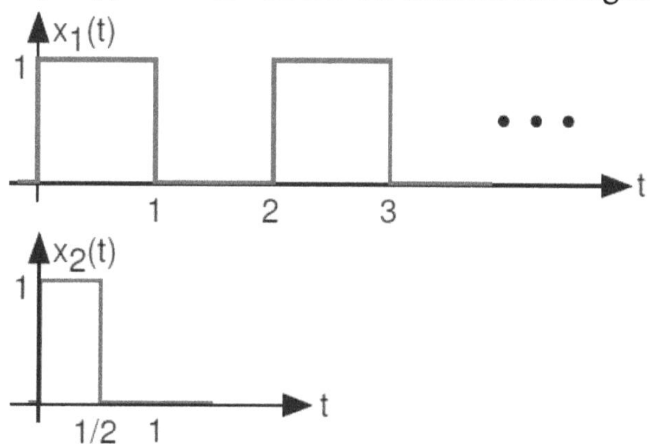


Figure 2.23

## Problem 2.10: Analog Computers

So-called analog computers use circuits to solve mathematical problems, particularly when they involve differential equations. Suppose we are given the following differential equation to solve.

$$\frac{dy(t)}{dt} + ay(t) = x(t)$$

In this equation, a is a constant.

- When the input is a unit step (x(t)=u(t)), the output is given by y(t)=(1-e-(at))u(t) What is the total energy expended by the input?
- Instead of a unit step, suppose the input is a unit pulse (unit-amplitude, unit-duration) delivered to the circuit at time t =10. what is the output voltage in this case? Sketch the waveform.

## Solutions to Exercises in Chapter 2

### Solution to Exercise 2.1.1

z + z*= a + jb + 2a = 2Re (z). Similarly, z– z* = a + jb–(a–jb) = 2jb = 2 (j; Im (z))

### Solution to Exercise 2.1.2

To convert 3 –2j to polar form, we _rst locate the number in the complex plane in the fourth quadrant. Thc distance from the origin to the complex number is the magnitude r, which equals $\sqrt{13} = \sqrt{3^2 + (-2)^2}$. The angle equals–arctan $\left(\frac{2}{3}\right)$ or –0:588 radians (–33:7 degrees). The final answer is $\sqrt{13}\angle(-33.7)$ degrees.

### Solution to Exercise 2.1.3

Zz* = (a + *jb*) (a– *jb*) = a² + b². Thus, zz* = r² = (|z|)².

### Solution to Exercise 2.3.1

$$\mathrm{sq}(t) = \sum_{n=-\infty}^{\infty} (-1)^n A p_{T/2}\left(t - n\frac{T}{2}\right)$$

### Solution to Exercise 2.6.1

In the _rst case, order does not matter; in the second it does. “Delay” means t→ t-r. “Time-reverse” means t→-t

**Case 1** y (t) = Gx(t–r ), and the way we apply the gain and delay the signal gives the same result.

**Case 2** Time-reverse then delay: y (t) = x (-(t–r )) = x (–t +r ). Delay then time-reverse: y (t) = x ((–t) –r ).

# Chapter 3 ANALOG SIGNAL PROCESSING

## 3.1 VOLTAGE, CURRENT, AND GENERIC CIRCUIT ELEMENTS

It is understood that information can be conveyed through signals. Now, we must comprehend the physical implementation of these signals. Electric signals have been identified as the most convenient to utilize over time. Voltage and currents are the physical manifestations of electrical signals. Therefore, it is necessary for us to explore the realm of electricity and electromagnetic. Circuits are the systems employed to directly alter electric signals. They enhance the representation of information or extract information from the voltage or current. Often, they serve as excellent illustrations of linear systems.

A generic circuit element establishes a relationship between the fundamental variables of a circuit: voltage and current. Voltage is the measure of electric potential, which signifies the force that propels electric charge to move between different locations. The movement of charge is caused by a physical separation between positive and negative charges. A battery produces an electric field by generating an excess of positive charge at one terminal and negative charge at the other terminal through electrochemical processes. Voltage is determined by the potential difference across a circuit element, where a positive sign indicates a positive voltage drop across the element. When a conductor establishes a connection between the positive and negative potentials, an electric current is generated. A positive current indicates that positive charges are moving from the positive terminal to the negative terminal. Electrons are responsible for the flow of electric current in numerous instances. Due to their negative charge, electrons exhibit a movement that is opposed to the direction of positive current flow. Specifically, when negative charge flows towards the right, it is comparable to positive charge traveling towards the left.

Comprehending the principles of current flow in conductors is crucial for recognizing the ingenuity behind novel electrical gadgets. Electric charge can originate from various sources, with the electron being the most basic and straightforward one. When we refer to the "flow of electrons through a conductor," we are describing the process in which the atoms that make up the conductor readily release electrons from their outer shells. "Flow" refers to the movement of electrons as they transfer from one atom to another, propelled by the applied electric potential. When one electron is absent, it might be considered as a virtual positive charge. Electrical engineers refer to these empty spaces as "holes," and in certain materials, especially specific semiconductors, the movement of electric current is really caused by these holes. Current flow is also present in the nerve cells located in your brain. Neurons communicate by transmitting voltage pulses that depend on the movement of positive ions (mostly potassium and sodium, and to a lesser extent calcium) across the outer membrane of the neuron. Hence, current can originate from several sources, and circuit theory can be employed to comprehend the manner in which current is influenced by electric fields.

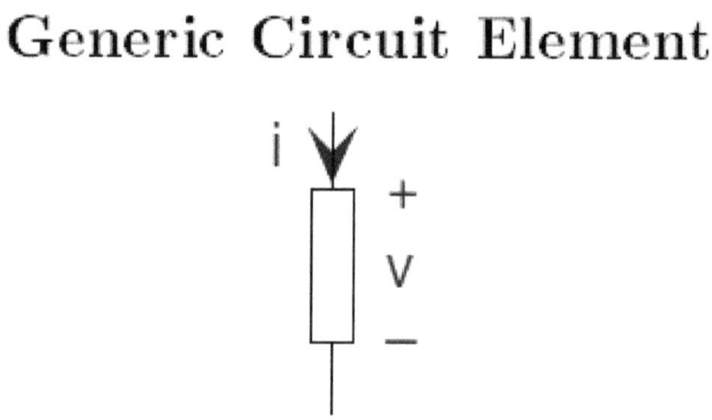


**Figure 3.1: The generic circuit element**

Electric current passes via circuit components, as shown in the diagram above, as well as through conductive materials, which are represented by lines in circuit diagrams. Each circuit element is associated with a specific voltage and current. The v-i relation of the element is determined by its inherent physical qualities. When defining the voltage-current relationship, we follow the convention that positive current flows from a higher voltage to a lower voltage. The unit of voltage is volts, and both the unit and the quantity are named after Volta. Current is measured in amperes, which is a unit named for the French physicist Ampère.

Both voltages and currents are responsible for transmitting electrical power. The convention illustrated in Figure 3.1 (Generic Circuit Element) is used to represent circuit elements. According to this convention, the instantaneous power consumed by an element at any given point is equal to the product of the voltage and current.

$$p(t) = v(t)\, i(t)$$

A positive value for power indicates that at time t the circuit element is consuming power; a negative value means it is producing power. With voltage expressed in volts and current in amperes, power defined this way has units of watts. Just as in all areas of physics and chemistry, power is the rate at which energy is consumed or produced. Consequently, energy is the integral of power.

$$E(t) = \int_{-\infty}^{t} p(\alpha)\, d\alpha$$

Positive energy is equivalent to energy that is used, whereas negative energy refers to energy that is produced. It should be noted that a circuit component with a power profile that alternates between positive and negative values within a specific time period can either consume or generate energy, depending on the sign of the power integral. The energy units are measured in joules, as a watt is equivalent to joules per second.

### Exercise 3.1.1

Residential energy bills typically state a home's energy usage in kilowatt-hours. Is this really a unit of energy? If so, how many joules equals one kilowatt-hour?

## 3.2 IDEAL CIRCUIT ELEMENTS

The elementary circuit elements—the resistor, capacitor, and inductor—impose linear relationships between voltage and current.

### 3.2.1 Resistor

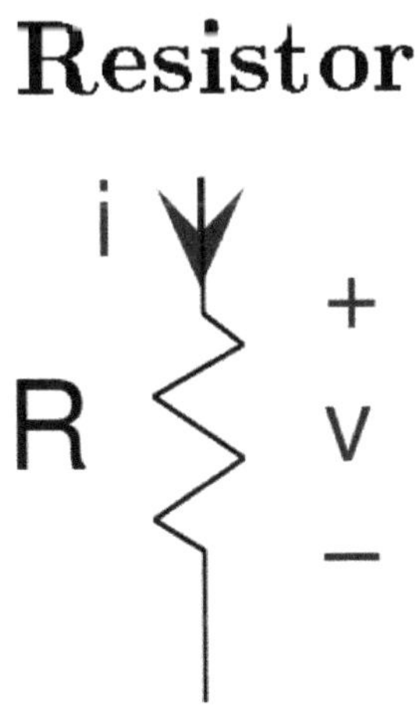


**Figure 3.2: Resistor v = Ri**

The resistor is far and away the simplest circuit element. In a resistor, the voltage is proportional to the current, with the constant of proportionality R, known as the resistance.

$$v(t) = Ri(t)$$

Resistance has units of ohms, denoted by Ω, named for the German electrical scientist Georg Ohm. Sometimes, the v-i relation for the resistor is written i = Gv, with G, the conductance, equal to $\frac{1}{R}$. Conductance has units of Siemens (S), and is named for the German electronics industrialist Werner von Siemens.

When resistance is positive, as it is in most cases, a resistor consumes power. A resistor's instantaneous power consumption can be written one of two ways.

$$p(t) = Ri^2(t) = \frac{1}{R}v^2(t)$$

As the resistance approaches infinity, we have what is known as an open circuit: No current flows but a non-zero voltage can appear across the open circuit. As the resistance becomes zero, the voltage goes to zero for a non-zero current flow. This situation corresponds to a short circuit. A superconductor physically realizes a short circuit.

### 3.2.2 Capacitor

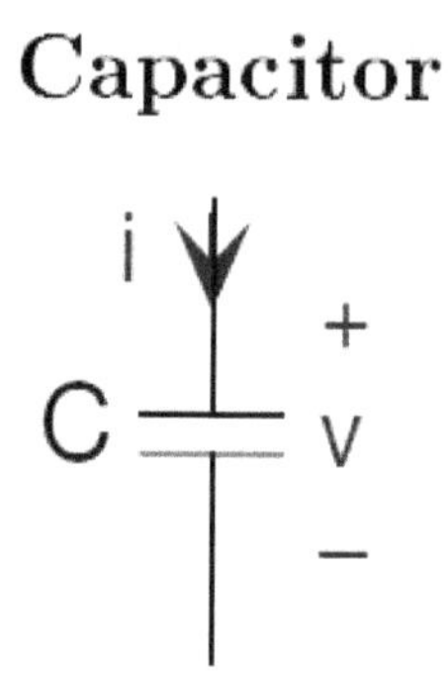


**Figure 3.3: Capacitor** $i = C\frac{dv(t)}{dt}$

The capacitor stores charge and the relationship between the charge stored and the resultant voltage is q = Cv. The constant of proportionality, the capacitance, has units of farads (F), and is named for the English experimental physicist Michael Faraday. As current is the rate of change of charge, the v-i relation can be expressed in differential or integral form.

$$i(t) = C\frac{dv(t)}{dt} \quad \text{or} \quad v(t) = \frac{1}{C}\int_{-\infty}^{t} i(\alpha)\,d\alpha \tag{3.1}$$

If the voltage across a capacitor is constant, then the current flowing into it equals zero. In this situation, the capacitor is equivalent to an open circuit. The power consumed/produced by a voltage applied to a capacitor depends on the product of the voltage and its derivative.

$$p(t) = Cv(t)\frac{dv(t)}{dt}$$

This result means that a capacitor's total energy expenditure up to time t is concisely given by

$$E(t) = \frac{1}{2}Cv^2(t)$$

This expression presumes the fundamental assumption of circuit theory: all voltages and currents in any circuit were zero in the far distant past (t = -∞).

### 3.2.3 Inductor

**Figure 3.4: Inductor** $v = L\frac{di(t)}{dt}$

The inductor stores magnetic flux, with larger valued inductors capable of storing more flux. Inductance has units of henries (H), and is named for the American physicist Joseph Henry. The differential and integral forms of the inductor's v-i relation are

$$v(t) = L\frac{di(t)}{dt} \quad \text{or} \quad i(t) = \frac{1}{L}\int_{-\infty}^{t} v(\alpha)\,d\alpha \tag{3.2}$$

The power consumed/produced by an inductor depends on the product of the inductor current and its derivative and its total energy expenditure up to time t is given by

$$p(t) = Li(t)\frac{di(t)}{dt}$$

$$E(t) = \frac{1}{2}Li^2(t)$$

### 3.2.4 Sources

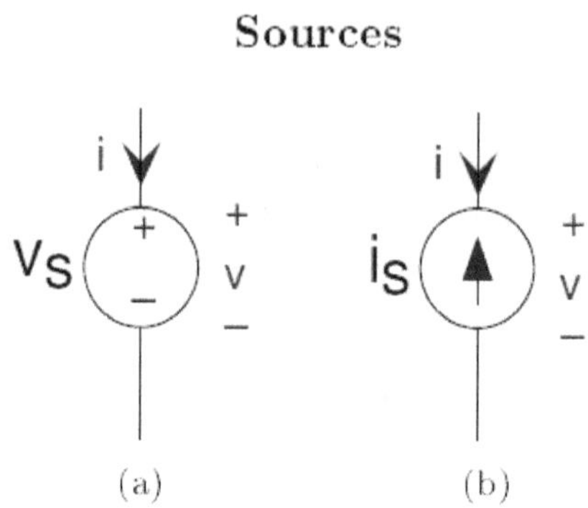


**Figure 3.5: The voltage source on the left and current source on the right are like all circuit elements in that they have a particular relationship between the voltage and current defined for them. For the voltage source, v=vs for any current i, for the current source, i=–is for any voltage v**

Sources of voltage and current are also circuit elements, but they are not linear in the strict sense of linear systems. For example, the voltage source's v-i relation is v=vs regardless of what the current might be. As for the current source, i=-is regardless of the voltage. Another name for a constant-valued voltage source is a battery, and can be purchased in any supermarket. Current sources, on the other hand, are much harder to acquire; we'll learn why later.

## 3.3 IDEAL AND REAL-WORLD CIRCUIT ELEMENTS

Source and linear circuit elements are ideal circuit elements. One central notion of circuit theory is combining the ideal elements to describe how physical elements operate in the real world. For example, the 1 kΩ resistor you can hold in your hand is not exactly an ideal 1 kΩ resistor. First of all, physical devices are manufactured to close tolerances (the tighter the tolerance, the more money you pay), but never have exactly their advertised values. The fourth band on resistors specifies their tolerance; 10% is common. More pertinent to the current discussion is another deviation from the ideal: If a sinusoidal voltage is placed across a physical resistor, the current will not be exactly proportional to it as frequency becomes high, say above 1 MHz. At very high frequencies, the way the resistor is constructed introduces inductance and capacitance effects. Thus, the smart engineer must be aware of the frequency ranges over which his ideal models match reality well.

On the other hand, physical circuit elements can be readily found that well approximate the ideal, but they will always deviate from the ideal in some way. For example, a flashlight battery, like a C-cell, roughly corresponds to a 1.5 V voltage source. However, it ceases to be modeled by a voltage source capable of supplying any current (that›s what ideal ones can do!) when the resistance of the light bulb is too small.

## 3.4 ELECTRIC CIRCUITS AND INTERCONNECTION LAWS

A circuit connects circuit elements together in a specific configuration designed to transform the source signal (originating from a voltage or current source) into another signal—the output—that corresponds to the current or voltage defined for a particular circuit element. A simple resistive circuit is shown in the Figure 3.6. below. This circuit is the electrical embodiment of a system having its input provided by a source system producing vin(t).

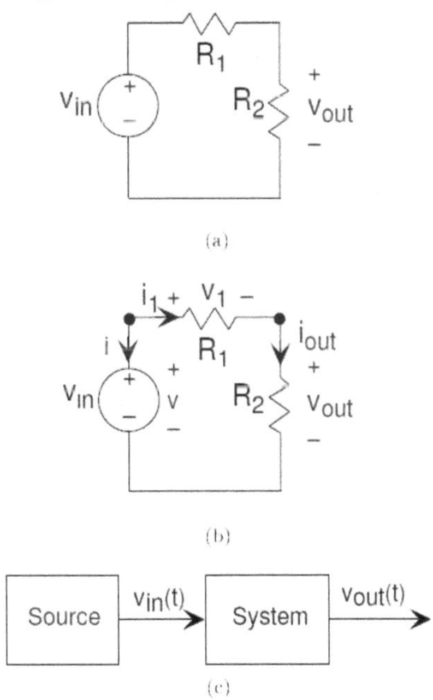


**Figure 3.6: The circuit shown in the top two figures is perhaps the simplest circuit that performs a signal processing function. On the bottom is the block diagram that corresponds to the circuit. The input is provided by the voltage source vin and the output is the voltage vout across the resistor label R2 As shown in the middle, we analyze the circuit—understand what it accomplishes—by defining currents and voltages for all circuit elements, and then solving the circuit and element equations.**

To understand what this circuit accomplishes, we want to determine the voltage across the resistor labeled by its value R2. Recasting this problem mathematically, we need to solve some set of equations so that we relate the output voltage vout to the source voltage. It would be simple—a little too simple at this point—if we could instantly write down the one equation that relates these two voltages. Until we have more knowledge about how circuits work, we must write a set of equations that allow us to find all the voltages and currents that can be defined for every circuit element. Because we have a three-element circuit, we have a total of six voltages and currents that must be either specified or determined. You can define the directions for positive current flow and positive voltage drop any way you like. Once the values for the voltages and currents are calculated, they may be positive or negative according to your definition. When two people define variables according to their individual preferences, the signs of their variables may not agree, but current flow and voltage drop values for each element will agree. Do recall in defining your voltage and current variables that the v-i relations for the elements presume that positive current flow is in the same direction as positive voltage drop. Once you define voltages and currents, we need six nonredundant equations to solve for the six unknown voltages and currents. By specifying the source, we have one; this amounts to providing the source›s v-i relation. The v-i relations for the resistors give us two more. We are only halfway there; where do we get the other three equations we need?

What we need to solve every circuit problem are mathematical statements that express how the circuit elements are interconnected. Said another way, we need the laws that govern the electrical connection of circuit elements. First of all, the places where circuit elements attach to each other are called nodes. Two nodes are explicitly indicated in Figure 3.6, a third is at the bottom where the voltage source and resistor R2 are connected. Electrical engineers tend to draw circuit diagrams—schematics— in a rectilinear fashion. Thus the long line connecting the bottom of the voltage source with the bottom of the resistor is intended to make the diagram look pretty. This line simply means that the two elements are connected together. Kirchhoff's Laws, one for voltage (Section 3.4.2: Kirchhoff's Voltage Law (KVL)) and one for current, determine what a connection among circuit elements means. These laws are essential to analyzing this and any circuit. They are named for Gustav Kirchhoff, a nineteenth century German physicist.

### 3.4.1 Kirchhoff's Current Law (KCL)

At every node, the sum of all currents entering or leaving a node must equal zero. What this law means physically is that charge cannot accumulate in a node; what goes in must come out. For the above example, we have a three-node circuit and thus have three KCL equations.

$$(-i) - i_1 = 0$$
$$i_1 - i_2 = 0$$
$$i + i_2 = 0$$

Note that the current entering a node is the negative of the current leaving the node.

Given any two of these KCL equations, we can find the other by adding or subtracting them. Thus, one of them is redundant and, in mathematical terms, we can discard any one of them. The convention is to discard the equation for the (unlabeled) node at the bottom of the circuit.

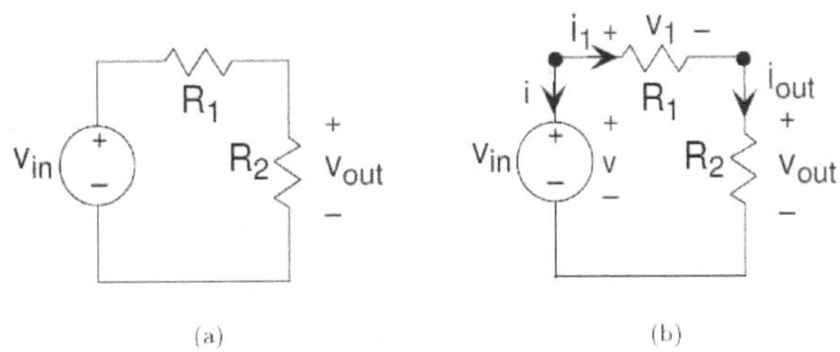


**Figure 3.7: The circuit shown is perhaps the simplest circuit that performs a signal processing function. The input is provided by the voltage source vin and the output is the voltage vout across the resistor labelled R2**

### Exercise 3.4.1

In writing KCL equations, you will find that in an n-node circuit, exactly one of them is always redundant. Can you sketch a proof of why this might be true? Hint: It has to do with the fact that charge won't accumulate in one place on its own.

### 3.4.2 Kirchhoff's Voltage Law (KVL)

The voltage law says that the sum of voltages around every closed loop in the circuit must equal zero. A closed loop has the obvious definition: Starting at a node, trace a path through the circuit that returns you to the origin node. KVL expresses the fact that electric fields are conservative: The total work performed in moving a test charge around a closed path is zero. The KVL equation for our circuit is:

$$v_1 + v_2 - v = 0$$

In writing KVL equations, we follow the convention that an element's voltage enters with a plus sign when traversing the closed path, we go from the positive to the negative of the voltage's definition.

For the circuit in Figure 3.7, we have three v-i relations, two KCL equations, and one KVL equation for solving for the circuit›s six voltages and currents.

$$\begin{aligned} \text{v-i:} \quad & v = v_{in} \\ & v_1 = R_1 i_1 \\ & v_{out} = R_2 i_{out} \\ \text{KCL:} \quad & (-i) - i_1 = 0 \\ & i_1 - i_{out} = 0 \\ \text{KVL:} \quad & -v + v_1 + v_{out} = 0 \end{aligned}$$

We have exactly the right number of equations! Eventually, we will discover shortcuts for solving circuit problems; for now, we want to eliminate all the variables but vout and determine how it depends on vin and on resistor values. The KVL equation can be rewritten as: vin=v1+vout. Substituting into it the resistor's v-i relation, we have: vin=R1i1+R2iout. Yes, we temporarily eliminate the quantity we seek. Though not obvious, it is the simplest way to solve the equations. One of the KCL equations says: i1=iout ,which means that: vin=R1iout+R2iout=(R1+R2) iout ..Solving for the current in the output resistor, we have: $i_{out} = \frac{v_{in}}{R_1+R_2}$. We have now solved the circuit: We have expressed one voltage or current in terms of sources and circuit-element values. To find any other circuit quantities, we can back substitute this answer into our original equations or ones we developed along the way. Using the v-i relation for the output resistor, we obtain the quantity we seek:

$$v_{out} = \frac{R_2}{R_1 + R_2} v_{in}$$

**Exercise 3.4.2**

Referring back to Figure 3.6, a circuit should serve some useful purpose. What kind of system does our circuit realize and, in terms of element values, what are the system's parameter(s)?

## 3.5 POWER DISSIPATION IN RESISTOR CIRCUITS

In circuits that consist of resistors and either voltage or current sources, it is possible to determine the values of voltages and currents. We need to assess whether the variables of these circuits adhere to the concept of Conservation of Power: as a closed system, a circuit should neither dissipate nor generate energy. For the moment, our approach is to investigate first a resistor circuit's power consumption/creation. Later, we will prove that because of KVL and KCL all circuits conserve power.

According to the definition provided on [link], the instantaneous power consumed or created by each circuit element is equal to the multiplication of its voltage and current. The total power consumed or generated by a circuit is equal to the sum of the power of each individual element.

$$P = \sum_k v_k i_k$$

Recall that each element's current and voltage must obey the convention that positive current is defined to enter the positive-voltage terminal. With this convention, a positive value of vkik corresponds to consumed power, a negative value to created power. Because the total power in a circuit must be zero (P = 0), some circuit elements must create power while others consume it.

Consider the simple series circuit should in [link]. In performing our calculations, we defined the current iout to flow through the positive-voltage terminals of both resistors and found it to equal: iout= $\frac{v_{\text{in}}}{R_1+R_2}$. The voltage across the resistor R2 is the output voltage and we found it to equal: $v_{\text{out}} = \frac{R_2}{R_1+R_2} v_{\text{in}}$.

Consequently, calculating the power for this resistor yields

$$P_2 = \frac{R_2}{(R_1+R_2)^2} {v_{\text{in}}}^2$$

Consequently, this resistor dissipates power because P2 is positive. This result should not be surprising since we showed that the power consumed by any resistor equals either of the following:

$$\frac{v^2}{R} \text{ or } i^2 R \tag{3.3}$$

Since resistors are positive-valued, resistors always dissipate power. But where does a resistor's power go? By Conservation of Power, the dissipated power must be absorbed somewhere. The answer is not directly predicted by circuit theory, but is by physics. Current flowing through a resistor makes it hot; its power is dissipated by heat.

Note: A physical wire has a resistance and hence dissipates power (it gets warm just like a resistor in a circuit). In fact, the resistance of a wire of length L and cross-sectional area A is given by:

$$R = \frac{\rho L}{A}$$

The quantity ρ is known as the resistivity and presents the resistance of a unit-length, unit cross-sectional area material constituting the wire. Resistivity has units of ohm-meters. Most materials have a positive value for ρ, which means the longer the wire, the greater the resistance and thus the power dissipated. The thicker the wire, the smaller the resistance. Superconductors have zero resistivity and hence do not dissipate power. If a room-temperature superconductor could be found, electric power could be sent through power lines without loss!

### Exercise 3.5.1

Calculate the power consumed/created by the resistor R1 in our simple circuit example.

We conclude that both resistors in our example circuit consume power, which points to the voltage source as the producer of power. The current flowing into the source›s positive terminal is -iout. Consequently, the power calculation for the source yields:

$$-\left(v_{\text{in}} i_{\text{out}}\right) = -\left(\frac{1}{R_1 + R_2} {v_{\text{in}}}^2\right)$$

We conclude that the source provides the power consumed by the resistors, no more, no less.

### Exercise 3.5.2

Confirm that the source produces exactly the total power consumed by both resistors.

This result is quite general: sources produce power and the circuit elements, especially resistors, consume it. But where do sources get their power? Again, circuit theory does not model how sources are constructed, but the theory decrees that all sources must be provided energy to work.

## 3.6 SERIES AND PARALLEL CIRCUITS

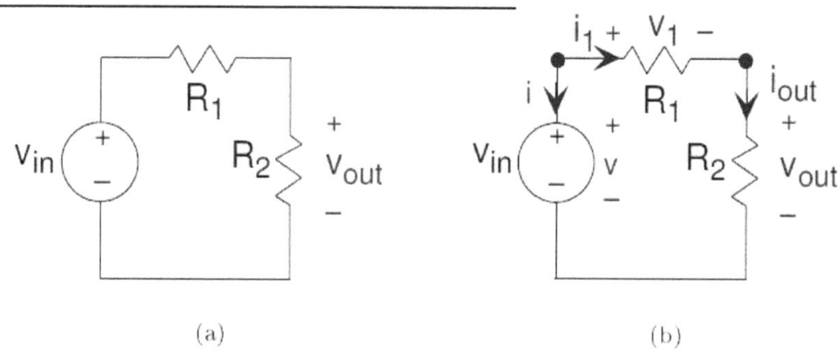


**Figure 3.8: The circuit shown is perhaps the simplest circuit that performs a signal processing function. The input is provided by the voltage source vin and the output is the voltage vout across the resistor labeled R2**

The results shown in other modules (circuit elements, (Section 3.4) KVL and KCL, interconnection laws (Section 3.4)) with regard to the above circuit (Figure

3.8) and the values of other currents and voltages in this circuit as well, have profound implications.

Resistors connected in such a way that current from one must flow only into another—currents in all resistors connected this way have the same magnitude—are said to be connected in series. For the two series-connected resistors in the example, the voltage across one resistor equals the ratio of that resistor's value and the sum of resistances times the voltage across the series combination. This concept is so pervasive it has a name: voltage divider.

The input-output relationship for this system, found in this particular case by voltage divider, takes the form of a ratio of the output voltage to the input voltage.

$$\frac{v_{\text{out}}}{v_{\text{in}}} = \frac{R_2}{R_1 + R_2}$$

In this way, we express how the components used to build the system affect the input-output relationship. Because this analysis was made with ideal circuit elements, we might expect this relation to break down if the input amplitude is too high (Will the circuit survive if the input changes from 1 volt to one million volts?) or if the source's frequency becomes too high. In any case, this important way of expressing input-output relationships—as a ratio of output to input—pervades circuit and system theory.

The current i1 is the current flowing out of the voltage source. Because it equals i2, we have that $\frac{v_{\text{in}}}{i_1} = R_1 + R_2$:

Resistors in Series: The series combination of two resistors acts, as far as the voltage source is concerned, as a single resistor having a value equal to the sum of the two resistances. This result is the first of several equivalent service ideas: In many cases, a complicated circuit when viewed from its terminals (the two places to which you might attach a source) appears to be a single circuit element (at best) or a simple combination of elements at worst. Thus, the equivalent circuit for a series combination of resistors is a single resistor having a resistance equal to the sum of its component resistances.

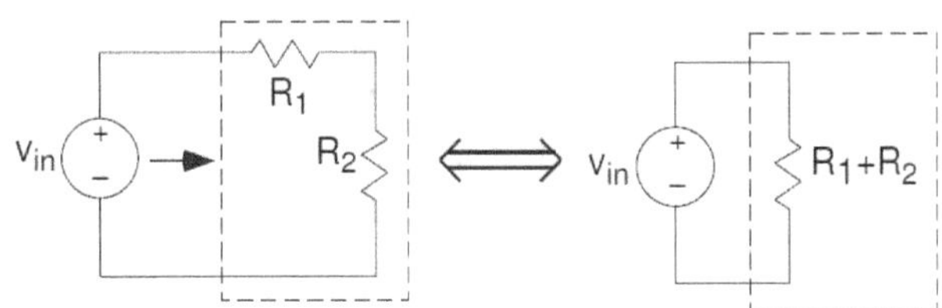


**Figure 3.9: The resistor (on the right) is equivalent to the two resistors (on the left) and has a resistance equal to the sum of the resistances of the other two resistors.**

Thus, the circuit the voltage source "feels" (through the current drawn from it) is a single resistor having resistance R1+R2. Note that in making this equivalent circuit, the output voltage can no longer be defined: The output resistor labeled R2 no longer appears. Thus, this equivalence is made strictly from the voltage source›s viewpoint.

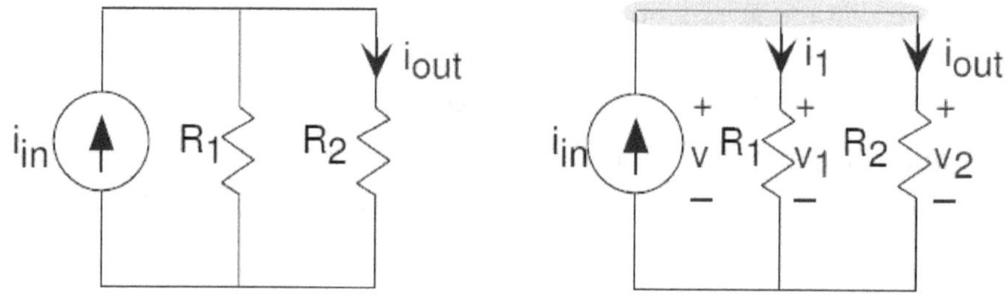


**Figure 3.10: A simple parallel circuit**

One interesting simple circuit in Figure 3.10 has two resistors connected side-by-side, what we will term a parallel connection, rather than in series. Here, applying KVL reveals that all the voltages are identical: v1=v and v2=v. This result typifies parallel connections. To write the KCL equation, note that the top node consists of the entire upper interconnection section. The KCL equation is: iin–i1–i2=0. Using the v-i relations, we find that

$$i_{\text{out}} = \frac{R_1}{R_1 + R_2} i_{\text{in}}$$

## Exercise 3.6.1

Suppose that you replaced the current source in Figure 3.10 by a voltage source. How would iout be related to the source voltage? Based on this result, what purpose does this revised circuit have?

This result does not depend on the resistor R1, which means that we simply have a resistor (R2) across a voltage source. The two-resistor circuit has no apparent use.

This circuit highlights some important properties of parallel circuits. You can easily show that the parallel combination of $R_1$ and $R_2$ has the v-i relation of a resistor having resistance : $\left(\frac{1}{R_1} + \frac{1}{R_2}\right)^{-1} = \frac{R_1 R_2}{R_1 + R_2}$. A shorthand notation for this quantity is $R_1 \| R_2$. As the reciprocal of resistance is conductance, we can say that for a parallel combination of resistors, the equivalent conductance is the sum of the conductances. (Section 3.2.1: Resistor), we can say that for a parallel combination of resistors, the equivalent conductance is the sum of the conductances.

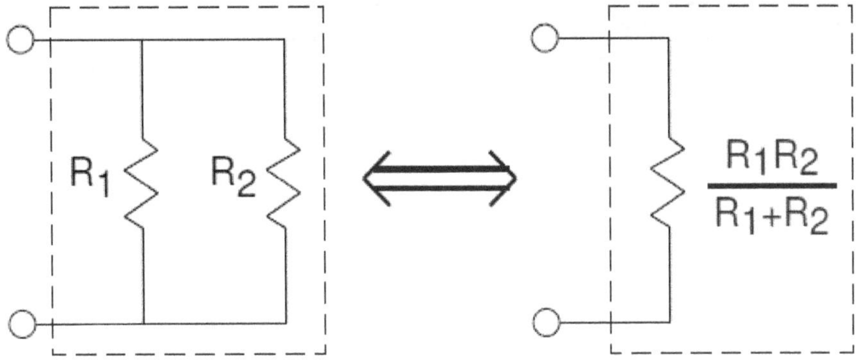


**Figure 3.11: Parallel Resistors**

Similar to voltage divider (p. 48) for series resistances, we have current divider for parallel resistances. The current through a resistor in parallel with another is the ratio of the conductance of the first to the sum of the conductances. Thus, for the depicted circuit, $i_2 = \frac{G_2}{G_1+G_2}i$. Expressed in terms of resistances, current divider takes the form of the resistance of the other resistor divided by the sum of resistances: $i_2 = \frac{R_1}{R_1+R_2}i$.

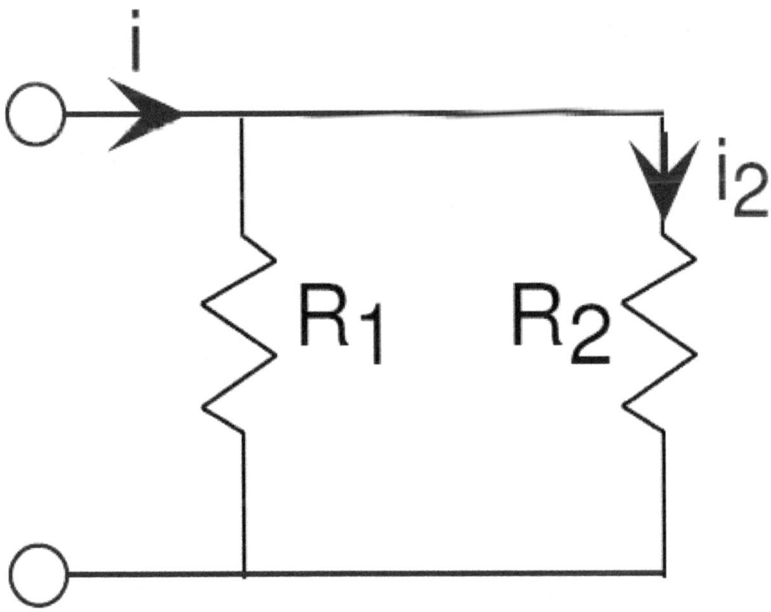


**Figure 3.12**

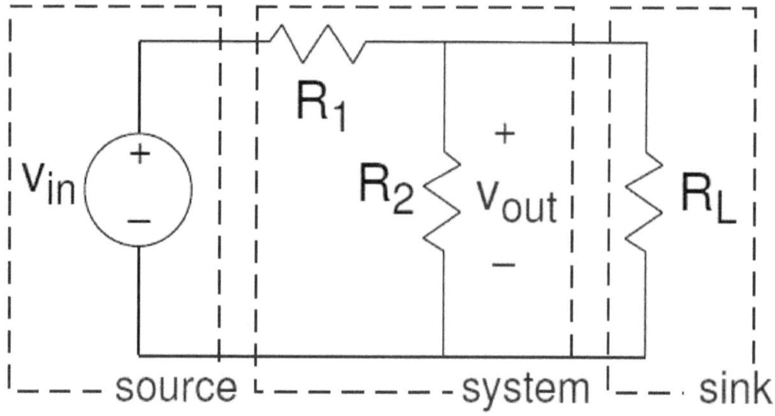


**Figure 3.13: The Simple Attenuator Circuit is attached to an oscilloscope's input. The input-output relation for the above circuit without a load is:** $v_{\text{out}} = \frac{R_2}{R_1+R_2}v_{\text{in}}$.

Suppose we want to pass the output signal into a voltage measurement device, such as an oscilloscope or a voltmeter. In system-theory terms, we want to pass our circuit's output to a sink. For most applications, we can represent these measurement devices as a resistor, with the current passing through it driving the measurement device through some type of display. In circuits, a sink is called a load; thus, we describe a system-theoretic sink as a load resistance RL. Thus, we have a complete system built from a cascade of three systems: a source, a signal processing system (simple as it is), and a sink.

We must analyze afresh how this revised circuit, shown in Figure 3.13 works. Rather than defining eight variables and solving for the current in the load resistor, let's take a hint from other analysis (series rules (p. 48), parallel rules (p. 49)). Resistors $R_2$ and $R_L$ are in a parallel configuration: The voltages across each resistor are the same while the currents are not. Because the voltages are the same, we can find the current through each from their v-i relations: $i_2 = \frac{v_{\text{out}}}{R_2}$ and $i_L = \frac{v_{\text{out}}}{R_L}$. Considering the node where all three resistors join, KCL says that the sum of the three currents must equal zero. Said another way, the current entering the node through $R_1$ must equal the sum of the other two currents leaving the node. Therefore, i1 = i2 +iL, which means that $i_1 = v_{\text{out}}\left(\frac{1}{R_2}+\frac{1}{R_L}\right)$.

Let Req denote the equivalent resistance of the parallel combination of $R_2$ and $R_L$. Using $R_1$ ‹s v-i relation, the voltage across it is $v_1 = \frac{R_1 v_{\text{out}}}{R_{\text{eq}}}$. The KVL equation written around the leftmost loop has $v_{in}=v_1+v_{out}$; Substituting for $v_1$, we find

$$v_{\text{in}} = v_{\text{out}}\left(\frac{R_1}{R_{\text{eq}}}+1\right)$$

Or

$$\frac{v_{\text{out}}}{v_{\text{in}}} = \frac{R_{\text{eq}}}{R_1 + R_{\text{eq}}}$$

Thus, we have the input-output relationship for our entire system having the form of voltage divider, but it does not equal the input-output relation of the circuit without the voltage measurement device. We can not measure voltages reliably unless the measurement device has little effect on what we are trying to measure. We should look more carefully to determine if any values for the load resistance would lessen its impact on the circuit. Comparing the input-output relations before and after, what we need is $R_{eq} \approx R_2$.

As $R_{\text{eq}} = \left(\frac{1}{R_2} + \frac{1}{R_L}\right)^{-1}$, The approximation would apply if $\frac{1}{R_2} \gg \frac{1}{R_L}$ or $R_2 \ll R_L$. This is the condition we seek:

Voltage measurement : Voltage measurement devices must have large resistances compared with that of the resistor across which the voltage is to be measured.

**Exercise 3.6.2**

Let's be more precise: How much larger would a load resistance need to be to affect the input-output relation by less than 10%? by less than 1%?

**Exercise 3.1**

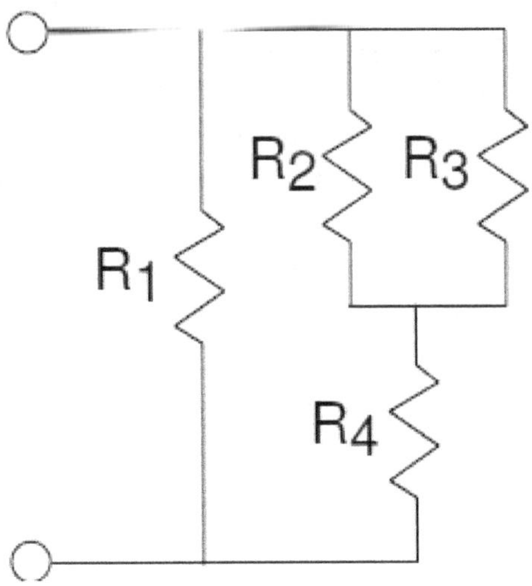


**Figure 3.14**

We want to find the total resistance of the example circuit. To apply the series and parallel combination rules, it is best to first determine the circuit's structure:

What is in series with what and what is in parallel with what at both small- and large-scale views. We have R2 in parallel with R3; this combination is in series with R4. This series combination is in parallel with R1. Note that in determining this structure, we started away from the terminals, and worked toward them. In most cases, this approach works well; try it first. The total resistance expression mimics the structure:

$$R_T = R_1 \parallel (R_2 \parallel R_3 + R_4)$$

$$R_T = \frac{R_1 R_2 R_3 + R_1 R_2 R_4 + R_1 R_3 R_4}{R_1 R_2 + R_1 R_3 + R_2 R_3 + R_2 R_4 + R_3 R_4}$$

Such complicated expressions typify circuit "simplifications." A simple check for accuracy is the units: Each component of the numerator should have the same units (here Ω3) as well as in the denominator(Ω2). The entire expression is to have units of resistance; thus, the ratio of the numerator's and denominator's units should be ohms. Checking units does not guarantee accuracy, but can catch many errors.

Another valuable lesson emerges from this example concerning the difference between cascading systems and cascading circuits. In system theory, systems can be cascaded without changing the input-output relation of intermediate systems. In cascading circuits, this ideal is rarely true unless the circuits are so designed. Design is in the hands of the engineer; he or she must recognize what have come to be known as loading effects. In our simple circuit, you might think that making the resistance RL large enough would do the trick. Because the resistors R1 and R2 can have virtually any value, you can never make the resistance of your voltage measurement device big enough. Said another way, a circuit cannot be designed in isolation that will work in cascade with all other circuits. Electrical engineers deal with this situation through the notion of specifications: Under what conditions will the circuit perform as designed? Thus, you will find that oscilloscopes and voltmeters have their internal resistances clearly stated, enabling you to determine whether the voltage you measure closely equals what was present before they were attached to your circuit. Furthermore, since our resistor circuit functions as an attenuator, with the attenuation (a fancy word for gains less than one) depending only on the ratio of the two resistor values: $\frac{R_2}{R_1+R_2} = \left(1 + \frac{R_1}{R_2}\right)^{-1}$, We can select any values for the two resistances we want to achieve the desired attenuation. The designer of this circuit must thus specify not only what the attenuation is, but also the resistance values employed so that integrators—people who put systems together from component systems—can combine systems together and have a chance of the combination working.

Figure 3.15 below, summarizes the series and parallel combination results. These results are easy to remember and very useful. Keep in mind that for series combinations, voltage and resistance are the key quantities, while for parallel combinations current and conductance are more important. In series combinations, the currents through each element are the same; in parallel ones, the voltages are the same.

series and parallel combination rules

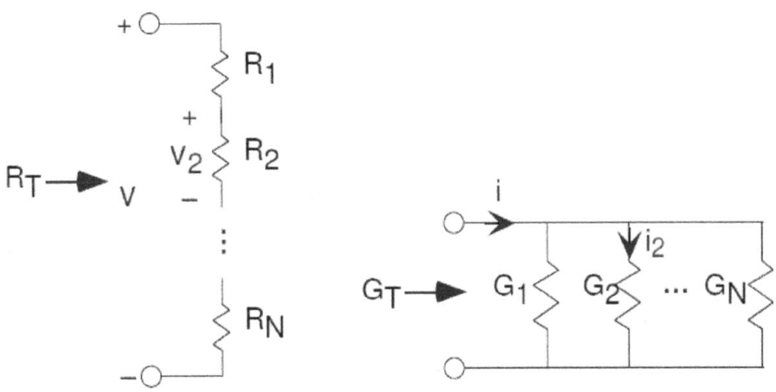


(a) series combination rule (b) parallel combination rule

**Figure 3.15: Series and parallel combination rules** (a) $R_T = \sum_{n=1}^{N} R_n$ $v_n = \frac{R_n}{R_T} v$ (b) $G_T = \sum_{n=1}^{N} G_n$ $i_n = \frac{G_n}{G_T} i$

**Exercise 3.6.3**

Contrast a series combination of resistors with a parallel one. Which variable (voltage or current) is the same for each and which differs? What are the equivalent resistances? When resistors are placed in series, is the equivalent resistance bigger, in between, or smaller than the component resistances? What is this relationship for a parallel combination?

## 3.7 EQUIVALENT CIRCUITS: RESISTORS AND SOURCES

We have found that the way to think about circuits is to locate and group parallel and series resistor combinations. Those resistors not involved with variables of interest can be collapsed into a single resistance. This result is known as an equivalent circuit: from the viewpoint of a pair of terminals, a group of resistors functions as a single resistor, the resistance of which can usually be found by applying the parallel and series rules.

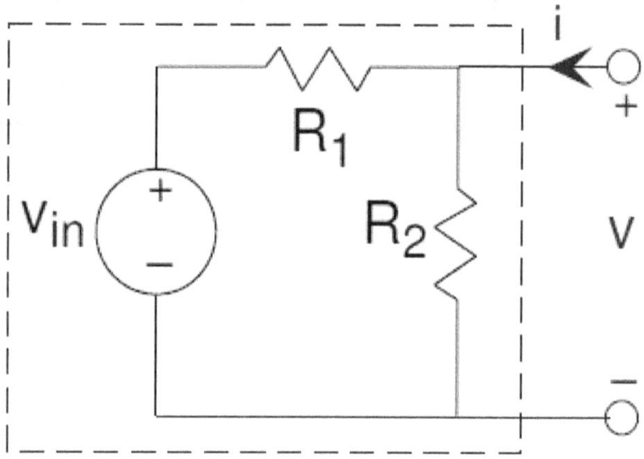


**Figure 3.16: Simple Attenuator Circuit**

This result generalizes to include sources in a very interesting and useful way. Let's consider our simple attenuator circuit (Shown in figure (Figure 3.16) from the viewpoint of the output terminals. We want to find the v-i relation for the output terminal pair, and then find the equivalent circuit for the boxed circuit. To perform this calculation, use the circuit laws and element relations, but do not attach anything to the output terminals. We seek the relation between v and i that describes the kind of element that lurks within the dashed box. The result is

$$v = (R_1 \parallel R_2)\, i + \frac{R_2}{R_1 + R_2} v_{\text{in}} \tag{3.4}$$

If the source were zero, it could be replaced by a short circuit, which would confirm that the circuit does indeed function as a parallel combination of resistors. However, the source's presence means that the circuit is not well modeled as a resistor.

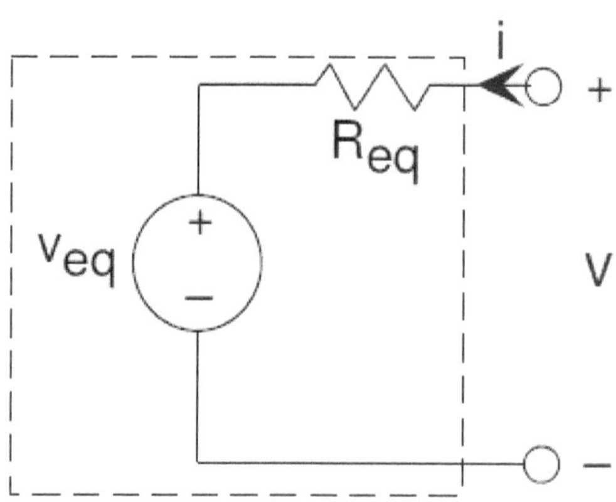


**Figure 3.17: The Thévenin equivalent circuit**

If we consider the simple circuit of Figure 3.17, we find it has the v-i relation at its terminals of

$$v = R_{\text{eq}} i + v_{\text{eq}} \tag{3.5}$$

Comparing the two v-i relations, we find that they have the same form. In this case the Thévenin equivalent resistance is: Req=$R_1 \| R_2$ and the Thévenin equivalent source has voltage: veq= $\frac{R_2}{R_1+R_2} v_{\text{in}}$. Thus, from viewpoint of the terminals, you cannot distinguish the two circuits. Because the equivalent circuit has fewer elements, it is easier to analyze and understand than any other alternative.

For any circuit containing resistors and sources, the v-i relation will be of the form and the Thévenin equivalent circuit for any such circuit is that of Figure 3.17. This equivalence applies no matter how many sources or resistors may be present in the circuit. In the example below, we know the circuit›s construction and element values, and derive the equivalent source and resistance. Because Thévenin's theorem applies in general, we should be able to make measurements or calculations only from the terminals to determine the equivalent circuit.

$$v = R_{\text{eq}} i + v_{\text{eq}} \tag{3.6}$$

To be more specific, consider the equivalent circuit in Figure 3.17. Let the terminals be open-circuited, which has the effect of setting the current i to zero. Because no current flows through the resistor, the voltage across it is zero (remember, Ohm's Law says that (v=iR). Consequently, by applying KVL we have that the so-called open-circuit voltage voc equals the Thévenin equivalent voltage. Now consider the situation when we set the terminal voltage to zero (short-circuit it) and measure the resulting current. Referring to the equivalent circuit, the source voltage now appears entirely across the resistor, leaving the short-circuit current to be: $i_{\text{sc}} = -\frac{v_{\text{eq}}}{R_{\text{eq}}}$. From this property, we can determine the equivalent resistance.

$$v_{\text{eq}} = v_{\text{oc}} \tag{3.7}$$

$$R_{\text{eq}} = -\frac{v_{\text{oc}}}{i_{\text{sc}}} \tag{3.8}$$

**Exercise 3.7.1**

Use the open/short-circuit approach to derive the Thévenin equivalent of the circuit shown in Figure 3.18.

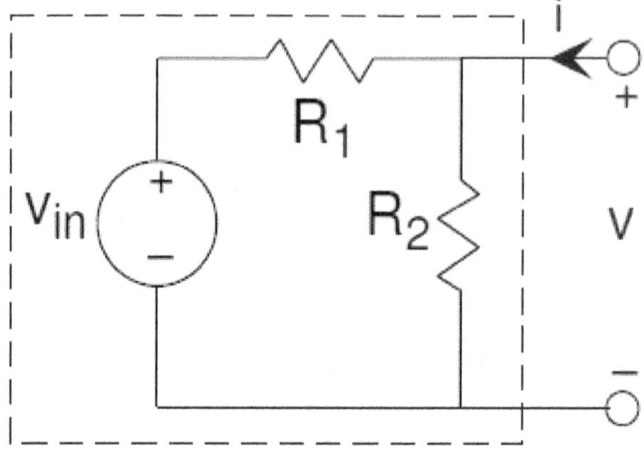


Figure 3.18

**Example 3.2**

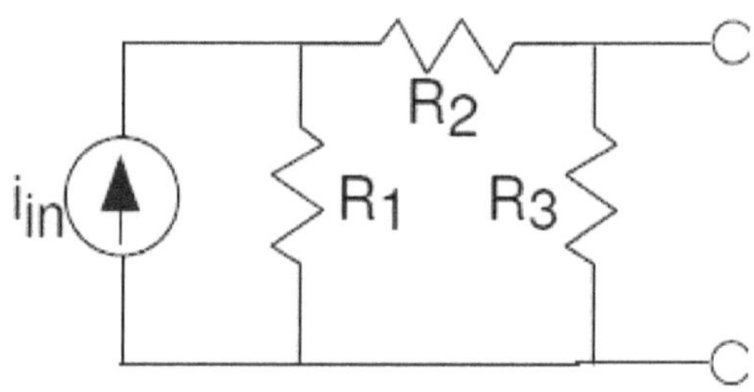


Figure 3.19

For the circuit depicted in the above figure, let's derive its Thévenin equivalent two different ways. Starting with the open/short-circuit approach, let's first find the open-circuit voltage voc. We have a current divider relationship as $R_1$ is in parallel with the series combination of $R_2$ and $R_3$. Thus $v_{oc} = \frac{i_{in} R_3 R_1}{R_1 + R_2 + R_3}$. When we short-circuit the terminals, no voltage appears across $R_3$, and thus no current flows through it. In short, $R_3$ does not affect the short-circuit current, and can be eliminated. We again have a current divider relationship: $i_{sc} = -\frac{i_{in} R_1}{R_1 + R_2}$. Thus, the Thévenin equivalent resistance is $\frac{R_3 (R_1 + R_2)}{R_1 + R_2 + R_3}$.

To verify, let's find the equivalent resistance by reaching inside the circuit and setting the current source to zero. Because the current is now zero, we can replace the current source by an open circuit. From the viewpoint of the terminals, resistor R3 is now in parallel with the series combination of $R_1$ and $R_2$. Thus $R_{eq} = R_3 \| R_1 + R_2$, and we obtain the same result.

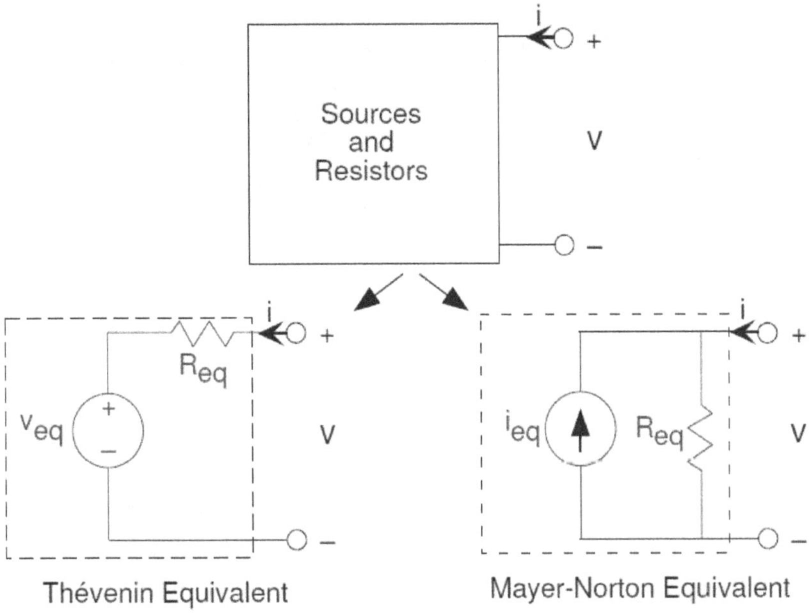


**Figure 3.20: All circuits containing sources and resistors can be described by simpler equivalent circuits. Choosing the one to use depends on the application, not on what is actually inside the circuit.**

As you might expect, equivalent circuits come in two forms: the voltage-source oriented Thévenin equivalent and the current-source oriented Mayer-Norton equivalent (Figure 3.20). To derive the latter, the v-i relation for the Thévenin equivalent can be written as:

$$v = R_{eq}i + v_{eq} \tag{3.9}$$

or

$$i = \frac{v}{R_{eq}} - i_{eq} \tag{3.10}$$

Where $i_{eq} = \frac{v_{eq}}{R_{eq}}$ is the Mayer-Norton equivalent source. The Mayer-Norton equivalent shown in Figure 3.20 can be easily shown to have this v-i relation. Note that both variations have the same equivalent resistance. The short-circuit current equals the negative of the Mayer-Norton equivalent source.

## Exercise 3.7.2

Find the Mayer-Norton equivalent circuit for the circuit below.

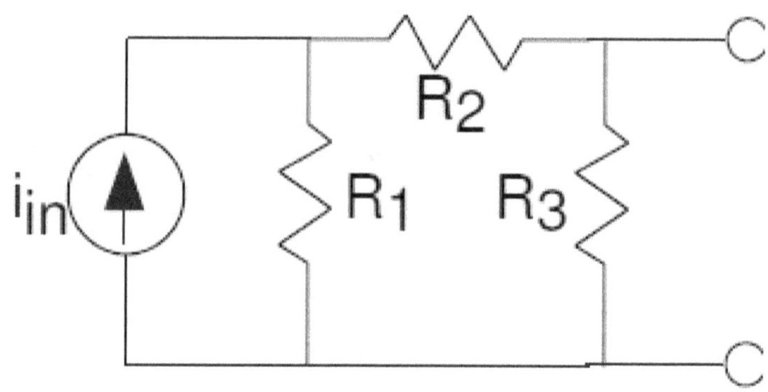


**Figure 3.21**

Equivalent circuits can be used in two basic ways. The first is to simplify the analysis of a complicated circuit by realizing the any portion of a circuit can be described by either a Thévenin or Mayer-Norton equivalent. Which one is used depends on whether what is attached to the terminals is a series configuration (making the Thévenin equivalent the best) or a parallel one (making Mayer-Norton the best).

Another application is modeling. When we buy a flashlight battery, either equivalent circuit can accurately describe it. These models help us understand the limitations of a battery. Since batteries are labeled with a voltage specification, they should serve as voltage sources and the Thévenin equivalent serves as the natural choice. If a load resistance RL is placed across its terminals, the voltage output can be found using voltage divider: $v = \frac{v_{eq}R_L}{R_L+R_{eq}}$. If we have a load resistance much larger than the battery's equivalent resistance, then, to a good approximation, the battery does serve as a voltage source. If the load resistance is much smaller, we certainly don't have a voltage source (the output voltage depends directly on the load resistance). Consider now the Mayer-Norton equivalent; the current through the load resistance is given by current divider, and equals: $i = -\frac{i_{eq}R_{eq}}{R_L+R_{eq}}$. For a current that does not vary with the load resistance, this resistance should be much smaller than the equivalent resistance. If the load resistance is comparable to the equivalent resistance, the battery serves neither as a voltage source or a current course. Thus, when you buy a battery, you get a voltage source if its equivalent resistance is much smaller than the equivalent resistance of the circuit to which you attach it. On the other hand, if you attach it to a circuit having a small equivalent resistance, you bought a current source.

Leon Charles Thevenin: He was an engineer with France's Postes, Télégraphe et Téléphone. In 1883, he published (twice!) a proof of what is now called the Thévenin equivalent while developing ways of teaching electrical engineering concepts at the École Polytechnique. He did not realize that the same result had been published by Hermann Helmholtz, the renowned nineteenth century physicist, thiry years earlier.

Hans Ferdinand Mayer: After earning his doctorate in physics in 1920, he turned to communications engineering when he joined Siemens & Halske in 1922. In 1926, he published in a German technical journal the Mayer-Norton equivalent. During his interesting career, he rose to lead Siemen's Central Laboratory in 1936, surruptiously leaked to the British all he knew of German warfare capabilities a month after the Nazis invaded Poland, was arrested by the Gestapo in 1943 for listening to BBC radio broadcasts, spent two years in Nazi concentration camps, and went to the United States for four years working for the Air Force and Cornell University before returning to Siemens in 1950. He rose to a position on Siemen's Board of Directors before retiring.

Edward L Norton: Edward Norton was an electrical engineer who worked at Bell Laboratory from its inception in 1922. In the same month when Mayer›s paper appeared, Norton wrote in an internal technical memorandum a paragraph describing the current-source equivalent. No evidence suggests Norton knew of Mayer›s publication.

## 3.8 CIRCUITS WITH CAPACITORS AND INDUCTORS

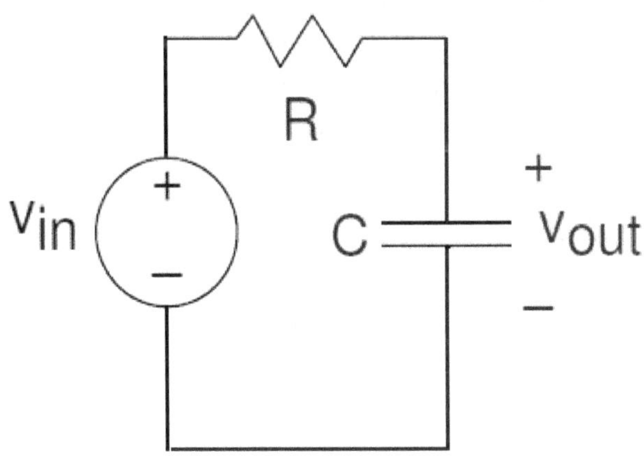


**Figure 3.22: A simple RC circuit**

Let's consider a circuit having something other than resistors and sources. Because of KVL, we know that: vin=vR+vout.The current through the capacitor is given by:$i = C\frac{dv_{out}}{dt}$,and this current equals that passing through the resistor. Substituting: vR=Ri into the KVL equation and using the v-i relation for the capacitor, we arrive at

$$RC\frac{dv_{\text{out}}}{dt} + v_{\text{out}} = v_{\text{in}} \tag{3.11}$$

The input-output relation for circuits involving energy storage elements takes the form of an ordinary differential equation, which we must solve to determine what the output voltage is for a given input. In contrast to resistive circuits, where we obtain an explicit input-output relation, we now have an implicit relation that requires more work to obtain answers.

At this point, we could learn how to solve differential equations. Note first that even finding the differential equation relating an output variable to a source is often very tedious. The parallel and series combination rules that apply to resistors don't directly apply when capacitors and inductors occur. We would have to slog our way through the circuit equations, simplifying them until we finally found the equation that related the source(s) to the output. At the turn of the twentieth century, a method was discovered that not only made finding the differential equation easy, but also simplified the solution process in the most common situation. Although not original with him, Charles Steinmetz presented the key paper describing the impedance approach in 1893. It allows circuits containing capacitors and inductors to be solved with the same methods we have learned to solved resistor circuits. To use impedances, we must master complex numbers. Though the arithmetic of complex numbers is mathematically more complicated than with real numbers, the increased insight into circuit behavior and the ease with which circuits are solved with impedances is well worth the diversion. But more importantly, the impedance concept is central to engineering and physics, having a reach far beyond just circuits.

## 3.9 THE IMPEDANCE CONCEPT

Rather than solving the differential equation that arises in circuits containing capacitors and inductors, let's pretend that all sources in the circuit are complex exponentials having the same frequency. Although this pretense can only be mathematically true, this fiction will greatly ease solving the circuit no matter what the source really is.

Simple Circuit

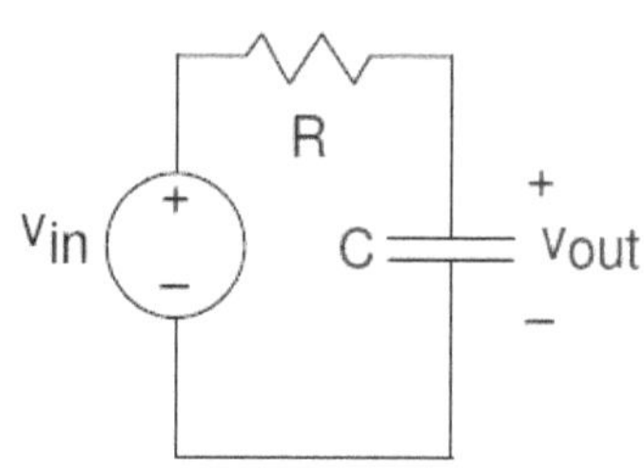


**Figure 3.23: A simple RC circuit**

For the above example RC circuit, (Figure 3.23 (Simple Circuit) let vin=Vinei2πft The complex amplitude Vin determines the size of the source and its phase. The critical consequence of assuming that sources have this form is that all voltages and currents in the circuit are also complex exponentials, having amplitudes governed by KVL, KCL, and the v-i relations and the same frequency as the source. To appreciate why this should be true, let›s investigate how each circuit element behaves when either the voltage or current is a complex exponential. For the resistor, v=Ri. When v=Vei2πft;Then $i = \frac{V}{R}e^{j2\pi ft}$. Thus, if the resistor's voltage is a complex exponential, so is the current, with an amplitude $I = \frac{V}{R}$ (determined by the resistor's v-i relation) and a frequency the same as the voltage. Clearly, if the current were assumed to be a complex exponential, so would the voltage. For a capacitor, $i = C\frac{dv}{dt}$. Letting the voltage be a complex exponential, we have: i=CVj2πfei2πft . The amplitude of this complex exponential is: I=CVj2πf.Finally, for the inductor, where $v = L\frac{di}{dt}$, Assuming the current to be a complex exponential results in the voltage having the form: v=LIj2πfei2πft, making its complex amplitude V=LIj2πf.

The major consequence of assuming complex exponential voltage and currents is that the ratio $Z = \frac{V}{I}$for each element does not depend on time, but does depend on source frequency. This quantity is known as the element's impedance.

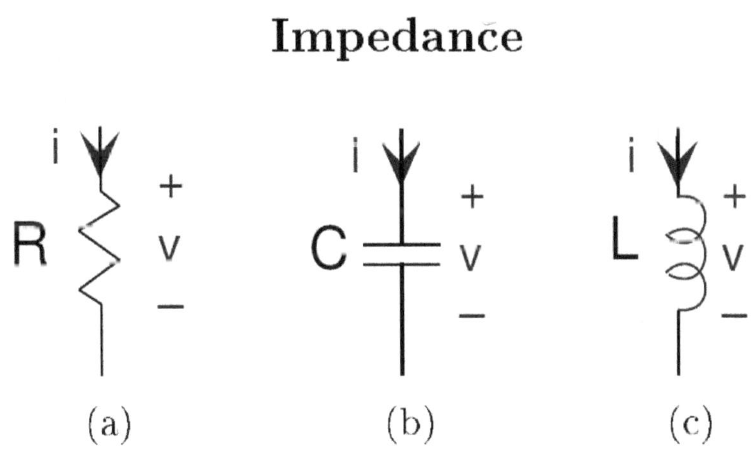


**Figure 3.24: (a) Resistor: ZR=R (b) Capacitor:** $Z_C = \frac{1}{j2\pi fC}$ **(c) Inductor: ZL=j2πf L**

The impedance is, in general, a complex-valued, frequency-dependent quantity. For example, the magnitude of the capacitor's impedance is inversely related to frequency, and has a phase of $-\frac{\pi}{2}$. This observation means that if the current is a complex exponential and has constant amplitude, the amplitude of the voltage decreases with frequency.

Let's consider Kirchoff's circuit laws. When voltages around a loop are all complex exponentials of the same frequency, we have:

$$\begin{aligned} \sum_{nn} v_n &= \sum_{nn} V_n e^{j2\pi ft} \\ &= 0 \end{aligned} \tag{3.12}$$

which means the complex amplitudes of the voltages obey KVL. We can easily imagine that the complex amplitudes of the currents obey KCL.

$$\sum_{nn} V_n = 0 \tag{3.13}$$

What we have discovered is that source(s) equaling a complex exponential of the same frequency forces all circuit variables to be complex exponentials of the same frequency. Consequently, the ratio of voltage to current for each element equals the ratio of their complex amplitudes, which depends only on the source's frequency and element values.

This situation occurs because the circuit elements are linear and time-invariant. For example, suppose we had a circuit element where the voltage equaled the square of the current: v(t)=Ki2(t).If i(t)=Iej2πft,v(t)=KI2ej2π2ft ,meaning that voltage and current no longer had the same frequency and that their ratio was time-dependent.

Because for linear circuit elements the complex amplitude of voltage is proportional to the complex amplitude of current: V=ZI assuming complex exponential sources means circuit elements behave as if they were resistors, where instead of resistance, we use impedance. Because complex amplitudes for voltage and current also obey Kirchoff's laws, we can solve circuits using voltage and current divider and the series and parallel combination rules by considering the elements to be impedances.

## 3.10 TIME AND FREQUENCY DOMAINS

When we find the differential equation relating the source and the output, we are faced with solving the circuit in what is known as the time domain. What we emphasize here is that it is often easier to find the output if we use impedances. Because impedances depend only on frequency, we find ourselves in the frequency domain. A common error in using impedances is keeping the time-dependent part, the complex exponential, in the fray. The entire point of using impedances is to get rid of time and concentrate on frequency. Only after we find the result in the frequency domain do we go back to the time domain and put things back together again.

To illustrate how the time domain, the frequency domain and impedances fit together, consider the time domain and frequency domain to be two work rooms. Since you can't be two places at the same time, you are faced with solving your circuit problem in one of the two rooms at any point in time. Impedances and complex exponentials are the way you get between the two rooms. Security guards make sure you don't try to sneak time domain variables into the frequency domain room and vice versa. Figure 3.25 (Two Rooms) shows how this works.

Two Rooms

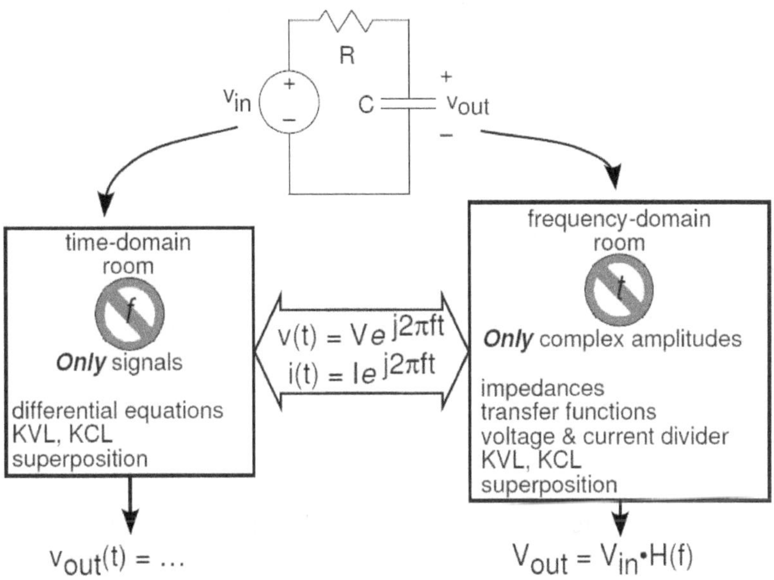


**Figure 3.25: The time and frequency domains are linked by assuming signals are complex exponentials. In the time domain, signals can have any form. Passing into the frequency domain "work room," signals are represented entirely by complex amplitudes.**

As we unfold the impedance story, we'll see that the powerful use of impedances suggested by Steinmetz greatly simplifies solving circuits, alleviates us from solving differential equations, and suggests a general way of thinking about circuits. Because of the importance of this approach, let's go over how it works.

- Even though it's not, pretend the source is a complex exponential. We do this because the impedance approach simplifies finding how input and output are related. If it were a voltage source having voltage vin = p(t) (a pulse), still let: vin=Vinej2πft. We'll learn how to "get the pulse back" later.
- With a source equaling a complex exponential, all variables in a linear circuit will also be complex exponentials having the same frequency. The circuit›s only remaining «mystery» is what each variable›s complex amplitude might be. To find these, we consider the source to be a complex number (Vin here) and the elements to be impedances.

- We can now solve using series and parallel combination rules how the complex amplitude of any variable relates to the sources complex amplitude.

### Example 3.3

To illustrate the impedance approach, we refer to the RC circuit (Figure 3.26 (Simple Circuits)) below, and we assume that vin=Vinej2πft

Simple Circuits

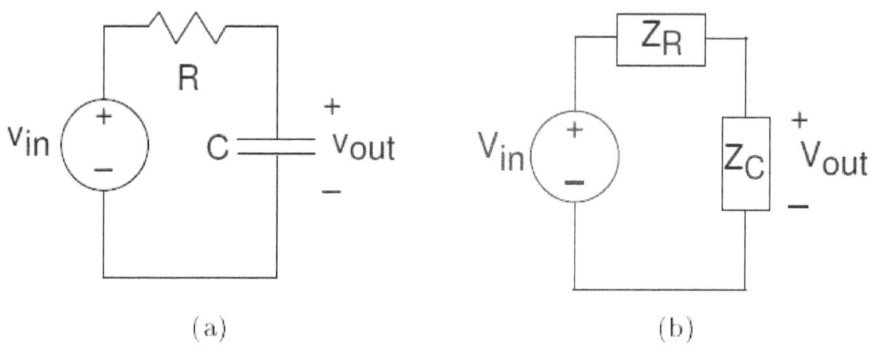


**Figure 3.26: (a) A simple RC circuit. (b) The impedance counterpart for the RC circuit. Note that the source and output voltage are now complex amplitudes.**

Using impedances, the complex amplitude of the output voltage Vout can be found using voltage divider:

$$V_{\text{out}} = \frac{Z_C}{Z_C + Z_R} V_{\text{in}}$$

$$V_{\text{out}} = \frac{\frac{1}{j2\pi fC}}{\frac{1}{j2\pi fC} + R} V_{\text{in}}$$

$$V_{\text{out}} = \frac{1}{j2\pi fRC + 1} V_{\text{in}}$$

If we refer to the differential equation for this circuit as shown in Circuits with Capacitors and Inductors (Section 3.8) to be: $RC\frac{dv_{\text{out}}}{dt} + v_{\text{out}} = v_{\text{in}}$), letting the output and input voltages be complex exponentials, we obtain the same relationship between their complex amplitudes. Thus, using impedances is equivalent to using the differential equation and solving it when the source is a complex exponential.

In fact, we can find the differential equation directly using impedances. If we cross-multiply the relation between input and output amplitudes,

$$V_{\text{out}} (j2\pi fRC + 1) = V_{\text{in}}$$

and then put the complex exponentials back in, we have

$$RCj2\pi fV_{\text{out}}e^{j2\pi ft} + V_{\text{out}}e^{j2\pi ft} = V_{\text{in}}e^{j2\pi ft}$$

In the process of defining impedances, note that the factor i2πf arises from the derivative of a complex exponential. We can reverse the impedance process, and revert back to the differential equation.

$$RC\frac{dv_{\text{out}}}{dt} + v_{\text{out}} = v_{\text{in}}$$

This is the same equation that was derived much more tediously in Circuits with Capacitors and Inductors. Finding the differential equation relating output to input is far simpler when we use impedances than with any other technique.

**Exercise 3.10.1**

Suppose you had an expression where a complex amplitude was divided by i2πf. What time-domain operation corresponds to this division?

## 3.11 POWER IN THE FREQUENCY DOMAIN

Recalling that the instantaneous power consumed by a circuit element or an equivalent circuit that represents a collection of elements equals the voltage times the current entering the positive-voltage terminal: p(t)=v(t)i(t),what is the equivalent expression using impedances? The resulting calculation reveals more about power consumption in circuits and the introduction of the concept of average power.

When all sources produce sinusoids of frequency f, the voltage and current for any circuit element or collection of elements are sinusoids of the same frequency.

$$v(t) = |V|\cos(2\pi ft + \phi)$$
$$i(t) = |I|\cos(2\pi ft + \theta)$$

Here, the complex amplitude of the V equals |V|ejφ and that of the current is |I|ejθ We can also write the voltage and current in terms of their complex amplitudes using Euler's formula (Section 2.1.2: Euler's Formula).

$$v(t) = \frac{1}{2}\left(Ve^{j2\pi ft} + V^*e^{-(j2\pi ft)}\right)$$
$$i(t) = \frac{1}{2}\left(Ie^{j2\pi ft} + I^*e^{-(j2\pi ft)}\right)$$

Multiplying these two expressions and simplifying gives

$$\begin{aligned} p(t) &= \frac{1}{4}\left(VI^* + V^*I + VIe^{j4\pi ft} + V^*I^*e^{-(j4\pi ft)}\right) \\ &= \frac{1}{2}Re(VI^*) + \frac{1}{2}Re\left(VIe^{j4\pi ft}\right) \\ &= \frac{1}{2}Re(VI^*) + \frac{1}{2}|V||I|\cos(4\pi ft + \phi + \theta) \end{aligned}$$

We define $\frac{1}{2}VI^*$ to be complex power. The real-part of complex power is the first term and since it does not change with time, it represents the power consistently consumed/produced by the circuit. The second term varies with time at a frequency twice that of the source. Conceptually, this term details how power "sloshes" back and forth in the circuit because of the sinusoidal source.

From another viewpoint, the real-part of complex power represents long-term energy consumption/production. Energy is the integral of power and, as the integration interval increases, the first term appreciates while the time-varying term "sloshes." Consequently, the most convenient definition of the average power consumed/produced by any circuit is in terms of complex amplitudes:

$$P_{\text{ave}} = \frac{1}{2} Re\left(VI^*\right)$$

## Exercise 3.11.1

Suppose the complex amplitudes of the voltage and current have fixed magnitudes. What phase relationship between voltage and current maximizes the average power? In other words, how are φ and Φ related for maximum power dissipation?

Because the complex amplitudes of the voltage and current are related by the equivalent impedance, average power can also be written as

$$P_{\text{ave}} = \frac{1}{2} Re\left(Z\right)\left(|I|\right)^2 = \frac{1}{2} Re\left(\frac{1}{Z}\right)\left(|V|\right)^2$$

These expressions generalize the results we obtained for resistor circuits. We have derived a fundamental result: Only the real part of impedance contributes to long-term power dissipation. Of the circuit elements, only the resistor dissipates power. Capacitors and inductors dissipate no power in the long term. It is important to realize that these statements apply only for sinusoidal sources. If you turn on a constant voltage source in an RC-circuit, charging the capacitor does consume power.

## Exercise 3.11.2

In an earlier problem, (Section 1.5.1:RWS Values), we found that the rms value of a sinusoid was its amplitude divided by $\sqrt{2}$.What is average power expressed in terms of the rms values of the voltage and current (Vrms and Irms respectively)?

## 3.12 EQUIVALENT CIRCUITS: IMPEDANCES AND SOURCES

When we have circuits with capacitors and/or inductors as well as resistors and sources, Thévenin and Mayer-Norton equivalent circuits can still be defined by using impedances and complex amplitudes for voltage and currents. For any circuit containing sources, resistors, capacitors, and inductors, the input-output relation for the complex amplitudes of the terminal voltage and current is :

$$V = Z_{\text{eq}} I + V_{\text{eq}}$$

$$I = \frac{V}{Z_{\text{eq}}} - I_{\text{eq}}$$

with Veq=Zeq Ieq. Thus, we have Thévenin and Mayer-Norton equivalent circuits as shown in Figure 3.27 (Equivalent Circuits).

Equivalent Circuits

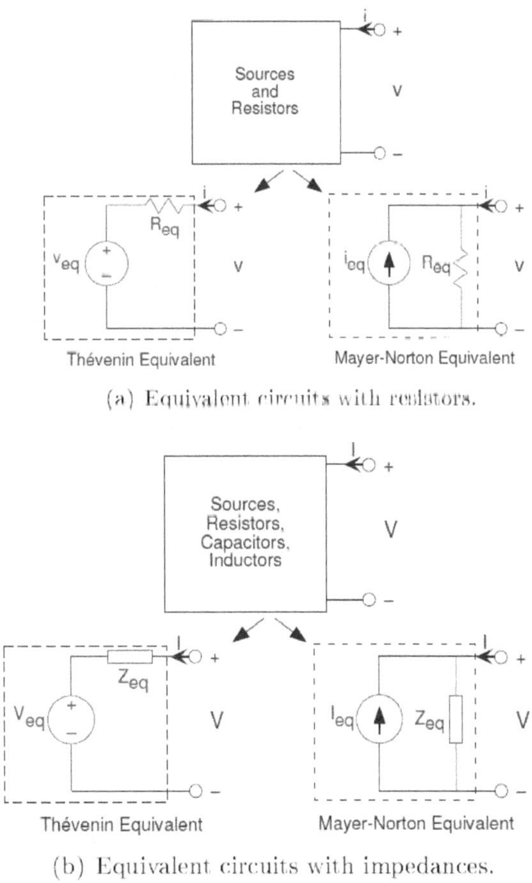


**Figure 3.27: Comparing the first, simpler, figure with the slightly more complicated second figure, we see two differences. First of all, more circuits (all those containing linear elements in fact) have equivalent circuits that contain equivalents. Secondly, the terminal and source variables are now complex amplitudes, which carries the implicit assumption that the voltages and currents are single complex exponentials, all having the same frequency.**

**Example 3.4**

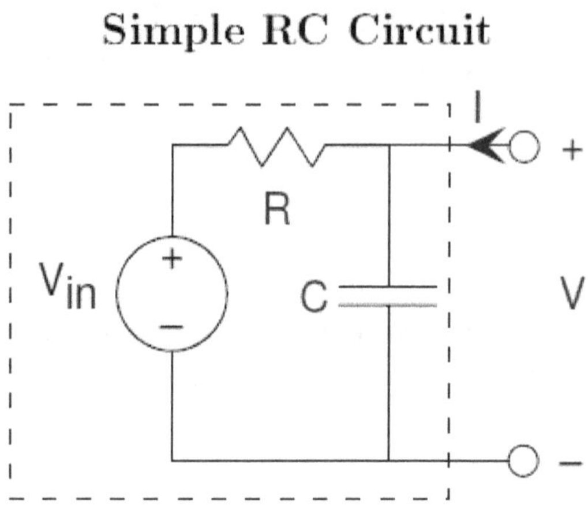


**Figure 3.28: Simple RC Circuit**

Let's find the Thévenin and Mayer-Norton equivalent circuits for the above Figure 3.28 (Simple RC Circuit) The open-circuit voltage and short-circuit current techniques still work, except we use impedances and complex amplitudes. The open-circuit voltage corresponds to the transfer function we have already found. When we short the terminals, the capacitor no longer has any effect on the circuit, and the short-circuit current $I_{sc}$ equals $\frac{V_{out}}{R}$. The equivalent impedance can be found by setting the source to zero, and finding the impedance using series and parallel combination rules. In our case, the resistor and capacitor are in parallel once the voltage source is removed (setting it to zero amounts to replacing it with a short-circuit). Thus, $Z_{eq} = R \parallel \frac{1}{j2\pi fC} = \frac{R}{1+j2\pi fRC}$.

Consequently, we have

$$V_{eq} = \frac{1}{1+j2\pi fRC} V_{in}$$

$$I_{eq} = \frac{1}{R} V_{in}$$

$$Z_{eq} = \frac{R}{1+j2\pi fRC}$$

Again, we should check the units of our answer. Note in particular that i2πfRCmust be dimensionless. Is it?

## 3.13 TRANSFER FUNCTIONS

The ratio of the output and input amplitudes for the Figure 3.29, known as the transfer function or the frequency response, is given by

$$\begin{aligned}\frac{V_{out}}{V_{in}} &= H(f) \\ &= \frac{1}{j2\pi fRC+1}\end{aligned} \quad (3.15)$$

Implicit in using the transfer function is that the input is a complex exponential, and the output is also a complex exponential having the same frequency. The transfer function reveals how the circuit modifies the input amplitude in creating the output amplitude. Thus, the transfer function completely describes how the circuit processes the input complex exponential to produce the output complex exponential. The circuit›s function is thus summarized by the transfer function. In fact, circuits are often designed to meet transfer function specifications. Because transfer functions are complex-valued, frequency-dependent quantities, we can better appreciate a circuit›s function by examining the magnitude and phase of its transfer function (Figure 3.30 (Magnitude and phase of the transfer function)).

Simple Circuit

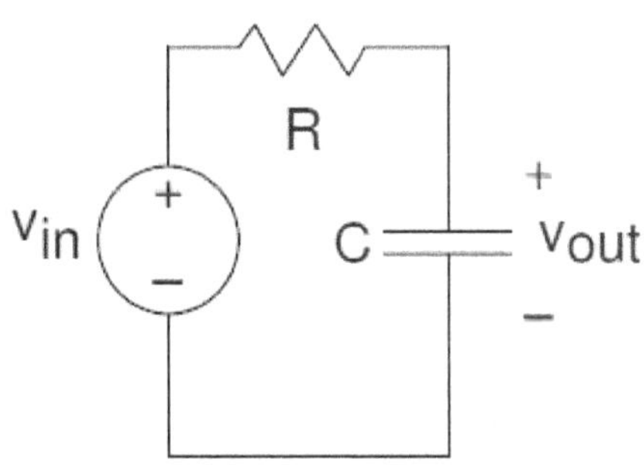


**Figure 3.29: A simple RC circuit.**

Magnitude and phase of the transfer function

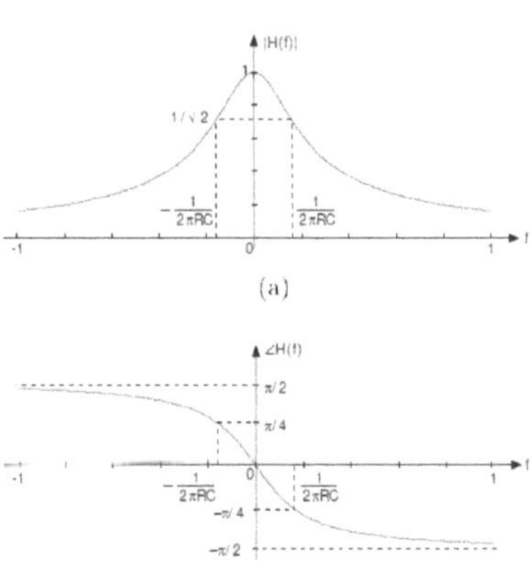


**Figure 3.13.2 Magnitude and phase of the transfer function of the RC circuit shown in Figure 3.29 (Simple Circuit)when RC=1. (a)** $|H(f)| = \frac{1}{\sqrt{(2\pi fRC)^2+1}}$ **(b) ∠ (H (f))= -arctan (2πfRC)**

This transfer function has many important properties and provides all the insights needed to determine how the circuit functions. First of all, note that we can compute the frequency response for both positive and negative frequencies. Recall that sinusoids consist of the sum of two complex exponentials, one having the negative frequency of the other. We will consider how the circuit acts on a sinusoid soon. Do note that the magnitude has even symmetry: The negative frequency portion is a mirror image of the positive frequency portion:|H(-f)|=|H(f)|. The phase has odd symmetry: ∠H(-f)=-∠H(f). These properties of this specific example apply for all transfer functions associated with circuits. Consequently, we don›t need to plot the negative frequency component; we know what it is from the positive frequency part.

The magnitude equals $\frac{1}{\sqrt{2}}$ of its maximum gain (1 at f=0) when 2πfRC=1 (the two terms in the denominator of the magnitude are equal). The frequency $f_c = \frac{1}{2\pi RC}$ defines the boundary between two operating ranges.

- For frequencies below this frequency, the circuit does not much alter the amplitude of the complex exponential source.
- For frequencies greater than fc the circuit strongly attenuates the amplitude. Thus, when the source frequency is in this range, the circuit's output has a much smaller amplitude than that of the source.

For these reasons, this frequency is known as the cutoff frequency. In this circuit the cutoff frequency depends only on the product of the resistance and the capacitance. Thus, a cutoff frequency of 1 kHz occurs when: $\frac{1}{2\pi RC} = 10^3$ or $RC = \frac{10^{-3}}{2\pi} = 1.59 \times 10^{-4}$. Thus resistance-capacitance combinations of 1.59 kΩ and 100 nF or 10 Ω and 1.59 μF result in the same cutoff frequency.

The phase shift caused by the circuit at the cutoff frequency precisely equals $-\frac{\pi}{4}$. Thus, below the cutoff frequency, phase is little affected, but at higher frequencies, the phase shift caused by the circuit becomes $-\frac{\pi}{2}$. This phase shift corresponds to the difference between a cosine and a sine.

We can use the transfer function to find the output when the input voltage is a sinusoid for two reasons. First of all, a sinusoid is the sum of two complex exponentials, each having a frequency equal to the negative of the other. Secondly, because the circuit is linear, superposition applies. If the source is a sine wave, we know that

$$\begin{aligned} v_{\text{in}}(t) &= A\sin(2\pi ft) \\ &= \frac{A}{2j}\left(e^{j2\pi ft} - e^{-(j2\pi ft)}\right) \end{aligned} \tag{3.16}$$

Since the input is the sum of two complex exponentials, we know that the output is also a sum of two similar complex exponentials, the only difference being that the complex amplitude of each is multiplied by the transfer function evaluated at each exponential's frequency.

$$v_{\text{out}}(t) = \frac{A}{2j} H(f) e^{j2\pi ft} - \frac{A}{2j} H(-f) e^{-(j2\pi ft)} \tag{3.17}$$

As noted earlier, the transfer function is most conveniently expressed in polar form: H(f)=|H(f)|ej∠H(f). Furthermore, |H(-f)|=|H(f)| (even symmetry of the magnitude) and ∠H(-f)=-∠H(f)) (odd symmetry of the phase). The output voltage expression simplifies to

$$\begin{aligned} v_{\text{out}}(t) &= \tfrac{A}{2j} |H(f)| e^{j2\pi ft + \angle(H(f))} - \tfrac{A}{2j} |H(f)| e^{(-(j2\pi ft)) - \angle(H(f))} \\ &= A|H(f)| \sin(2\pi ft + \angle(H(f))) \end{aligned} \tag{3.18}$$

The circuit's output to a sinusoidal input is also a sinusoid, having a gain equal to the magnitude of the circuit's transfer function evaluated at the source frequency and a phase equal to the phase of the transfer function at the source frequency. It will turn out that this input-output relation description applies to any linear circuit having a sinusoidal source.

### Exercise 3.13.1

This input-output property is a special case of a more general result. Show that if the source can be written as the imaginary part of a complex exponential – vin(t)=I(Vei2πft)– the output is given by vout(t)=I(VH(f)ej2πft). Show that a similar result also holds for the real part.

The notion of impedance arises when we assume the sources are complex exponentials. This assumption may seem restrictive; what would we do if the source were a unit step? When we use impedances to find the transfer function between the source and the output variable, we can derive from it the differential equation that relates input and output. The differential equation applies no matter what the source may be. As we have argued, it is far simpler to use impedances to find the differential equation (because we can use series and parallel combination rules) than any other method. In this sense, we have not lost anything by temporarily pretending the source is a complex exponential.

In fact we can also solve the differential equation using impedances! Thus, despite the apparent restrictiveness of impedances, assuming complex exponential sources is actually quite general.

## 3.14 DESIGNING TRANSFER FUNCTIONS

According to linear system theory, when the source contains several signals, the output voltage is equal to the sum of the individual outputs produced by each signal. Linear circuits, being a subset of linear systems, are subject to the principle of superposition. Specifically, assume that these constituent signals are complex exponentials, each with a distinct frequency from the others. The transfer function illustrates the impact of the circuit on the amplitude and phase of each component, enabling us to comprehend the circuit's behavior on a complex signal. Components with a frequency lower than the cutoff frequency are able to flow through the circuit with minimal alteration, whereas components with higher frequencies are suppressed. The circuit functions as a filter by selectively attenuating or amplifying different frequency components of the source signal, dependent on the frequency of each complex exponential component. The filter is referred to as a lowpass filter since it specifically allows low frequencies to pass through. In addition, we have discovered that the transfer function allows for convenient computation of the output for sinusoidal inputs. Once the transfer function is determined, we may directly express the output based on the circuit's response to a sinusoidal input (3.18).

### Example 3.5

RL circuit

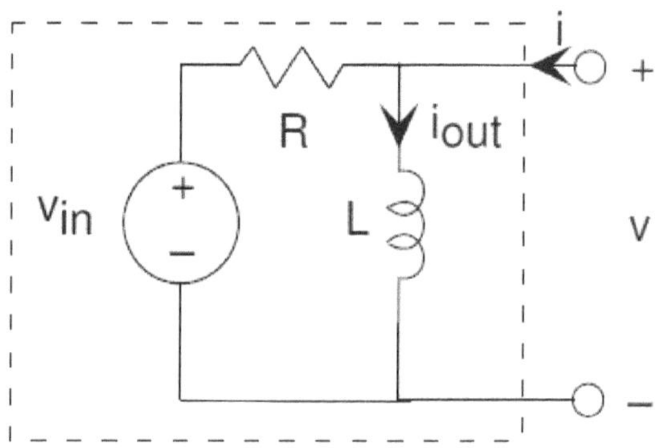


**Figure 3.31: RL circuit**

Let's apply these results to a final example, in which the input is a voltage source and the output is the inductor current. The source voltage equals: $V_{in}=2\cos(2\pi 60t)+3$. We want the circuit to pass constant (offset) voltage essentially unaltered (save for the fact that the output is a current rather than a voltage) and remove the 60 Hz term. Because the input is the sum of two sinusoids--a constant is a zero-frequency cosine--our approach is

- Find the transfer function using impedances
- Use it to find the output due to each input component
- Add the results
- Find element values that accomplish our design criteria

Because the circuit is a series combination of elements, let's use voltage divider to find the transfer function between Vin and V, then use the v-i relation of the inductor to find its current.

$$\begin{aligned} \frac{I_{\text{out}}}{V_{\text{in}}} &= \frac{j2\pi fL}{R+j2\pi fL}\frac{1}{j2\pi fL} \\ &= \frac{1}{j2\pi fL+R} \\ &= H(f) \end{aligned} \tag{3.19}$$

Where

$$\text{voltage divider} = \frac{j2\pi fL}{R+j2\pi fL}$$

[Do the units check?] The form of this transfer function should be familiar; it is a lowpass filter, and it will perform our desired function once we choose element values properly.

The constant term is easiest to handle. The output is given by: $3|H(0)| = \frac{3}{R}$. Thus, the value we choose for the resistance will determine the scaling factor of how voltage is converted into current. For the 60 Hz component signal, the output current is $2|H(60)|\cos(2\pi 60t + \angle(H(60)))$. The total output due to our source is:

$$i_{\text{out}} = 2|H(60)|\cos(2\pi 60t + \angle(H(60))) + 3\times H(0) \tag{3.20}$$

The cutoff frequency for this filter occurs when the real and imaginary parts of the transfer function's denominator equal each other. Thus, 2πfcL=R, which gives $f_c = \frac{R}{2\pi L}$. We want this cutoff frequency to be much less than 60 Hz. Suppose we place it at, say, 10 Hz. This specification would require the component values to be related by $\frac{R}{L} = 20\pi = 62.8$ The transfer function at 60 Hz would be

$$\left|\frac{1}{j2\pi 60L+R}\right| = \frac{1}{R}\left|\frac{1}{6j+1}\right| = \frac{1}{R}\frac{1}{\sqrt{37}} \simeq 0.16\times\frac{1}{R} \tag{3.21}$$

which yields an attenuation (relative to the gain at zero frequency) of about 1/6, and result in an output amplitude of $\frac{0.3}{R}$ relative to the constant term›s amplitude of $\frac{3}{R}$. A factor of 10 relative size between the two components seems reasonable. Having a 100 mH inductor would require a 6.28 Ω resistor. An easily available resistor value is 6.8 Ω; thus, this choice results in cheaply and easily purchased parts. To make the resistance bigger would require a proportionally larger inductor. Unfortunately, even a 1 H inductor is physically large; consequently low cutoff frequencies require small-valued resistors and large-valued inductors. The choice made here represents only one compromise.

The phase of the 60 Hz component will very nearly be $-\frac{\pi}{2}$, leaving it to be: $\frac{0.3}{R}\cos\left(2\pi 60t - \frac{\pi}{2}\right) = \frac{0.3}{R}\sin(2\pi 60t)$. The waveforms for the input and output are shown in Figure 3.32 (Waveforms).

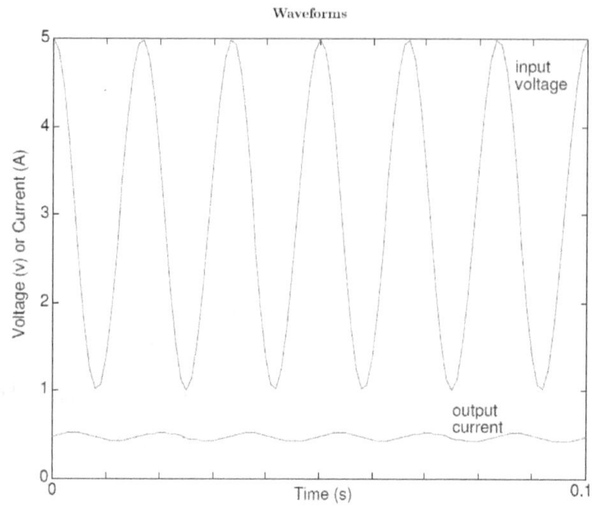


**Figure 3.32: Input and output waveforms for the example RL circuit when the element values are R=6.28Ω and L=100mH**

Note that the sinusoid's phase has indeed shifted; the lowpass filter not only reduced the 60 Hz signal's amplitude, but also shifted its phase by 90°.

## 3.15 FORMAL CIRCUIT METHODS: NODE METHOD

In some (complicated) cases, we cannot use the simplification techniques--such as parallel or series combination rules--to solve for a circuit's input-output relation. In other modules, we wrote v-i relations and Kirchoff›s laws haphazardly, solving them more on intuition than procedure. We need a formal method that produces a small, easy set of equations that lead directly to the input-output relation we seek. One such technique is the node method.

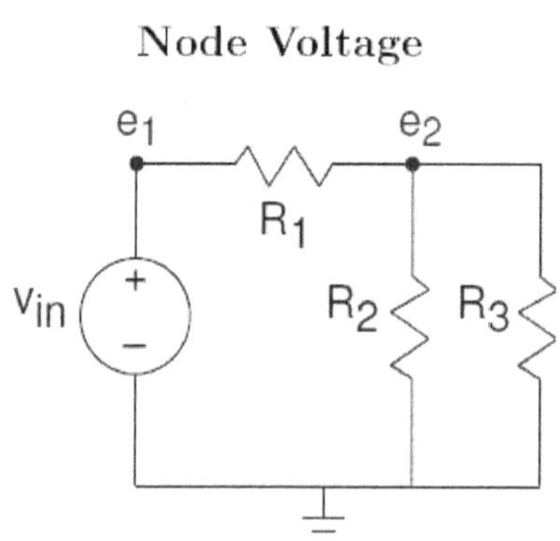


**Figure 3.33: Node Voltage Circuit**

The node method begins by finding all nodes--places where circuit elements attach to each other--in the circuit. We call one of the nodes the reference node; the choice of reference node is arbitrary, but it is usually chosen to be a point of symmetry or the "bottom" node. For the remaining nodes, we define node voltages en that represent the voltage between the node and the reference. These nodes voltages constitute the only unknowns; all we need is a sufficient number of equations to solve for them. In our example, we have two node voltages. The very act of defining nodes voltages is equivalent to using all the KVL equations at your disposal. The reason for this simple, but astounding, fact is that a node voltage is uniquely defined regardless of what path is traced between the node and the reference. Because two paths between a node and reference have the same voltage, the sum of voltages around the loop equals zero.

In some cases, a node voltage corresponds exactly to the voltage across a voltage source. In such cases, the node voltage is specified by the source and is not an unknown. For example, in our circuit, e1=vin; Thus, we need only to find one node voltage.

The equations governing the node voltages are obtained by writing KCL equations at each node having an unknown node voltage, using the v-i relations for each element. In our example, the only circuit equation is

$$\frac{e_2 - v_{\text{in}}}{R_1} + \frac{e_2}{R_2} + \frac{e_2}{R_3} = 0 \tag{3.22}$$

A little reflection reveals that when writing the KCL equations for the sum of currents leaving a node, that node's voltage will always appear with a plus sign, and all other node voltages with a minus sign. Systematic application of this procedure makes it easy to write node equations and to check them before solving them. Also remember to check units at this point: Every term should have units of current. In our example, solving for the unknown node voltage is easy:

$$e_2 = \frac{R_2 R_3}{R_1 R_2 + R_1 R_3 + R_2 R_3} v_{\text{in}} \tag{3.23}$$

Have we really solved the circuit with the node method? Along the way, we have used KVL, KCL, and the v-i relations. Previously, we indicated that the set of equations resulting from applying these laws is necessary and sufficient. This result guarantees that the node method can be used to «solve» any circuit. One fallout of this result is that we must be able to find any circuit variable given the node voltages and sources. All circuit variables can be found using the v-i relations and voltage divider. For example, the current through R3 equals $\frac{e_2}{R_3}$.

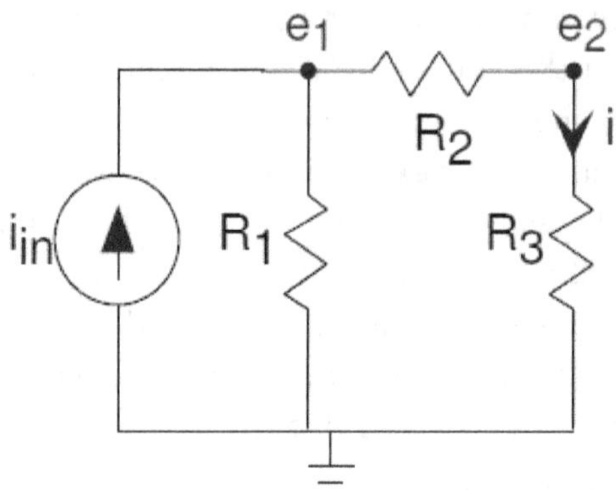


**Figure 3.34: Node Voltage Circuit**

The presence of a current source in the circuit does not affect the node method greatly; just include it in writing KCL equations as a current leaving the node. The circuit has three nodes, requiring us to define two node voltages. The node equations are

$$\frac{e_1}{R_1} + \frac{e_1 - e_2}{R_2} - i_{\text{in}} = 0 \quad \text{(Node 1)}$$

$$\frac{e_2 - e_1}{R_2} + \frac{e_2}{R_3} = 0 \quad \text{(Node 2)}$$

Note that the node voltage corresponding to the node that we are writing KCL for enters with a positive sign, the others with a negative sign, and that the units of each term is given in amperes. Rewrite these equations in the standard set-of-linear-equations form.

$$e_1\left(\frac{1}{R_1} + \frac{1}{R_2}\right) - e_2\frac{1}{R_2} = i_{\text{in}}$$

$$e_1\frac{1}{R_2} + e_2\left(\frac{1}{R_2} + \frac{1}{R_3}\right) = 0$$

Solving these equations gives

$$e_1 = \frac{R_2 + R_3}{R_3}e_2$$

$$e_2 = \frac{R_1 R_3}{R_1 + R_2 + R_3}i_{\text{in}}$$

To find the indicated current, we simply use: $i = \frac{e_2}{R_3}$.

## Example 3.6: Node Method Example

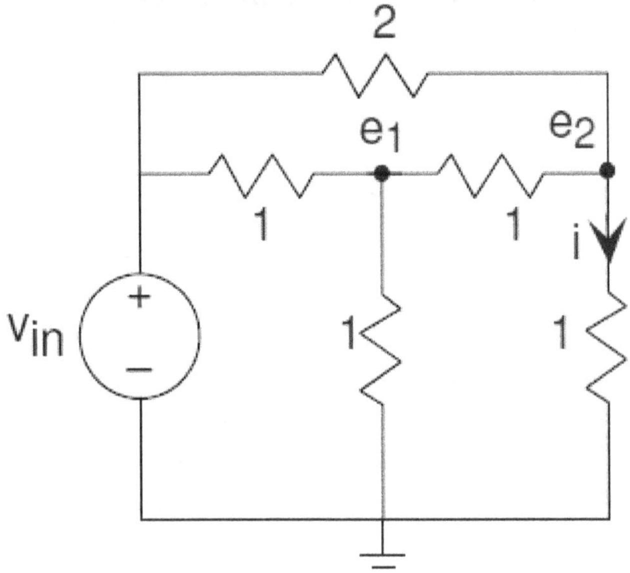


Figure 3.35

In the above circuit we cannot use the series/parallel combination rules: The vertical resistor at node 1 keeps the two horizontal 1 Ω resistors from being in series, and the 2 Ω resistor prevents the two 1 Ω resistors at node 2 from being in series. We really do need the node method to solve this circuit! Despite having six elements, we need only define two node voltages. The node equations are:

$$\frac{e_1 - v_{in}}{1} + \frac{e_1}{1} + \frac{e_1 - e_2}{1} = 0 \quad \text{(Node 1)}$$

$$\frac{e_2 - v_{in}}{2} + \frac{e_2}{1} + \frac{e_2 - e_1}{1} = 0 \quad \text{(Node 2)}$$

Solving these equations yields: $e_1 = \frac{6}{13}v_{in}$ and $e_2 = \frac{5}{13}v_{in}$. The output current equals: $\frac{e_2}{1} = \frac{5}{13}v_{in}$. One unfortunate consequence of using the element's numeric values from the outset is that it becomes impossible to check units while setting up and solving equations.

### Exercise 3.15.1

What is the equivalent resistance seen by the voltage source?

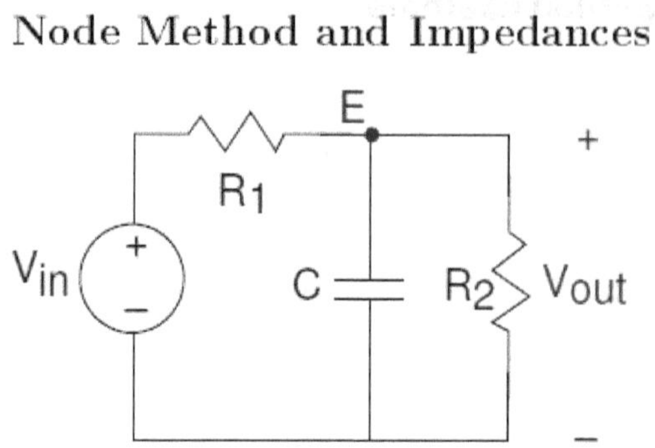


**Figure 3.36: Modification of the circuit shown on the left to illustrate the node method and the effect of adding the resistor R2**

The node method applies to RLC circuits, without significant modification from the methods used on simple resistive circuits, if we use complex amplitudes. We rely on the fact that complex amplitudes satisfy KVL, KCL, and impedance-based v-i relations. In the example circuit, we define complex amplitudes for the input and output variables and for the node voltages. We need only one node voltage here, and its KCL equation is:

$$\frac{E - V_{\text{in}}}{R_1} + Ej2\pi fC + \frac{E}{R_2} = 0$$

with the result

$$E = \frac{R_2}{R_1 + R_2 + j2\pi fR_1R_2C}V_{\text{in}}$$

To find the transfer function between input and output voltages, we compute the ratio E/Vin. The transfer function's magnitude and angle are:

$$|H(f)| = \frac{R_2}{\sqrt{(R_1 + R_2)^2 + (2\pi fR_1R_2C)^2}}$$

$$\angle(H(f)) = -\arctan\left(\frac{2\pi fR_1R_2C}{R_1 + R_2}\right)$$

This circuit differs from the one shown previously in that the resistor R2 has been added across the output. What effect has it had on the transfer function, which in the original circuit was a lowpass filter having cutoff frequency: $f_c = \frac{1}{2\pi R_1C}$? As shown in Figure 3.37 below, adding the second resistor has two effects: it lowers the gain in the passband (the range of frequencies for which the filter has little effect on the input) and increases the cutoff frequency.

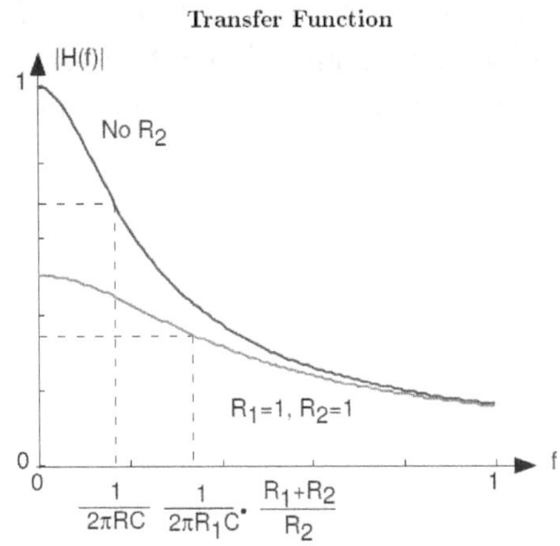


**Figure 3.37 Transfer functions of the circuits shown in Figure 3.36 (Node Method and Impedances). Here, $R_1 = 1$, $R_2 = 1$, and C = 1.**

When $R_2$=$R_1$ as shown on the plot, the passband gain becomes half of the original, and the cutoff frequency increases by the same factor. Thus, adding R2 provides a ‹knob› by which we can trade passband gain for cutoff frequency.

### Exercise 3.15.2

We can change the cutoff frequency without affecting passband gain by changing the resistance in the original circuit. Does the addition of the R2 resistor help in circuit design?

## 3.16 POWER CONSERVATION IN CIRCUITS

Now that we have a formal method---the node method---for solving circuits, we can use it to prove a powerful result: KVL and KCL are all that are required to show that all circuits conserve power, regardless of what elements are used to build the circuit.

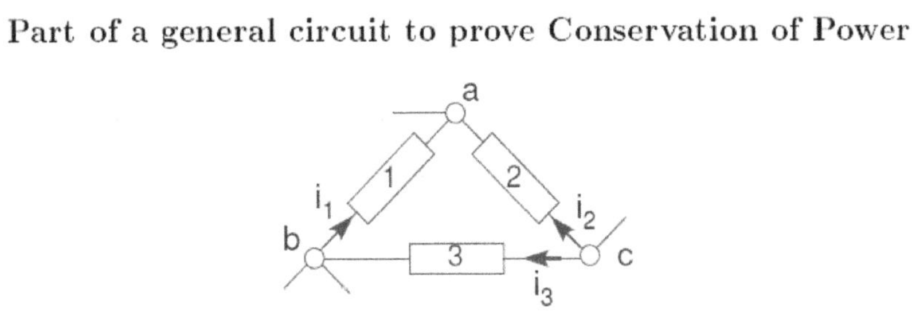


**Figure 3.38 Part of a general circuit to prove Conservation of Power**

First of all, define node voltages for all nodes in a given circuit. Any node chosen as the reference will do. For example, in the portion of a large circuit depicted above, we define node voltages for nodes a, b and c. With these node voltages, we can express the voltage across any element in terms of them. For example, the voltage across element 1 is given by: v1=eb–ea. The instantaneous power for element 1 becomes:

$$v_1 i_1 = (e_b - e_a) i_1 = e_b i_1 - e_a i_1$$

Writing the power for the other elements, we have

$$\begin{aligned} v_2 i_2 &= e_c i_2 - e_a i_2 \\ v_3 i_3 &= e_c i_3 - e_b i_3 \end{aligned}$$

When we add together the element power terms, we discover that once we collect terms involving a particular node voltage, it is multiplied by the sum of currents leaving the node minus the sum of currents entering. For example, for node b, we have: eb (i3–i1). We see that the currents will obey KCL that multiply each node voltage. Consequently, we conclude that the sum of element powers must equal zero in any circuit regardless of the elements used to construct the circuit.

$$\sum_k v_k i_k = 0$$

The simplicity and generality with which we proved this results generalizes to other situations as well. In particular, note that the complex amplitudes of voltages and currents obey KVL and KCL, respectively. Consequently, we have that: $\sum_k V_k I_k = 0$. Furthermore, the complex-conjugate of currents also satisfies KCL, which means we also have: $\sum_k V_k I_k{}^* = 0$. And finally, we know that evaluating the real-part of an expression is linear. Finding the real-part of this power conservation gives the result that average power is also conserved in any circuit.

$$\sum_k \frac{1}{2} Re\left(V_k I_k{}^*\right) = 0$$

Note: This proof of power conservation can be generalized in another very interesting way. All we need is a set of voltages that obey KVL and a set of currents that obey KCL. Thus, for a given circuit topology (the specific way elements are interconnected), the voltages and currents can be measured at different times and the sum of v-i products is zero.

$$\sum_k v_k(t_1)\, i_k(t_2) = 0$$

Even more interesting is the fact that the elements don't matter. We can take a circuit and measure all the voltages. We can then make element-for-element replacements and, if the topology has not changed, we can measure a set of currents. The sum of the product of element voltages and currents will also be zero!

## 3.17 ELECTRONICS

Thus far, we have examined and studied electrical circuits. The source signal possesses a greater magnitude of power compared to the output variable, whether it is a voltage or a current. The concept of power has not been clearly defined, however, this is not of great importance. Resistors, inductors, and capacitors, when used as separate components, do not contribute to power amplification. Similarly, circuits constructed solely with these components would not possess the ability to generate power. These circuits are referred to as electrical circuits, as opposed to electronic circuits, which do provide power gain. The process of amplifying power, such as when your stereo reads a CD and generates sound, is achieved through the use of semiconductor circuits that incorporate transistors. The fundamental concept behind the transistor is to allow a feeble input signal to modulate a robust current supplied by an electrical power source, known as the power supply, in order to generate a more potent signal. An example of a physical analogy is a water faucet: By manipulating the faucet in a reciprocating motion, the water flow fluctuates correspondingly, and possesses far greater force than the energy exerted in manipulating the handle. The waterpower is generated by the hydrostatic pressure of the water in your plumbing system, which is created by the water utility pumping water up to your local water tower. The power source can be likened to a water tower, while the transistor acts as a faucet that is controlled by the input signal. Similar to this example, a power supply functions as a consistent source of voltage, just as a water tower is designed to maintain a steady water pressure. The operational amplifier, sometimes referred to as the op-amp, is a device that offers greater convenience in terms of supplying gain and other useful qualities compared to the transistor. An operational amplifier (op-amp) is an integrated circuit consisting of several transistors arranged on a single chip. When connected to a power supply, it exhibits a significant amplification of voltage. A new circuit element, known as the dependent source, can be used to simulate the op-amp.

## 3.18 DEPENDENT SOURCES

A dependent source is an electrical component that can be either a voltage or current source. Its magnitude is directly related to the magnitude of another voltage or current in the circuit.

There are four distinct types of dependent sources. In order to explain an operational amplifier (op-amp), we require a voltage-dependent voltage source. Nevertheless, the conventional circuit-theoretical representation of a transistor includes a current-controlled current source. Dependent sources cannot be used as inputs to a circuit in the same way that independent sources may. Active circuits, which include electronic components, are modeled using them. The RLC circuits we have previously discussed are referred to as passive circuits.

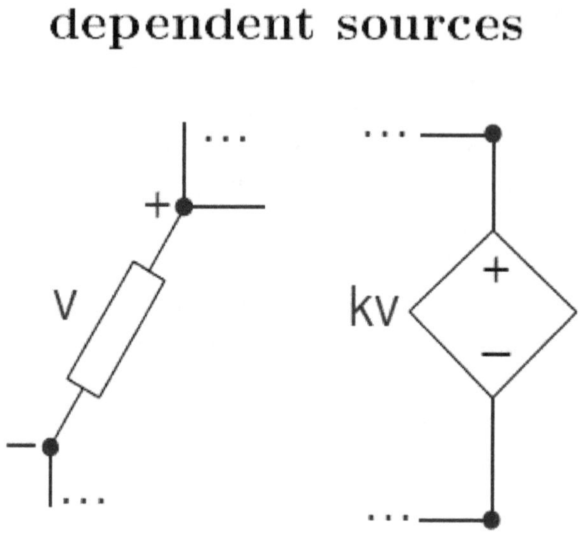


**Figure 3.39: Of the four possible dependent sources, depicted is a voltage-dependent voltage source in the context of a generic circuit.**

Figure 3.40 (op-amp) shows the circuit symbol for the op-amp and its equivalent circuit in terms of a voltage-dependent voltage source.

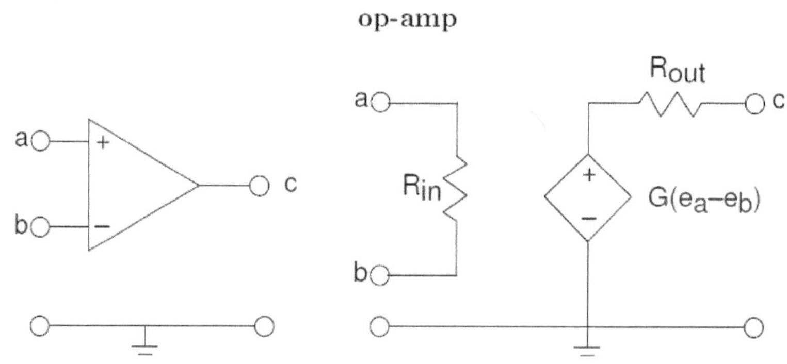


**Figure 3.40 The op-amp has four terminals to which connections can be made. Inputs attach to nodes a and b, and the output is node c. As the circuit model on the right shows, the op-amp serves as an amplifier for the difference of the input node voltages.**

The output voltage is equal to an amplified version of the voltage difference between the node voltages across its inputs. The dependent source model accurately depicts the functioning of the operational amplifier. Typically, the power supply is omitted in active circuit designs, however it is necessary for the circuit model to be precise. Proper functioning of most operational amplifiers necessitates the presence of both positive and negative supply voltages.

Due to the fact that dependent sources cannot be characterized as impedances and the dependent variable cannot be eliminated when applying parallel/series combining rules, it is often not advisable to employ circuit simplifications such as current and voltage dividers. When analyzing circuits that have dependent sources, it is necessary to employ formal approaches such as the node method, as explained in Section 3.15. Applying the node approach to these circuits is rather simple, where node voltages are assigned to the source and treated as if they were already known, similar to independent sources. Examine the circuit depicted at the top in Figure 3.41, which is a feedback operational amplifier.

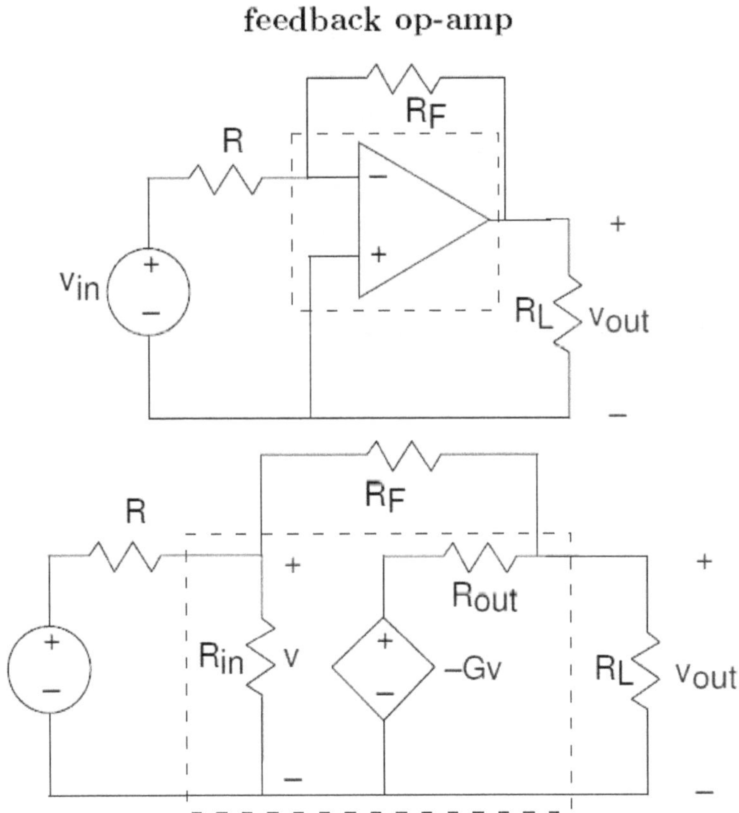


**Figure 3.41: The top circuit depicts an op-amp in a feedback amplifier configuration. On the bottom is the equivalent circuit, and integrates the op-amp circuit model into the circuit.**

Note that the op-amp is placed in the circuit "upside-down," with its inverting input at the top and serving as the only input. As we explore op-amps in more detail in the next section, this configuration will appear again and again and its usefulness demonstrated. To determine how the output voltage is related to the input voltage, we apply the node method. Only two node voltages - v and vout need be defined; the remaining nodes are across sources or serve as the reference. The node equations are:

$$\frac{v - v_{\text{in}}}{R} + \frac{v}{R_{\text{in}}} + \frac{v - v_{\text{out}}}{R_F} = 0 \tag{3.24}$$

$$\frac{v_{\text{out}} - (-G)\,v}{R_{\text{out}}} + \frac{v_{\text{out}} - v}{R_F} + \frac{v_{\text{out}}}{R_L} = 0 \tag{3.25}$$

Note that no special considerations were used in applying the node method to this dependent-source circuit. Solving these to learn how vout relates to vin yields:

$$\left(\frac{R_F R_{\text{out}}}{R_{\text{out}} - G R_F}\left(\frac{1}{R_{\text{out}}} + \frac{1}{R_{\text{in}}} + \frac{1}{R_L}\right)\left(\frac{1}{R} + \frac{1}{R_{\text{in}}} + \frac{1}{R_F}\right) - \frac{1}{R_F}\right) v_{\text{out}} = \frac{1}{R} v_{\text{in}} \tag{3.26}$$

This expression represents the general input-output relation for this circuit, known as the standard feedback configuration. Once we learn more about op-amps, in particular what its typical element values are, the expression will simplify greatly. Do note that the units check, and that the parameter G of the dependent source is a dimensionless gain.

## 3.19 OPERATIONAL AMPLIFIERS

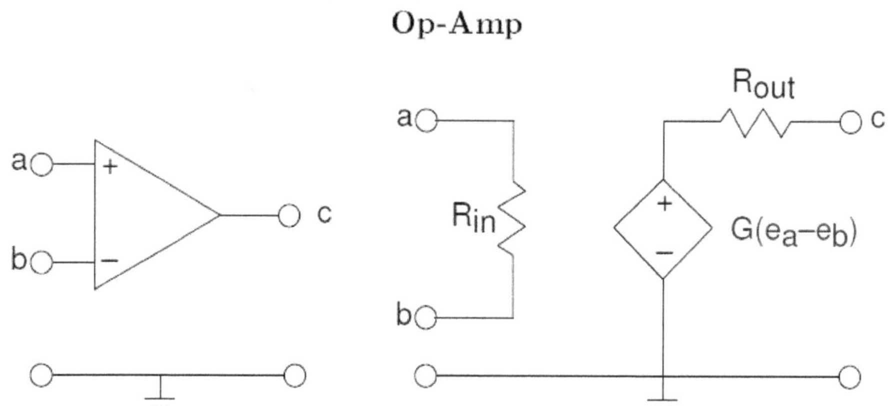


**Figure 3.42: The op-amp has four terminals to which connections can be made. Inputs attach to nodes a and b, and the output is node c. As the circuit model on the right shows, the op-amp serves as an amplifier for the difference of the input node voltages.**

Op-amps not only have the circuit model shown in Figure 3.42(Op-Amp), but their element values are very special.

- The input resistance, Rin, is typically large, on the order of 1 MΩ.
- The output resistance, Rout, is small, usually less than 100 Ω.
- The voltage gain, G, is large, exceeding 105.

The large gain catches the eye; it suggests that an op-amp could turn a 1 mV input signal into a 100 V one. If you were to build such a circuit--attaching a voltage source to node a, attaching node b to the reference, and looking at the output--you would be disappointed. In dealing with electronic components, you cannot forget the unrepresented but needed power supply.

Unmodelled Limitations Imposed By Power Supplies: It is impossible for electronic components to yield voltages that exceed those provided by the power supply or for them to yield currents that exceed the power supply's rating.

Typical power supply voltages required for op-amp circuits are + (15V). Attaching the 1 mv signal not only would fail to produce a 100 V signal, the resulting waveform would be severely distorted. While a desirable outcome if you are a rock & roll aficionado, high-quality stereos should not distort signals. Another consideration in designing circuits with op-amps is that these element values are typical: Careful control of the gain can only be obtained by choosing a circuit so that its element values dictate the resulting gain, which must be smaller than that provided by the op-amp.

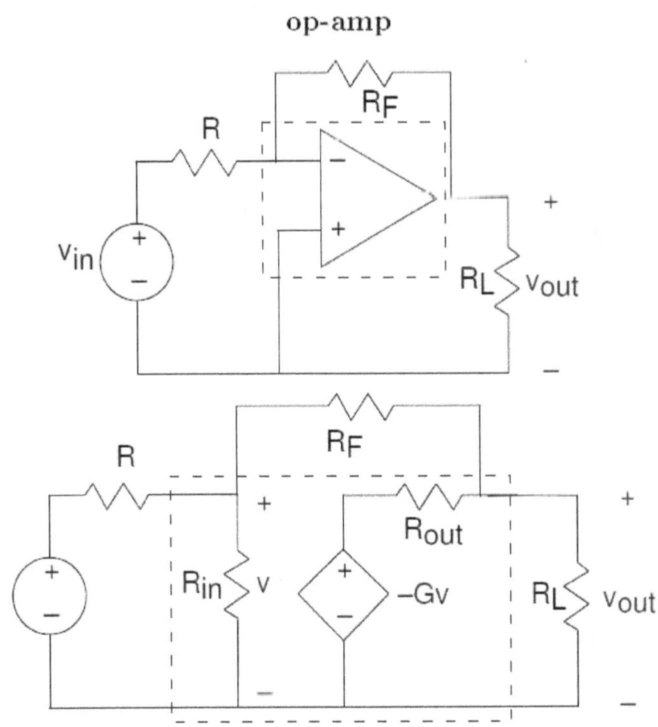


**Figure 3.43: The top circuit depicts an op-amp in a feedback amplifier configuration. On the bottom is the equivalent circuit, and integrates the op-amp circuit model into the circuit.**

### 3.19.1 Inverting Amplifier

The feedback configuration shown in Figure 3.19.2 is the most common op-amp circuit for obtaining what is known as an inverting amplifier.

$$\left(\frac{R_F R_{\text{out}}}{R_{\text{out}} - GR_F}\left(\frac{1}{R_{\text{out}}} + \frac{1}{R_{\text{in}}} + \frac{1}{R_L}\right)\left(\frac{1}{R} + \frac{1}{R_{\text{in}}} + \frac{1}{R_F}\right) - \frac{1}{R_F}\right) v_{\text{out}} = \frac{1}{R} v_{\text{in}} \tag{3.27}$$

provides the exact input-output relationship. In choosing element values with respect to op-amp characteristics, we can simplify the expression dramatically.

- Make the load resistance, RL, much larger than Rout. This situation drops the term 1/RL from the second factor of the above equation.
- Make the resistor, R, smaller than Rin, which means that the 1/Rin term in the third factor is negligible.

with these two design criteria, the expression ((3.27)) becomes

$$\left(\frac{R_F}{R_{\text{out}} - GR_F}\left(\frac{1}{R} + \frac{1}{R_F}\right) - \frac{1}{R_F}\right) v_{\text{out}} = \frac{1}{R} v_{\text{out}} \tag{3.28}$$

Because the gain is large and the resistance Rout is small, the first term becomes $-\frac{1}{G}$, leaving us with:

$$\left(\left(-\frac{1}{G}\right)\left(\frac{1}{R} + \frac{1}{R_F}\right) - \frac{1}{R_F}\right) v_{\text{out}} = \frac{1}{R} v_{\text{in}} \tag{3.29}$$

If we select the values of RF and R so that GR>>RF, this factor will no longer depend on the op-amp's inherent gain, and it will equal $-\frac{1}{R_F}$.

Under these conditions, we obtain the classic input-output relationship for the op-amp-based inverting amplifier.

$$v_{\text{out}} = -\left(\frac{R_F}{R} v_{\text{in}}\right) \tag{3.30}$$

Consequently, the gain provided by our circuit is entirely determined by our choice of the feedback resistor RF and the input resistor R. It is always negative, and can be less than one or greater than one in magnitude. It cannot exceed the op-amp's inherent gain and should not produce such large outputs that distortion results (remember the power supply!). Interestingly, note that this relationship does not depend on the load resistance. This effect occurs because we use load resistances large compared to the op-amp's output resistance. Thus observation means that, if careful, we can place op-amp circuits in cascade, without incurring the effect of succeeding circuits changing the behavior (transfer function) of previous ones; see this problem (Problem 3.44).

### 3.19.2 Active Filters

The input-output relationship for the inverting amplifier remains valid whether the feedback and input circuit parts are impedances, such as resistors, capacitors, and inductors, as long as the design constraints are satisfied.

op-amp

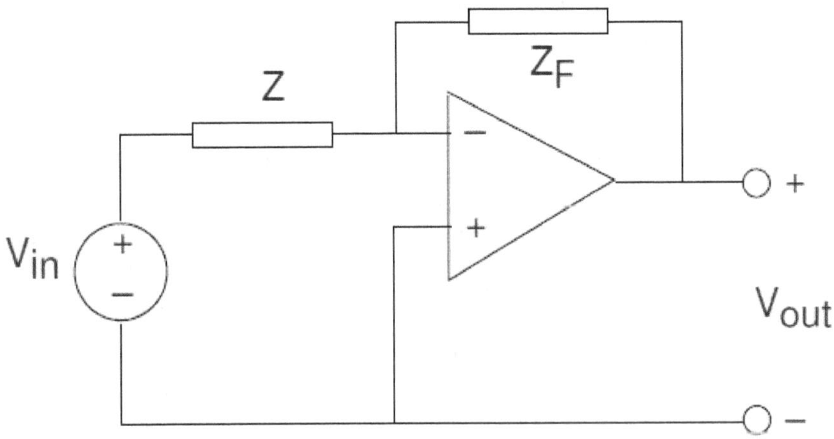


**Figure 3.44: Op-amp** $\frac{V_{out}}{V_{in}} = -\frac{Z_F}{Z}$

**Example 3.7**

Let's design an op-amp circuit that functions as a lowpass filter. We want the transfer function between the output and input voltage to be:

$$H(f) = \frac{K}{1 + \frac{if}{f_c}}$$

where K equals the passband gain and fc is the cutoff frequency. Let›s assume that the inversion (negative gain) does not matter. With the transfer function of the above op-amp circuit in mind, let›s consider some choices.

- $Z_F = K$, $Z = 1 + \frac{if}{f_c}$. This choice means the feedback impedance is a resistor and that the input impedance is a series combination of an inductor and a resistor. In circuit design, we try to avoid inductors because they are physically bulkier than capacitors.
- $Z_F = \frac{1}{1+\frac{if}{f_c}}$, $Z = \frac{1}{K}$. Consider the reciprocal of the feedback impedance (its admittance):

${Z_F}^{-1} = 1 + \frac{if}{f_c}$. Since this admittance is a sum of admittances, this expression suggests the parallel combination of a resistor (value = 1 Ω) and a capacitor (value $= \frac{1}{f_c}$ F). We have the right idea, but the values (like 1 Ω) are not right. Consider the general RC parallel combination; its admittance is $\frac{1}{R_F}+j2\pi fC$. Letting the input resistance equal R, the transfer function of the op-amp inverting amplifier now is $H(f)=-\frac{\frac{R_F}{R}}{1+j2\pi fR_FC}$ $\frac{R_F}{R}$ and the cutoff frequency $\frac{1}{R_FC}$.

Creating a specific transfer function with op-amps does not have a unique answer. As opposed to design with passive circuits, electronics is more flexible (a cascade of circuits can be built so that each has little effect on the others; see [link]) and gain (increase in power and amplitude) can result. To complete our example, let's assume we want a lowpass filter that emulates what the telephone companies do. Signals transmitted over the telephone have an upper frequency limit of about 3 kHz. For the second design choice, we require RFC=5.3x10-5. Thus, many choices for resistance and capacitance values are possible. A 1 μF capacitor and a 330 Ω resistor, 10 nF and 33 kΩ, and 10 pF and 33 MΩ would all theoretically work. Let's also desire a voltage gain of ten: $\frac{R_F}{R}=10$, which means $R=\frac{R_F}{10}$. Recall that we must have R<Rin. As the op-amp's input impedance is about 1 MΩ, we don't want R too large, and this requirement means that the last choice for resistor/capacitor values won't work. We also need to ask for less gain than the op-amp can provide itself. Because the feedback "element" is an impedance (a parallel resistor capacitor combination), we need to examine the gain requirement more carefully. We must have $\frac{|Z_F|}{R}<10^5$ for all frequencies of interest. Thus, $\frac{\frac{R_F}{|1+j2\pi fR_FC|}}{R}<10^5$. As this impedance decreases with frequency, the design specification of $\frac{R_F}{R}=10$ means that this criterion is easily met. Thus, the first two choices for the resistor and capacitor values (as well as many others in this range) will work well. Additional considerations like parts cost might enter into the picture. Unless you have a high-power application (this isn't one) or ask for high-precision components, costs don't depend heavily on component values as long as you stay close to standard values. For resistors, having values r10d, easily obtained values of r are 1, 1.4, 3.3, 4.7 and 6.8 and the decades span 0-8.

## Exercise 3.19.1

What is special about the resistor values; why these rather odd-appearing values for r?

## 3.19.3 Intuitive Way of Solving Op-Amp Circuits

When we meet op-amp design specifications, we can simplify our circuit calculations greatly, so much so that we don't need the op-amp's circuit model to determine the transfer function. Here is our inverting amplifier.

op-amp

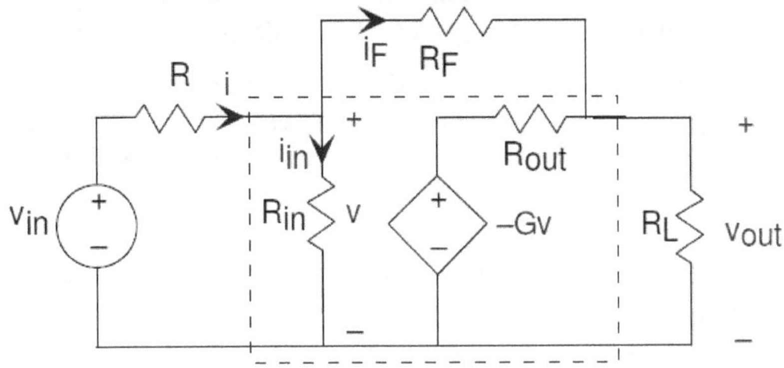


**Figure 3.45**

op-amp

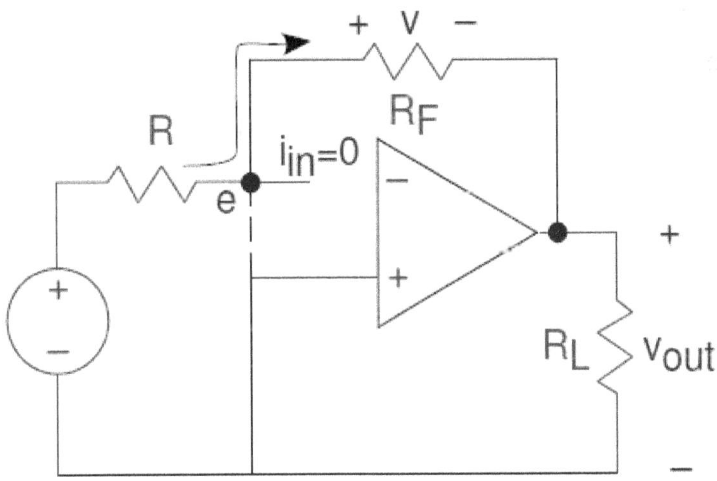


**Figure 3.46**

When we take advantage of the op-amp's characteristics—large input impedance, large gain, and small output impedance—we note the two following important facts.

- The current iin must be very small. The voltage produced by the dependent source is 105 times the voltage v. Thus, the voltage v must be shall, which means that $i_{in} = \frac{v}{R_{in}}$ must be tiny. For example, if the output is about 1 V, the voltage v=10-5 V, making the current iin=10-11 A. Consequently, we can ignore im in our calculations and assume it to be zero.

- Because of this assumption—essentially no current flow through Rin —the voltage v must also be essentially zero. This means that in op-amp circuits, the voltage across the op-amp's input is basically zero.

Armed with these approximations, let's return to our original circuit as shown in Figure 3.46 (op-amp). The node voltage e is essentially zero, meaning that it is essentially tied to the reference node. Thus, the current through the resistor R equals $\frac{v_{\text{in}}}{R}$. Furthermore, the feedback resistor appears in parallel with the load resistor. Because the current going into the op-amp is zero, all of the current flowing through R flows through the feedback resistor (iF=i)! The voltage across the feedback resistor R equals: $\frac{v_{\text{in}} R_F}{R}$. Because the left end of the feedback resistor is essentially attached to the reference node, the voltage across it equals the negative of that across the output resistor: $v_{\text{out}} = -v = -\frac{v_{\text{in}} R_F}{R}$. Using this approach makes analyzing new op-amp circuits much easier. When using this technique, check to make sure the results you obtain are consistent with the assumptions of essentially zero current entering the op-amp and nearly zero voltage across the op-amp's inputs.

### Example 3.8

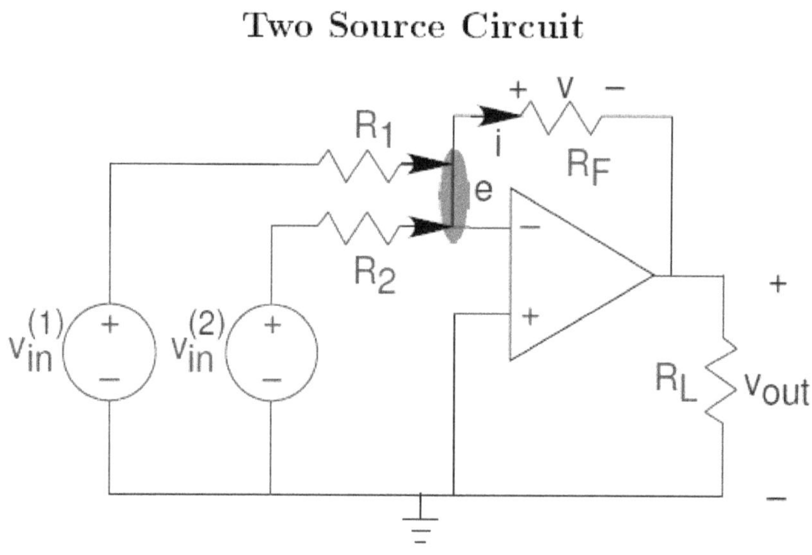


**Figure 3.47: Two-source, single-output op-amp circuit example.**

Let's try this analysis technique on a simple extension of the inverting amplifier configuration shown in the above Figure 3.47 (Two Source Circuit). If either of the source-resistor combinations were not present, the inverting amplifier remains, and we know that transfer function. By superposition, we know that the input-output relation is:

$$v_{\text{out}} = \left(-\left(\frac{R_F}{R_1}v_{\text{in}}^{(1)}\right)\right) - \frac{R_F}{R_2}v_{\text{in}}^{(2)} \tag{3.31}$$

When we start from scratch, the node joining the three resistors is at the same potential as the reference, e≃0 and the sum of currents flowing into that node is zero. Thus, the current i flowing in the resistor RF equals $\frac{v_{\text{in}}^{(1)}}{R_1} + \frac{v_{\text{in}}^{(2)}}{R_2}$. Because the feedback resistor is essentially in parallel with the load resistor, the voltages must satisfy v=-vout. In this way, we obtain the input-output relation given above.

What utility does this circuit have? Can the basic notion of the circuit be extended without bound?

## 3.20 THE DIODE

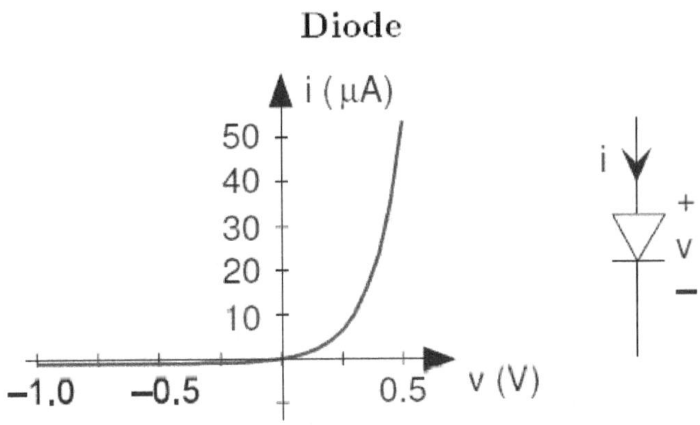


**Figure 3.48: v-i relation and schematic symbol for the diode. Here, the diode parameters were room temperature and I0=1μA**

The resistor, capacitor, and inductor are linear circuit elements because their voltage-current relations follow a linear mathematical pattern. Voltage and current sources are nonlinear devices, meaning that increasing the current through a voltage source does not result in a proportional increase in voltage. A diode is a very conspicuous and highly functional nonlinear circuit device (read more). The input-output relation follows an exponential pattern.

$$i(t) = I_0\left(e^{\frac{q}{kT}v(t)} - 1\right) \tag{3.32}$$

Here, the quantity q represents the charge of a single electron in coulombs, k is Boltzmann›s constant, and T is the diode's temperature in K. At room temperature, the ratio $\frac{kT}{q} = 25\text{ mV}$. The constant I0 is the leakage current, and is usually very small. Viewing this v-i relation in Figure 3.95, the nonlinearity becomes obvious. When the voltage is positive, current flows easily through the diode. This situation is known as forward biasing. When we apply a negative voltage, the current is quite small, and equals I0, known as the leakage or reverse-bias current. A less detailed model for the diode has any positive current flowing through the diode when it is forward biased, and no current when negative biased. Note that the diode›s schematic symbol looks like an arrowhead; the direction of current flow corresponds to the direction the arrowhead points.

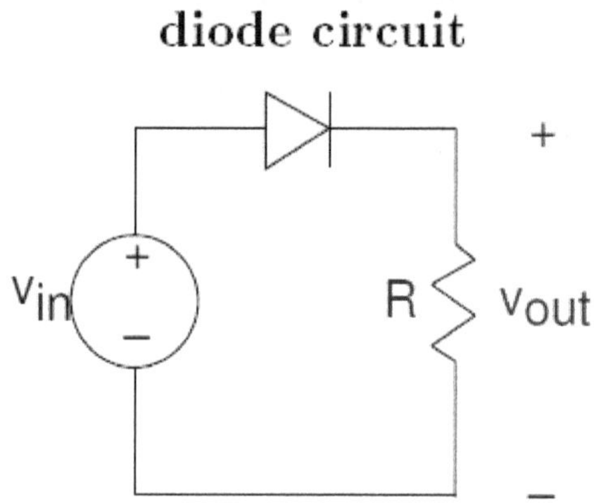


**Figure 3.49: The Diode circuit**

Because of the diode's nonlinear nature, we cannot use impedances nor series/parallel combination rules to analyze circuits containing them. The reliable node method can always be used; it only relies on KVL for its application, and KVL is a statement about voltage drops around a closed path regardless of whether the elements are linear or not. Thus, for this simple circuit we have

$$\frac{v_{\text{out}}}{R} = I_0\left(e^{\frac{q}{kT}(v_{\text{in}}-v_{\text{out}})} - 1\right) \tag{3.33}$$

This equation cannot be solved in closed form. We must understand what is going on from basic principles, using computational and graphical aids. As an approximation, when vin is positive, current flows through the diode so long as the voltage vout is smaller than vin (so the diode is forward biased). If the source is negative or vout «tries» to be bigger than vin the diode is reverse-biased, and the reverse-bias current flows through the diode. Thus, at this level of analysis, positive input voltages result in positive output voltages with negative ones resulting in: vout= – (RI0).

We need to detail the exponential nonlinearity to determine how the circuit distorts the input voltage waveform. We can of course numerically solve Figure 3.96 to determine the output voltage when the input is a sinusoid. To learn more, let's express this equation graphically. We plot each term as a function of vout for various values of the input voltage vin where they intersect gives us the output voltage. The left side, the current through the output resistor, does not vary itself with vin, and thus we have a fixed straight line. As for the right side, which expresses the diode's v-i relation, the point at which the curve crosses the vout axis gives us the value of vin. Clearly, the two curves will always intersect just once for any value of vin, and for positive vin the intersection occurs at a value for vout smaller than vin. This reduction is smaller if the straight line has a shallower slope, which corresponds to using a bigger output resistor. For negative vin, the diode is reverse-biased and the output voltage equals – (RI0).

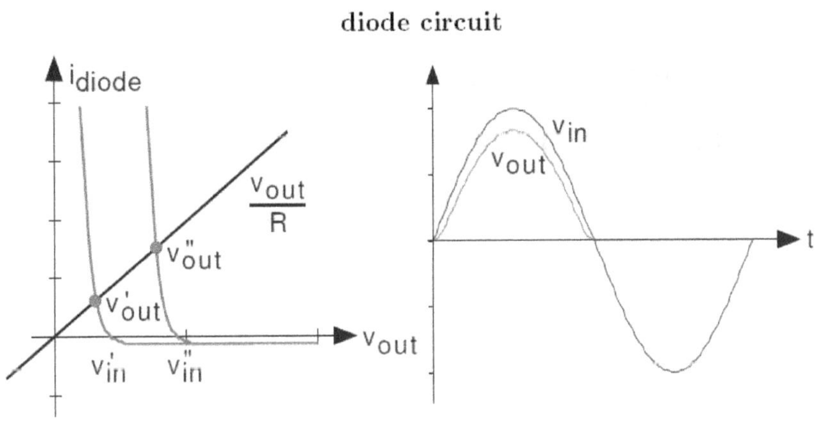


**Figure 3.50**

What utility might this simple circuit have? The diode's nonlinearity cannot be escaped here, and the clearly evident distortion must have some practical application if the circuit were to be useful. This circuit, known as a half-wave rectifier, is present in virtually every AM radio twice and each serves very different functions! We›ll learn what functions later.

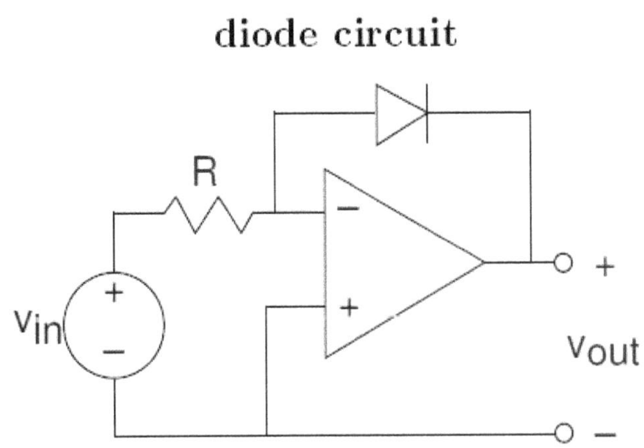


**Figure 3.51**

Here is a circuit involving a diode that is actually simpler to analyze than the previous one. We know that the current through the resistor must equal that through the diode. Thus, the diode's current is proportional to the input voltage. As the voltage across the diode is related to the logarithm of its current, we see that the input-output relation is:

$$v_{\text{out}} = -\left(\frac{kT}{q}\ln\left(\frac{v_{\text{in}}}{RI_0}+1\right)\right) \tag{3.34}$$

Clearly, the name logarithmic amplifier is justified for this circuit.

## 3.21 ANALOG SIGNAL PROCESSING PROBLEMS

### Problem 3.1: Simple Credit Analysis

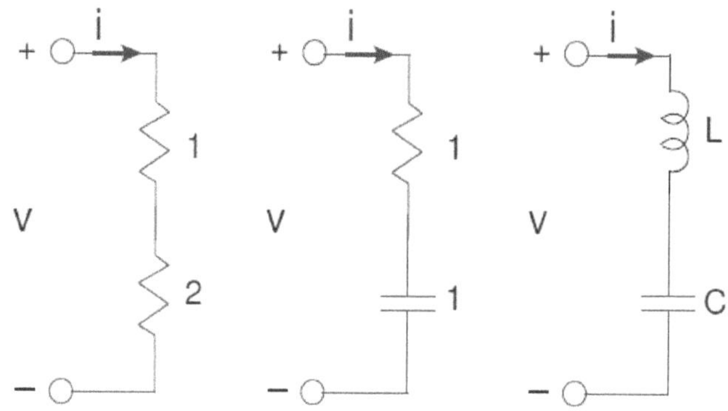


(a) Circuit a (b) Circuit b (c) Circuit c

**Figure 3.52**

For each circuit shown in Figure 3.52, the current (2πt).

- What is the voltage across each element and what is the voltage v in each case?
- For the last circuit, are there element values that make the voltage v equal zero for all time? If so, what element values work?
- Again, for the last circuit, if zero voltage were possible, what circuit element could substitute for the capacitor-inductor series combination that would yield the same voltage?

### Problem 3.2 Solving Simple Circuits

- Write the set of equations that govern Circuit A's behavior.
- Solve these equations for i1: In other words, express this current in terms of element and source values by eliminating non-source voltages and currents.

- For Circuit B, find the value for RL that results in a current of 5A passing through it.
- What is the power dissipated by the load resistor RL in this case?

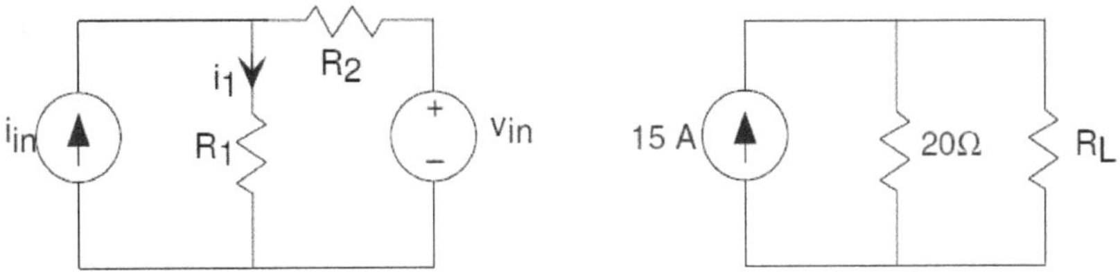


(a) Circuit A (b) Circuit B

**Figure 3.53**

## Problem 3.3 Equivalent Resistance

For each of the following circuits (Figure 3.54), find the equivalent resistance using series and parallel combination rules.

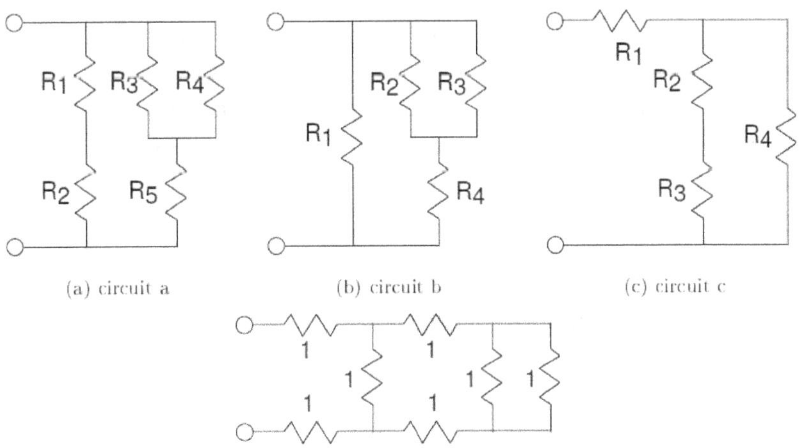


(a) circuit a (b) circuit b (c) circuit c (d) circuit d

**Figure 3.54**

Calculate the conductance seen at the terminals for circuit (c) in terms of each element's conductance. Compare this equivalent conductance formula with the equivalent resistance formula you found for circuit (b). How is the circuit (c) derived from circuit (b)?

## Problem 3.4: Superposition Principle

One of the most important consequences of circuit laws is the Superposition Principle: The current or voltage defined for any element equals the sum of the currents or voltages produced in the element by the independent sources. This Principle has important consequences in simplifying the calculation of ciruit variables in multiple source circuits.

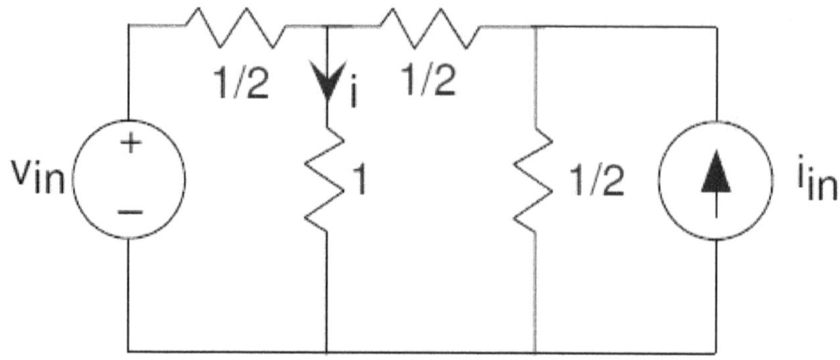


Figure 3.55

- For the depicted circuit (Figure 3.55), find the indicated current using any technique you like (you should use the simplest).
- You should have found that the current i is a linear combination of the two source values: i=C1 vin + C2in. This result means that we can think of the current as a superposition of two components, each of which is due to a source. We can find each component by setting the other sources to zero. Thus, to find the voltage source component, you can set the current source to zero (an open circuit) and use the usual tricks. To find the current source component, you would set the voltage source to zero (a short circuit) and find the resulting current. Calculate the total current i using the Superposition Principle. Is applying the Superposition Principle easier than the technique you used in part (1)?

## Problem 3.5: Current and Voltage Divider

Use current or voltage divider rules to calculate the indicated circuit variables in Figure 3.56.

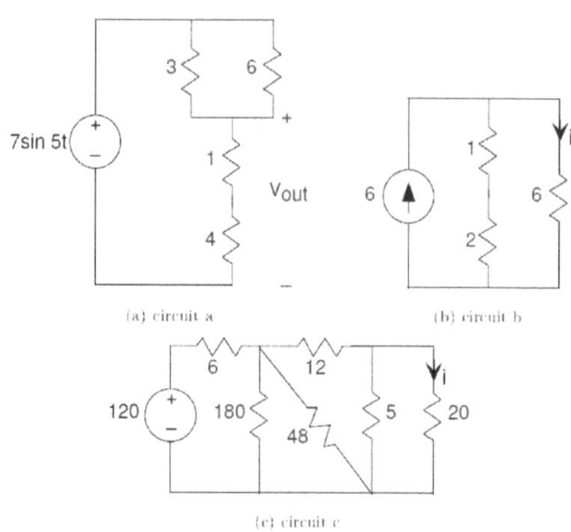


Figure 3.56

## Problem 3.6: Thévenin and Mayer-Norton Equivalents

Find the Thévenin and Mayer-Norton equivalent circuits for the following circuits (Figure 3.57).

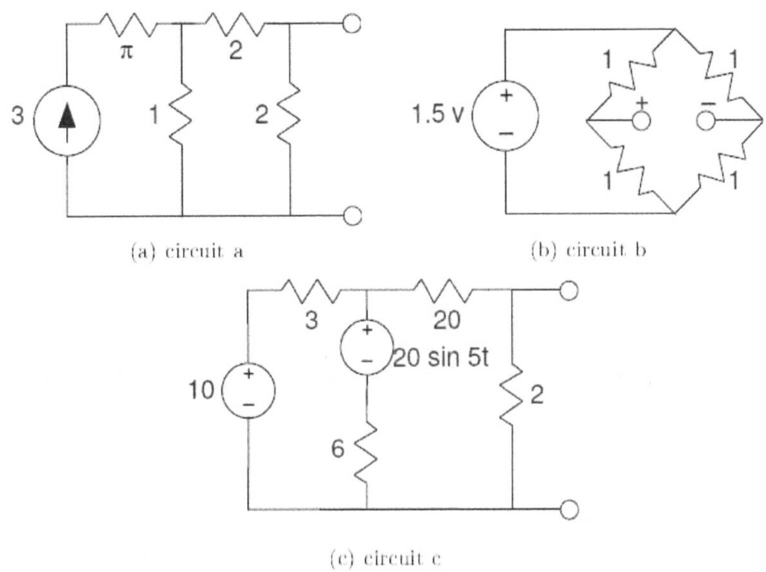


(a) circuit a

(b) circuit b

(c) circuit c

**Figure 3.57**

## Problem 3.7: Detective Work

In the depicted circuit (Figure 3.58), the circuit N1 has the v-i relation v1=3i1+7 when is=2.

- a) Find the Thévenin equivalent circuit for circuit N2.
- b) With is=2 determine R such that i1=-1.

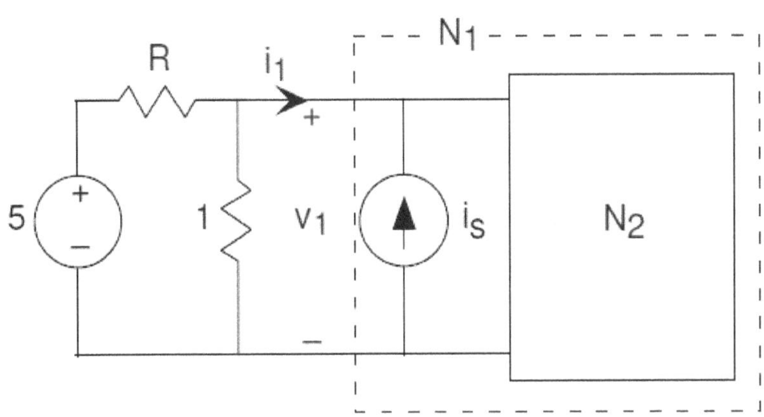


**Figure 3.58**

## Problem 3.8: Bridge Circuits

Circuits having the form of Figure 3.59 are termed bridge circuits.

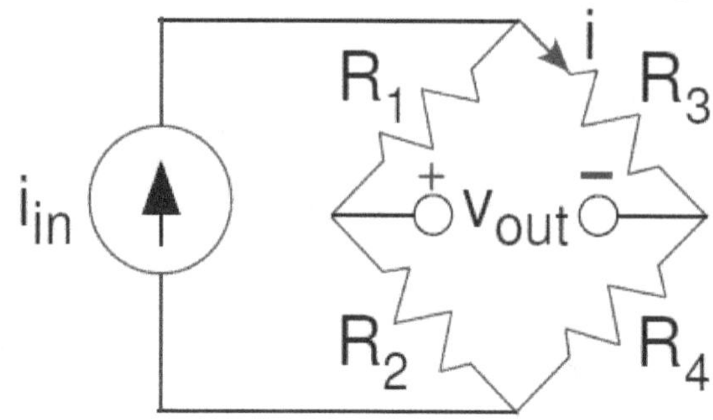


**Figure 3.59**

- What resistance does the current source see when nothing is connected to the output terminals?
- What resistor values, if any, will result in a zero voltage for vout?
- Assume R1=1Ω, R2=2Ω, R3=2Ω, R4=4Ω. Find the current i when the current source iin is Im ((4+2i)ej2π20t) Express your answer as a sinusoid.

## Problem 3.9: Cartesian to Polar Conversion

Convert the following expressions into polar form. Plot their location in the complex plane.

- $\left(1+\sqrt{-3}\right)^2$
- $\frac{2-j\frac{6}{\sqrt{3}}}{2+j\frac{6}{\sqrt{3}}}$
- $\left(4-j^3\right)\left(1+j\frac{1}{2}\right)$
- $3e^{j\pi}+4e^{j\frac{\pi}{2}}$
- $\left(\sqrt{3}+j\right)2\times\sqrt{2}e^{-\left(j\frac{\pi}{4}\right)}$
- $\frac{3}{1+j3\pi}$

## Problem 3.10: The Complex Plane

The complex variable z is related to the real variable u according to

$$z=1+e^{ju}$$

- Sketch the contour of values z takes on in the complex plane.
- What are the maximum and minimum values attainable by |z|?
- Sketch the contour the rational function $\frac{z-1}{z+1}$ traces in the complex plane.

## Problem 3.11: Cool Curves

In the following expressions, the variable x runs from zero to infinity. What geometric shapes do the following trace in the complex plane?

- ejx
- 1+ejx
- e-x ejx
- $e^{jx} + e^{j\left(x+\frac{\pi}{4}\right)}$

## Problem 3.12: Trigonometric Identities and Complex Exponentials

Show the following trigonometric identities using complex exponentials. In many cases, they were derived using this approach.

- sin(2u)=2sin(u)cos(u)
- $\cos^2(u) = \frac{1+\cos(2u)}{2}$
- cos2(u)+sin2(u)=1
- $\frac{d}{du}(\sin(u)) = \cos(u)$

## Problem 3.13: Transfer Functions

Find the transfer function relating the complex amplitudes of the indicated variable and the source shown in the Figure 3.60. Plot the magnitude and phase of the transfer function.

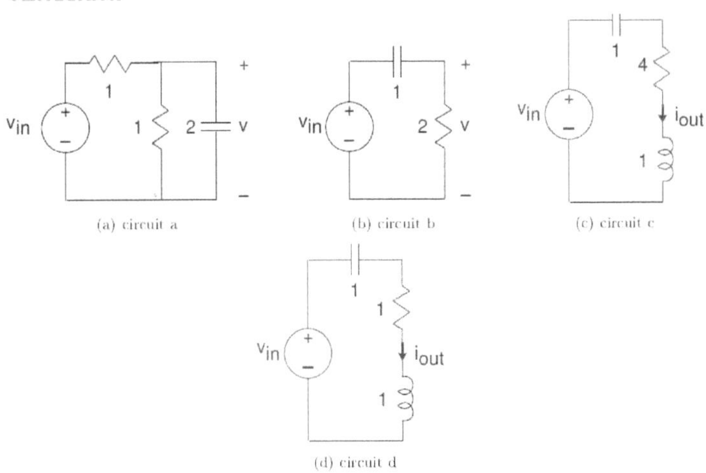


**Figure 3.60**

## Problem 3.14: Using Impedances

Find the differential equation relating the indicated variable to the source(s) using impedances for each circuit shown in Figure 3.61.

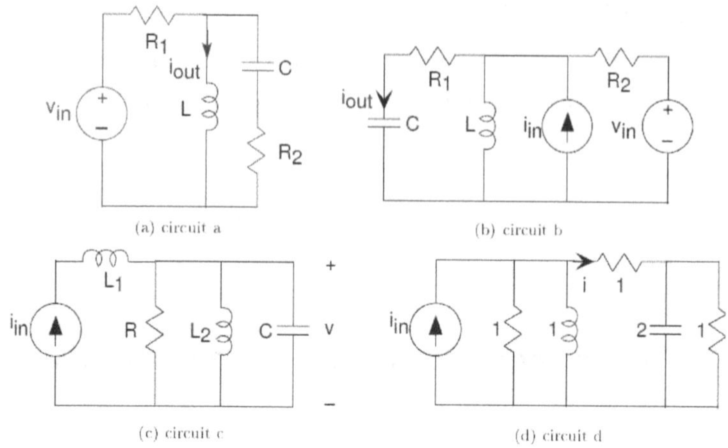


Figure 3.61

## Problem 3.15: Measurement Chaos

The simple circuit in (Figure 3.62) was constructed but the signal measurements were made haphazardly. When the source was: (2πf0t), the current: i(t) equalled $\frac{\sqrt{2}}{3}\sin\left(2\pi f_0 t + \frac{\pi}{4}\right)$ and the voltage $v_2(t) = \frac{1}{3}\sin(2\pi f_0 t)$.

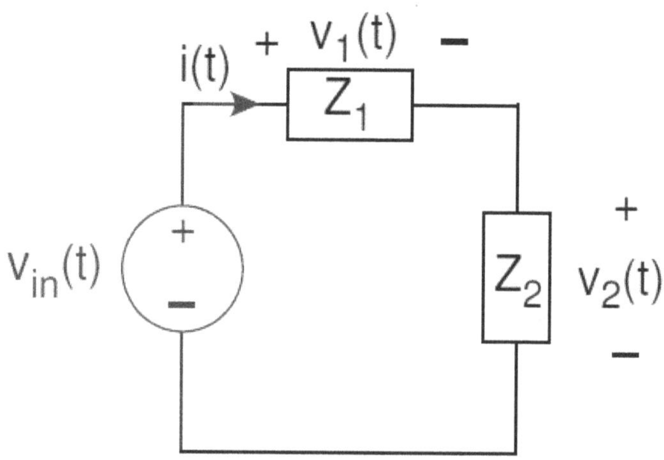


Figure 3.62

- What is the voltage v1(t)?
- Find the impedances Z1 and Z2.
- Construct these impedances from elementary circuit elements.

## Problem 3.16: Transfer Functions

In the circuit in (Figure3.63), the voltage source equals : $v_{\text{in}}(t) = 10\sin\left(\frac{t}{2}\right)$.

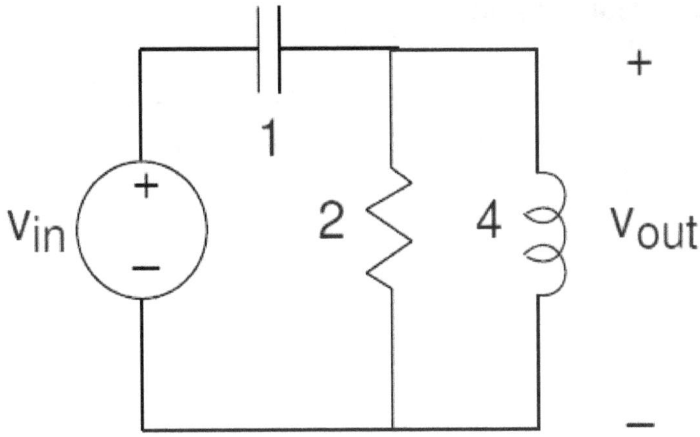


**Figure 3.63**

- Find the transfer function between the source and the indicated output voltage.
- For the given source, find the output voltage.

## Problem 3.17: A Simple Circuit

You are given this simple circuit (Figure 3.64).

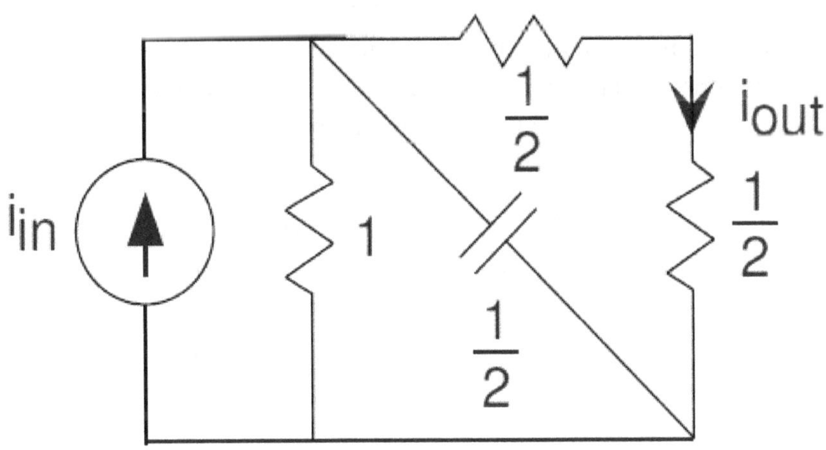


**Figure 3.64**

- What is the transfer function between the source and the indicated output current?
- If the output current is measured to be Cos(2t), what was the source?

## Problem 3.18: Circuit Design

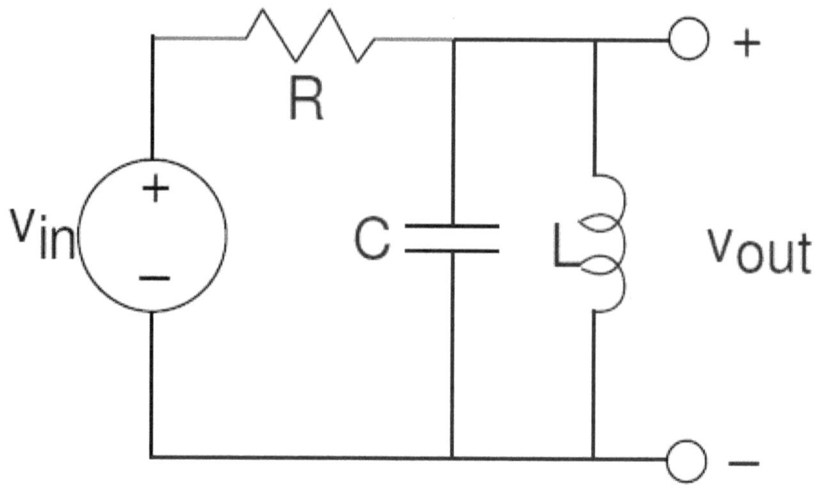


Figure 3.65

- Find the transfer function between the input and the output voltages for the circuits shown in Figure 3.65.
- At what frequency does the transfer function have a phase shift of zero? What is the circuit's gain at this frequency?
- Specifications demand that this circuit have an output impedance (its equivalent impedance) less than 8Ω for frequencies above 1 kHz, the frequency at which the transfer function is maximum. Find element values that satisfy this criterion.

## Problem 3.19: Equivalent Circuits and Power

Suppose we have an arbitrary circuit of resistors that we collapse into an equivalent resistor using the series and parallel rules. Is the power dissipated by the equivalent resistor equal to the sum of the powers dissipated by the actual resistors comprising the circuit? Let's start with simple cases and build up to a complete proof.

- Suppose resistors R1 and R2 are connected in parallel. Show that the power dissipated by R1||R2 equals the sum of the powers dissipated by the component resistors.
- Now suppose R1 and R2 are connected in series. Show the same result for this combination.
- Use these two results to prove the general result we seek.

## Problem 3.20: Power Transmission

The network shown in the figure represents a simple power transmission system. The generator produces 60 Hz and is modeled by a simple Thévenin equivalent. The transmission line consists of a long length of copper wire and can be accurately described as a 50Ω resistor.

- Determine the load current RL and the average power the generator must produce so that the load receives 1,000 watts of average power.
- Why does the generator need to generate more than 1,000 watts of average power to meet this requirement?
- Suppose the load is changed to that shown in the figure for Modified load circuit. Now how much power must the generator produce to meet the same power requirement? Why is it more than it had to produce to meet the requirement for the resistive load?
- The load can be compensated to have a unity power factor (see exercise (Exercise 3.11.2)) so that the voltage and current are in phase for maximum power efficiency. The compensation technique is to place a circuit in parallel to the load circuit. What element works and what is its value?
- With this compensated circuit, how much power must the generator produce to deliver 1,000 average power to the load?

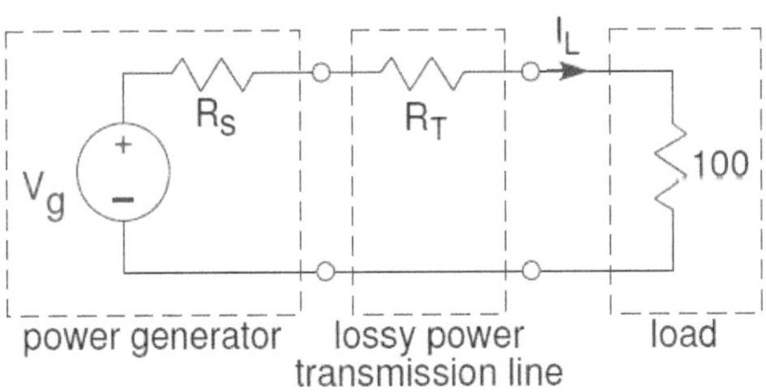


(a) Simple power transmission system

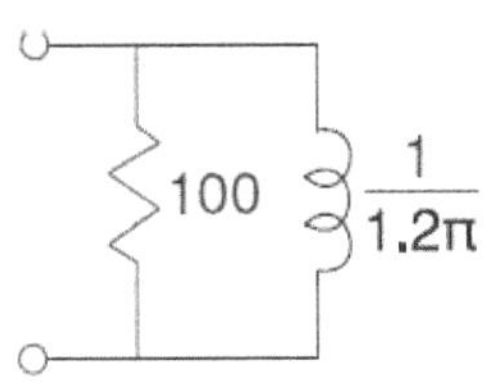


(b) Modified load circuit

**Figure 3.66**

## Problem 3.21: Optimal Power Transmission

The following figure (Figure 3.67) shows a general model for power transmission. The power generator is represented by a Thévinin equivalent and the load by a simple impedance. In most applications, the source components are fixed while there is some latitude in choosing the load.

- Suppose we wanted the maximize "voltage transmission:" make the voltage across the load as large as possible. What choice of load impedance creates the largest load voltage? What is the largest load voltage?
- If we wanted the maximum current to pass through the load, what would we choose the load impedance to be? What is this largest current?
- What choice for the load impedance maximizes the average power dissipated in the load? What is most power the generator can deliver?

Note: One way to maximize a function of a complex variable is to write the expression in terms of the variable's real and imaginary parts, evaluate derivatives with respect to each, set both derivatives to zero and solve the two equations simultaneously.

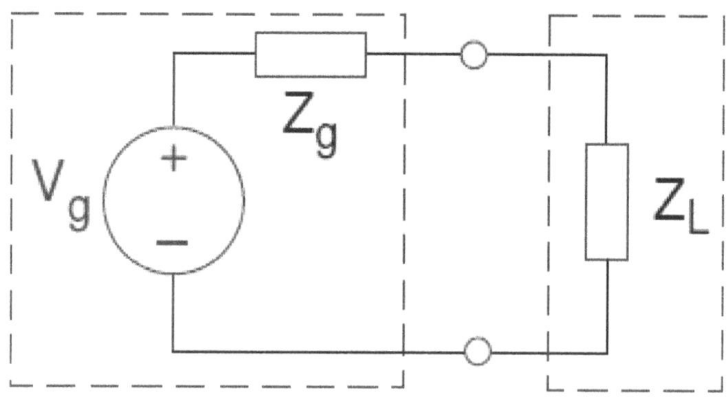


**Figure 3.67**

## Problem 3.22: Big is Beautiful

Sammy wants to choose speakers that produce very loud music. He has an amplifier and notices that the speaker terminals are labeled «8Ω source».

- What does this mean in terms of the amplifier's equivalent circuit?
- Any speaker Sammy attaches to the terminals can be well-modeled as a resistor. Choosing a speaker amounts to choosing the values for the resistor. What choice would maximize the voltage across the speakers?
- Sammy decides that maximizing the power delivered to the speaker might be a better choice. What values for the speaker resistor should be chosen to maximize the power delivered to the speaker?

## Problem 3.23: Sharing a Channel

Two transmitter-receiver pairs want to share the same digital communications channel. The transmitter signals will be added together by the channel. Receiver design is greatly simplified if first we remove the unwanted transmission (as much as possible). Each transmitter signal has the form:

$$x_i(t) = A\sin(2\pi f_i t) \quad , \quad 0 \leq t \leq T$$

where the amplitude is either zero or A and each transmitter uses its own frequency fi. Each frequency is harmonically related to the bit interval duration T, where the transmitter 1 uses the the frequency 1/T. The datarate is 10Mbps.

- Draw a block diagram that expresses this communication scenario.
- Find circuits that the receivers could employ to separate unwanted transmissions. Assume the received signal is a voltage and the output is to be a voltage as well.
- Find the second transmitter's frequency so that the receivers can suppress the unwanted transmission by at least a factor of ten.

### Problem 3.24: Circuit Detective Work

In the lab, the open-circuit voltage measured across an unknown circuit's terminals equals sin(t). When a 1Ω resistor is place across the terminals, a voltage of $\frac{1}{\sqrt{2}}\sin\left(t+\frac{\pi}{4}\right)$appears.

- What is the Thévenin equivalent circuit?
- What voltage will appear if we place a 1F capacitor across the terminals?

### Problem 3.25: Mystery Circuit

We want to determine as much as we can about the circuit lurking in the impenetrable box shown in Figure 3.68. A voltage source vin=2V has been attached to the left-hand terminals, leaving the right terminals for tests and measurements.

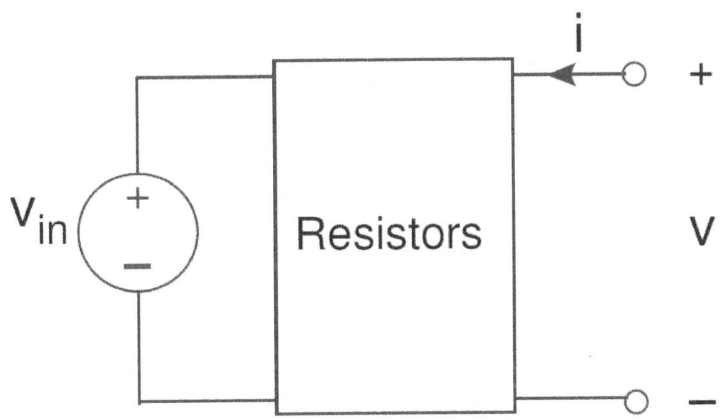


**Figure 3.68**

- Sammy measures v=10V when a 1 Ω resistor is attached to the terminals. Samantha says he is wrong. Who is correct and why?

- When nothing is attached to the right-hand terminals, a voltage of v=1V is measured. What circuit could produce this output?
- When a current source is attached so that i=2A, the voltage v is now 3 V. What resistor circuit would be consistent with this and the previous part?

## Problem 3.26: More Circuit Detective Work

The left terminal pair of a two terminal-pair circuit is attached to a testing circuit. The test source vin (t) equals sin (t) (Figure 3.69).

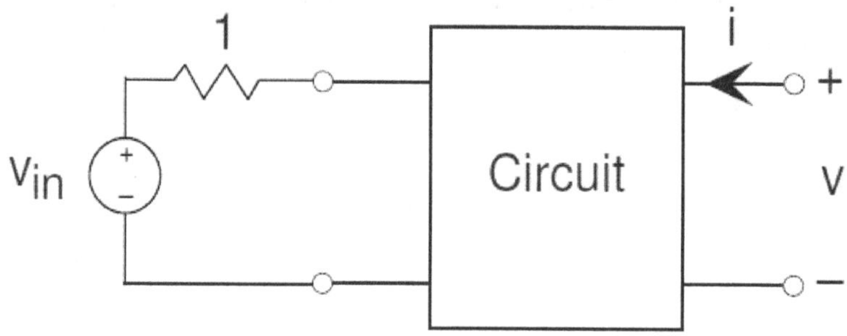


**Figure 3.69**

We make the following measurements:

- With nothing attached to the terminals on the right, the voltage v(t) equals $\frac{1}{\sqrt{2}}\cos\left(t+\frac{\pi}{4}\right)$.
- When a wire is placed across the terminals on the right, the current i(t) was –sin(t).
  - a) What is the impedance "seen" from the terminals on the right?
  - b) Find the voltage v(t) if a current source is attached to the terminals on the right so that i(t) = sin (t).

## Problem 3.27: Linear, Time-Invariant Systems

For a system to be completely characterized by a transfer function, it needs not only be linear, but also to be time-invariant. A system is said to be time-invariant if delaying the input delays the output by the same amount. Mathematically, if S(x(t))=y(t),meaning y(t) is the output of a system S(•) when x(t) is the input, S(•) is the time-invariant if: S(x(t–τ))=y(t–τ) for all delays τ and all inputs x(t). Note that both linear and nonlinear systems have this property. For example, a system that squares its input is time-invariant.

- Show that if a circuit has fixed circuit elements (their values don't change over time), its input-output relationship is time-invariant. Hint: Consider the differential equation that describes a circuit's input-output relationship. What is its general form? Examine the derivative(s) of delayed signals.

- Show that impedances cannot characterize time-varying circuit elements (R, L, and C). Consequently, show that linear, time-varying systems do not have a transfer function.
- Determine the linearity and time-invariance of the following. Find the transfer function of the linear, time-invariant (LTI) one(s).
  - diode
  - $y(t) = x(t) \sin(2\pi f0t)$
  - $y(t) = x(t-\tau 0)$
  - $y(t) = x(t) + N(t)$

## Problem 3.28: Long and Sleepless Nights

Sammy went to lab after a long, sleepless night, and constructed the circuit shown in Figure 3.70. He cannot remember what the circuit, represented by the impedance Z was. Clearly, this forgotten circuit is important as the output is the current passing through it.

- What is the Thévenin equivalent circuit seen by the impedance?
- In searching his notes, Sammy finds that the circuit is to realize the transfer function:

$$H(f) = \frac{1}{j10\pi f + 2}$$

Find the impedance Z as well as values for the other circuit elements.

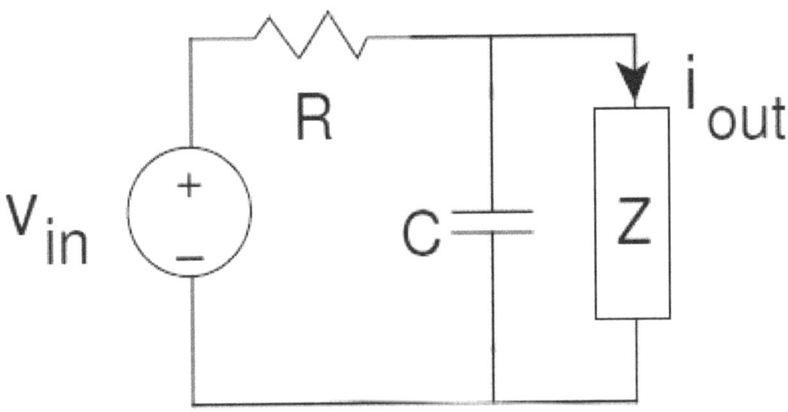


**Figure 3.70**

## Problem 3.29: A Testing Circuit

The simple circuit as shown in (Figure 3.71) was given on a test.

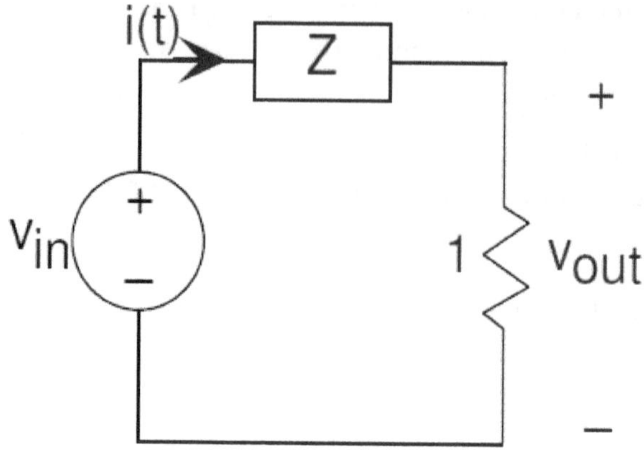


**Figure 3.71**

When the voltage source is: $\sqrt{5}\sin(t)$, the current $i(t) = \sqrt{2}\cos\left(t - \arctan(2) - \frac{\pi}{4}\right)$.

- What is voltage vout(t)?
- What is the impedance Z at the frequency of the source?

## Problem 3.30: Black-Box Circuit

You are given a circuit shown in (Figure 3.72) that has two terminals for attaching circuit elements.

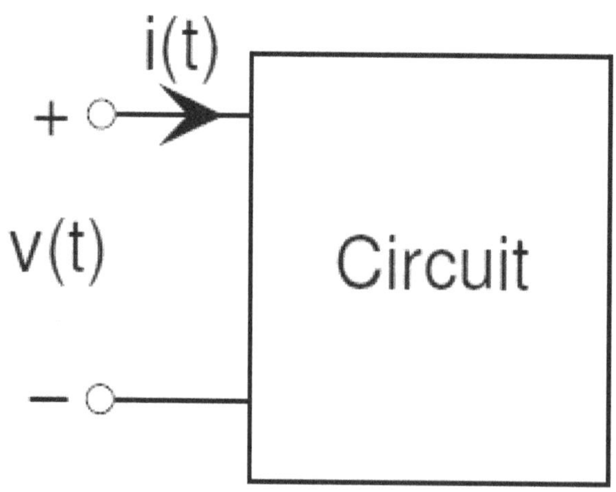


**Figure 3.72**

When you attach a voltage source equaling sin(t) to the terminals, the current through the source equals: $4\sin\left(t + \frac{\pi}{4}\right) - 2\sin(4t)$. When no source is attached (open-circuited terminals), the voltage across the terminals has the form: $A\sin(4t + \phi)$.

- What will the terminal current be when you replace the source by a short circuit?
- If you were to build a circuit that was identical (from the viewpoint of the terminals) to the given one, what would your circuit be?
- For your circuit, what are A and φ?

## Problem 3.31: Solving a Mystery Circuit

Sammy must determine as much as he can about a mystery circuit by attaching elements to the terminal and measuring the resulting voltage. When he attaches a 1Ω resistor to the circuit's terminals, he measures the voltage across the terminals to be 3sin(t). When he attaches a 1F capacitor across the terminals, the voltage is now: $3 \times \sqrt{2}\sin\left(t - \frac{\pi}{4}\right)$.

- What voltage should he measure when he attaches nothing to the mystery circuit?
- What voltage should Sammy measure if he doubled the size of the capacitor to 2 F and attached it to the circuit?

## Problem 3.32: Find the Load Impedance

The circuit in (Figure 3.73) has a transfer function between the output voltage and the source equal to

$$H(f) = \frac{-8\pi^2 f^2}{8\pi^2 f^2 + 4 + j6\pi f}$$

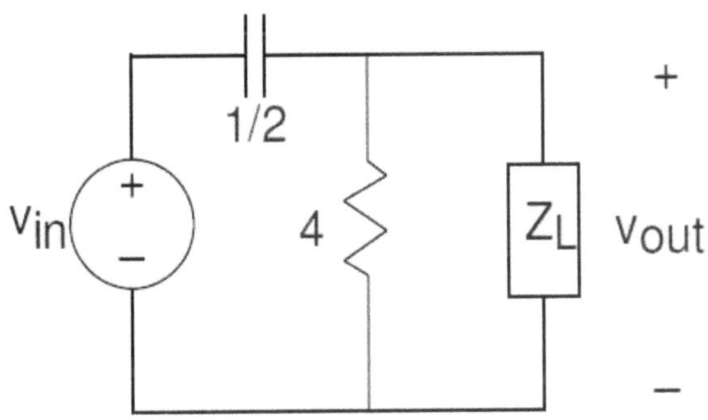


**Figure 3.73**

- Sketch the magnitude and phase of the transfer function.
- At what frequency does the phase equal $\frac{\pi}{2}$?
- Find a circuit that corresponds to this load impedance. Is your answer unique? If so, show it to be so; if not, give another example.

## Problem 3.33: Analog "Hum" Rejection

"Hum" refers to corruption from wall socket power that frequently sneaks into circuits. "Hum" gets its name because it sounds like a persistent humming sound. We want to find a circuit that will remove hum from any signal. A Rice engineer suggests using a simple voltage divider circuit (Figure 3.74) consisting of two series impedances.

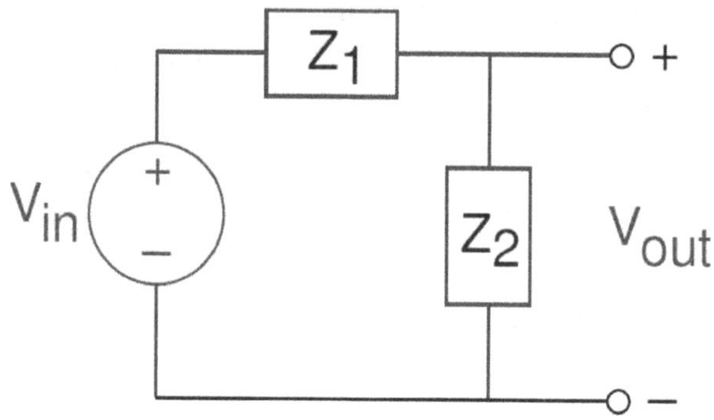


Figure 3.74

- The impedance Z1 is a resistor. The engineer must decide between the two circuits shown in the other figure for the impedance Z2. Which of these will work?
- Picking one circuit that works, choose circuit element values that will remove hum.
- Sketch the magnitude of the resulting frequency response.

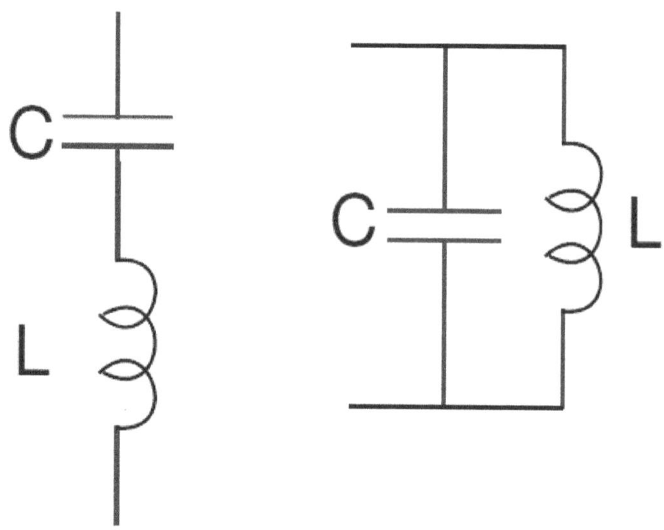


Figure 3.75

**Problem 3.34: An Interesting Circuit**

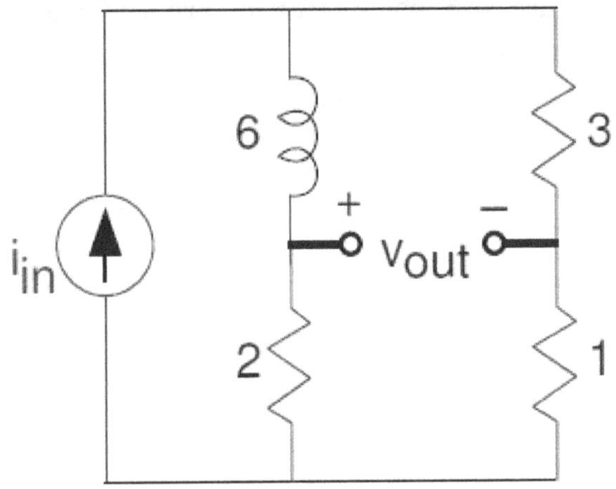


Figure 3.76

- For the circuit shown in Figure 3.76, find the transfer function.
- What is the output voltage when the input has the form: iin=5sin(2000πt)

**Problem 3.35: A Simple Circuit**

You have given the depicted circuit (Figure 3.77).

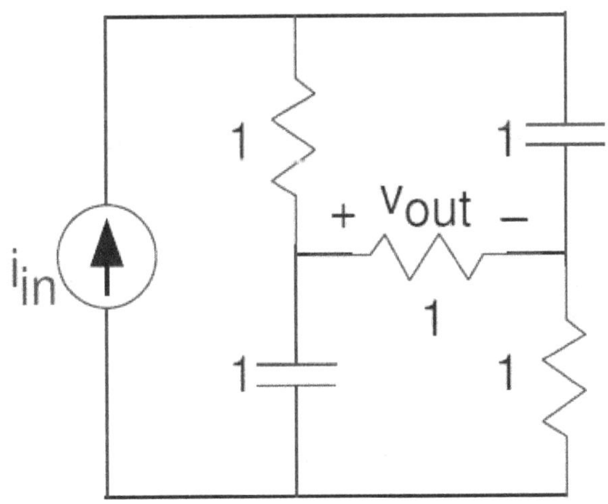


Figure 3.77

- What is the transfer function between the source and the output voltage?
- What will the voltage be when the source equals sin(t)?
- Many function generators produce a constant offset in addition to a sinusoid. If the source equals 1+sin(t), what is the output voltage?

## Problem 3.36: An Interesting and Useful Circuit

The circuit in (Figure 3.78) has interesting properties, which are exploited in high-performance oscilloscopes.

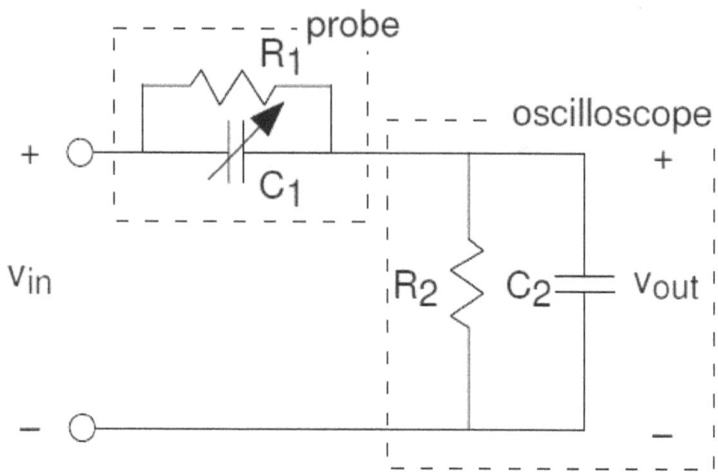


**Figure 3.78**

The portion of the circuit labeled "Oscilloscope" represents the scope's input impedance. R2=1MΩ and C2=30pF (note the label under the channel 1 input in the lab's oscilloscopes). A probe is a device to attach an oscilloscope to a circuit, and it has the indicated circuit inside it.

- Suppose for a moment that the probe is merely a wire and that the oscilloscope is attached to a circuit that has a resistive Thévenin equivalent impedance. What would be the effect of the oscilloscope's input impedance on measured voltages?
- Using the node method, find the transfer function relating the indicated voltage to the source when the probe is used.
- Plot the magnitude and phase of this transfer function when R1=9Ω and C1=2pF.
- For a particular relationship among the element values, the transfer function is quite simple. Find that relationship and describe what is so special about it.
- The arrow through C1 indicates that its value can be varied. Select the value for this capacitor to make the special relationship valid. What is the impedance seen by the circuit being measured for this special value?

## Problem 3.37: A Circuit Problem

You are given a circuit as shown in (Figure 3.79).

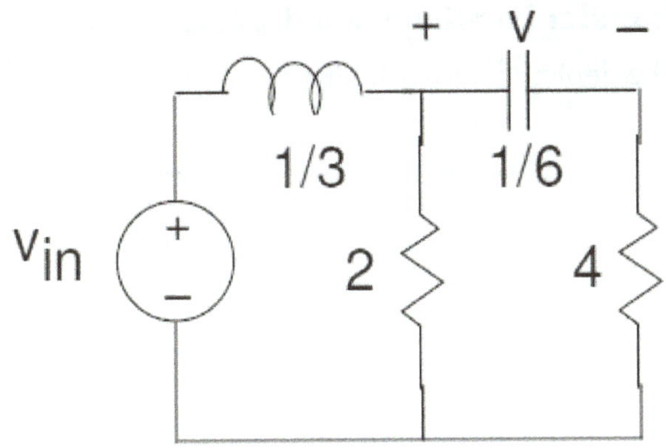


Figure 3.79

- Find the differential equation relating the output voltage to the source.
- What is the impedance "seen" by the capacitor?

## Problem 3.38: Analog Computers

Because the differential equations arising in circuits resemble those that describe mechanical motion, we can use circuit models to describe mechanical systems. An ELEC 241 student wants to understand the suspension system on his car. Without a suspension, the car's body moves in concert with the bumps in the raod. A well-designed suspension system will smooth out bumpy roads, reducing the car's vertical motion. If the bumps are very gradual (think of a hill as a large but very gradual bump), the car's vertical motion should follow that of the road. The student wants to find a simple circuit that will model the car's motion. He is trying to decide between two circuit models (Figure 3.80).

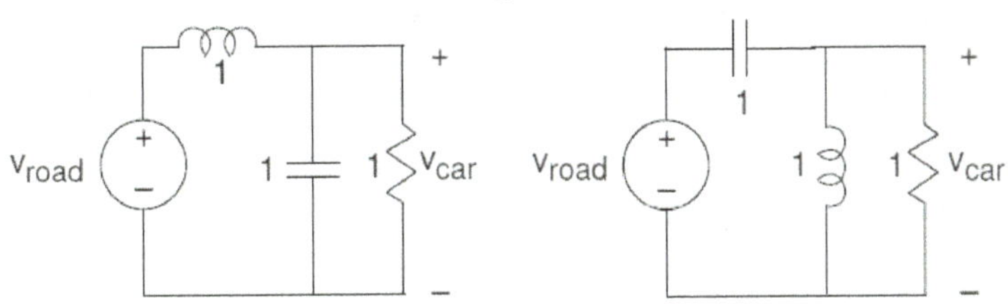


Figure 3.80

Here, road and car displacements are represented by the voltages vroad (t) and vcar (t) respectively.

- Which circuit would you pick? Why?
- For the circuit you picked, what will be the amplitude of the car's motion if the road has a displacement given by: vroad(t)=1+sin(2t)?

## Problem 3.39: Transfer Functions and Circuits

You are given the depicted network (Figure 3.81).

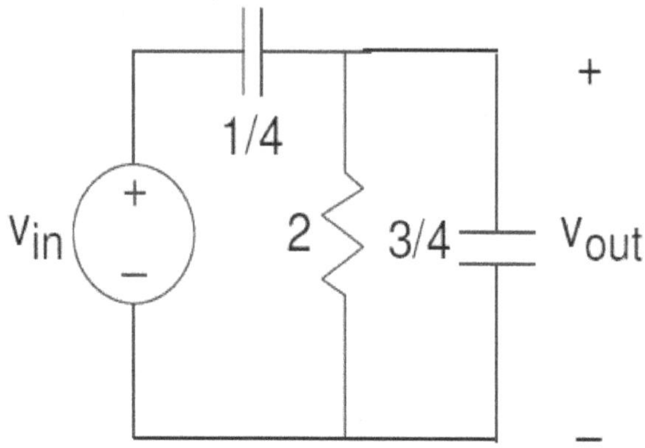


**Figure 3.81**

- Find the transfer function between Vin and Vout.
- Sketch the magnitude and phase of your transfer function. Label important frequency, amplitude and phase values.
- Find vout(t) when $v_{in}(t) = \sin\left(\frac{t}{2} + \frac{\pi}{4}\right)$.

## Problem 3.40: Fun in the Lab

You are given an unopenable box that has two terminals sticking out. You assume the box contains a circuit. You measure the voltage $\sin\left(t + \frac{\pi}{4}\right)$ across the terminals when nothing is connected to them and the current $\sqrt{2}\cos(t)$ when you place a wire across the terminals.

- Find a circuit that has these characteristics.
- You attach a 1 H inductor across the terminals. What voltage do you measure?

## Problem 3.41: Dependent Sources

Find the voltage vout in each of the circuits in (Figure 3.82).

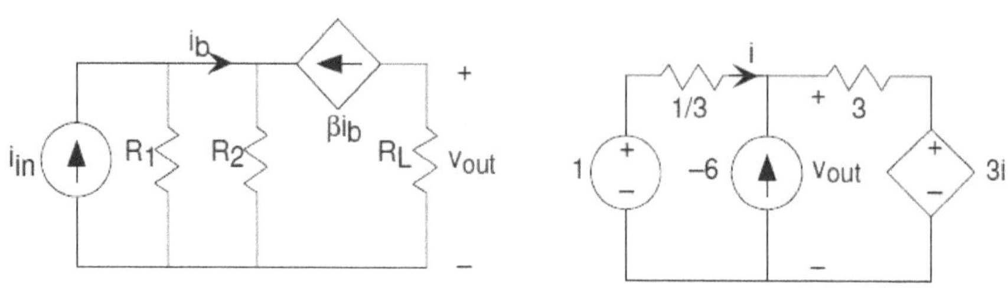


(a) circuit a

(b) circuit b

**Figure 3.82**

## Problem 3.42: Operational Amplifiers

Find the transfer function between the source voltage(s) and the indicated output voltage for the circuits shown in Figure 3.83.

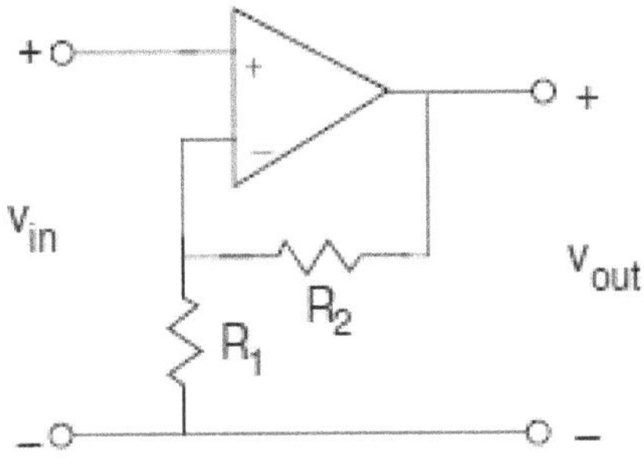


(a) op amp a

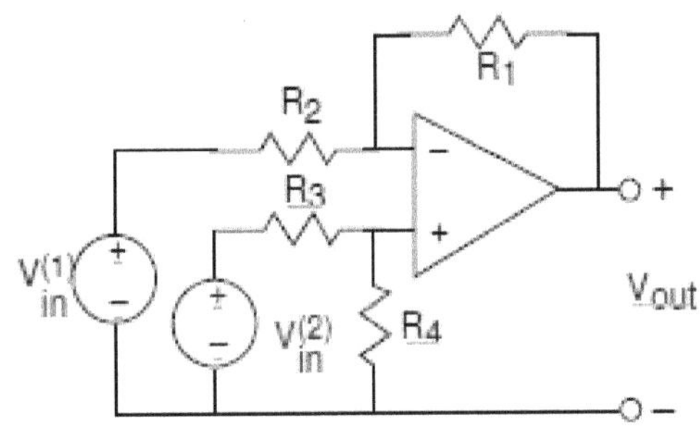


(b) op amp b

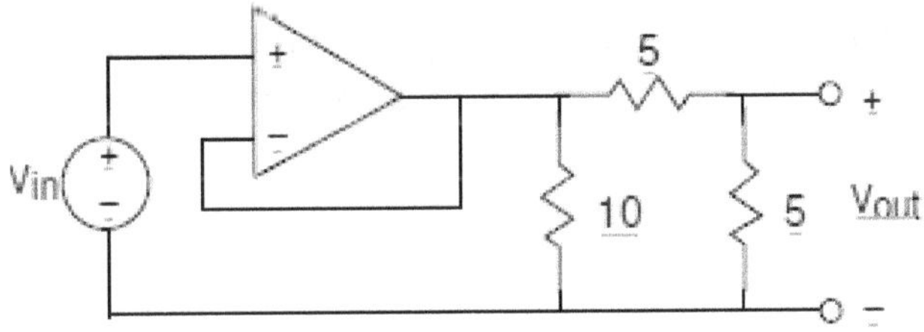


(c) op amp c

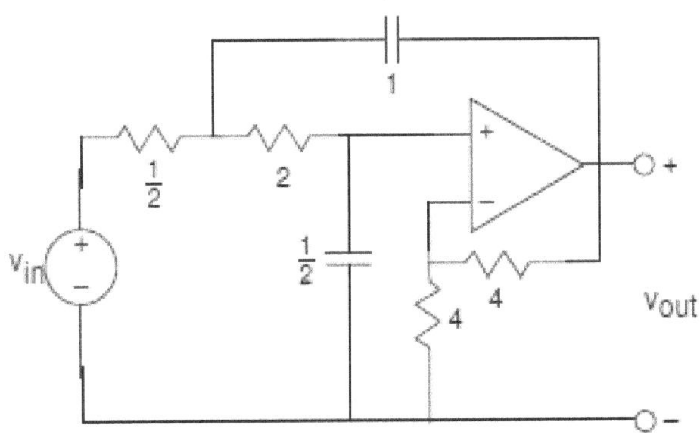


(d) op amp d

**Figure 3.83**

## Problem 3.43: Op-Amp Circuit

The following circuit (Figure 3.84) is claimed to serve a useful purpose.

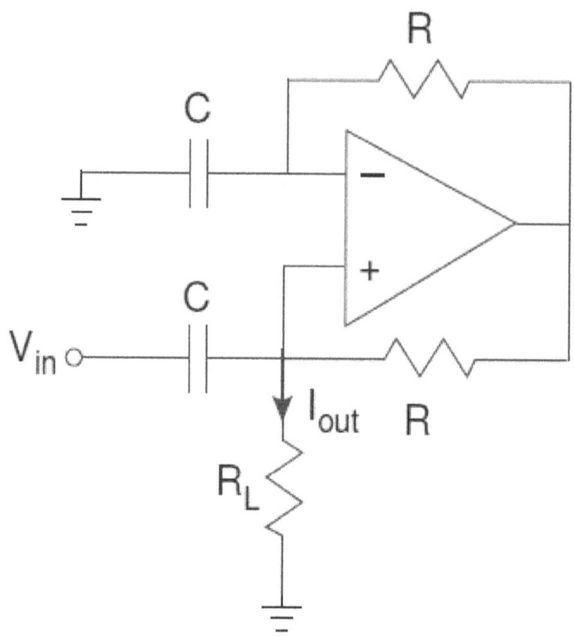


**Figure 3.84**

- What is the transfer function relating the complex amplitude of the output signal, the current Iout, to the complex amplitude of the input, the voltage Vin?
- What equivalent circuit does the load resistor RL see?
- Find the output current when $v_{in} = V_0 e^{-\frac{t}{\tau}}$.

## Problem 3.44: Why Op-Amps are Useful

The circuit (Figure 3.85) of a cascade of op-amp circuits illustrate the reason why op-amp realizations of transfer functions are so useful.

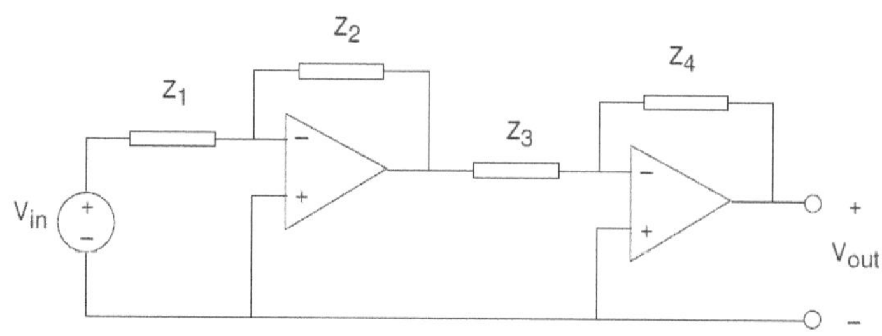


**Figure 3.85**

- Find the transfer function relating the complex amplitude of the voltage vout(t) to the source. Show that this transfer function equals the product of each stage's transfer function.
- What is the load impedance appearing across the first op-amp's output?
- Figure 3.86 below illustrates that sometimes "designs" can go wrong. Find the transfer function for this op-amp circuit, and then show that it can't work! Why can't it?

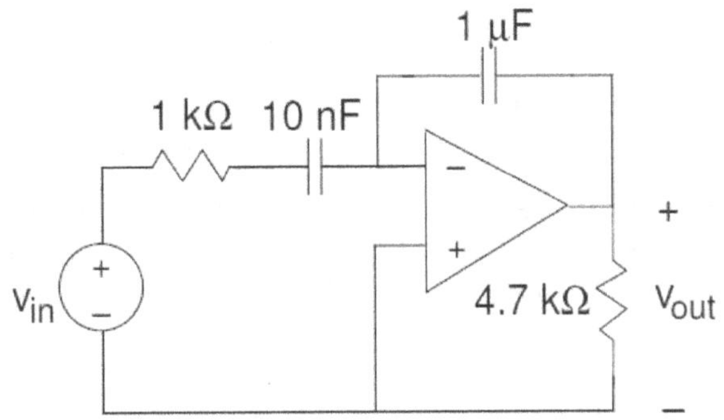


Figure 3.86

## Problem 3.45: Operational Amplifiers

Consider the circuit in (Figure 3.87).

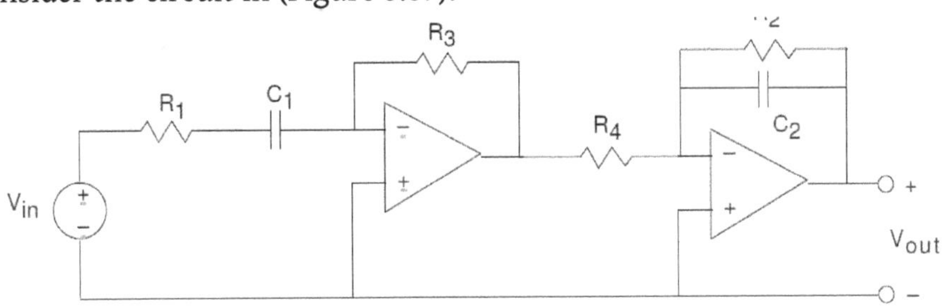


Figure 3.87

- Find the transfer function relating the voltage vout(t) to the source.
- R1=530Ω, C1=1μF, R2=5.3KΩ, C2=0.01μF, R3=R4=5.3KΩ. Characterize the resulting transfer function and determine what use this circuit might have.

## Problem 3.46: Designing a Bandpass Filter

We want to design a bandpass filter that has transfer the function

$$H(f) = 10\frac{j2\pi f}{\left(j\frac{f}{f_l}+1\right)\left(j\frac{f}{f_h}+1\right)}$$

Here, fl is the cutoff frequency of the low-frequency edge of the passband and fh is the cutoff frequency of the high-frequency edge. We want fl = 1KHz and fh = 10KHz.

- Plot the magnitude and phase of this frequency response. Label important amplitude and phase values and the frequencies at which they occur.
- Design a bandpass filter that meets these specifications. Specify component values.

## Problem 3.47: Pre-emphasis or De-emphasis?

In audio applications, prior to analog-to-digital conversion signals are passed through what is known as a pre-emphasis circuit that leaves the low frequencies alone but provides increasing gain at increasingly higher frequencies beyond some frequency f0. De-emphasis circuits do the opposite and are applied after digital-to-analog conversion. After pre-emphasis, digitization, conversion back to analog and de-emphasis, the signal's spectrum should be what it was.

The op-amp circuit here (Figure 3.88) has been designed for pre-emphasis or de-emphasis (Samantha can't recall which).

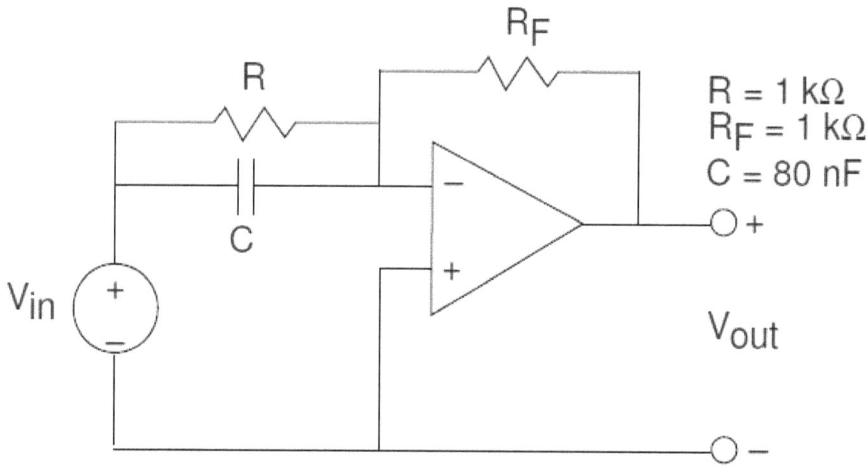


**Figure 3.88**

- Is this a pre-emphasis or de-emphasis circuit? Find the frequency f0 that defines the transition from low to high frequencies.
- What is the circuit's output when the input voltage is $\sin(2\pi ft)$, with f=4kHz?
- What circuit could perform the opposite function to your answer for the first part?

## Problem 3.48: Active Filter

Find the transfer function of the depicted active filter (Figure 3.89).

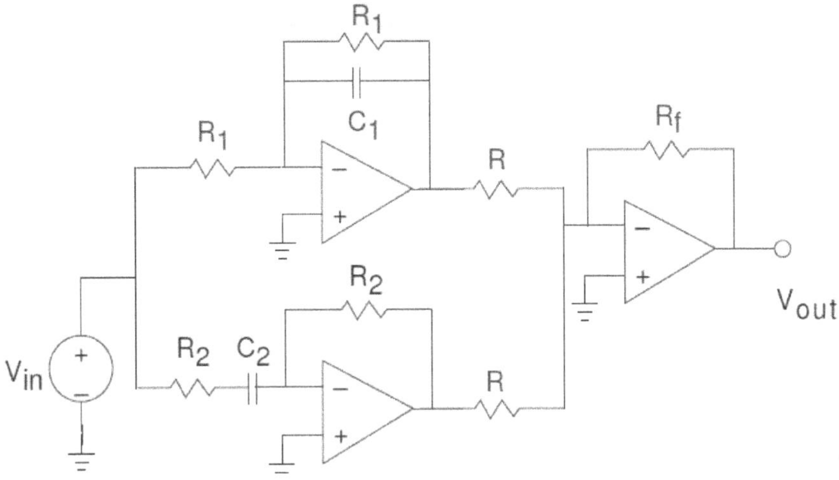


Figure 3.89

## Problem 3.49: This is a filter?

You are given a circuit (Figure 3.90).

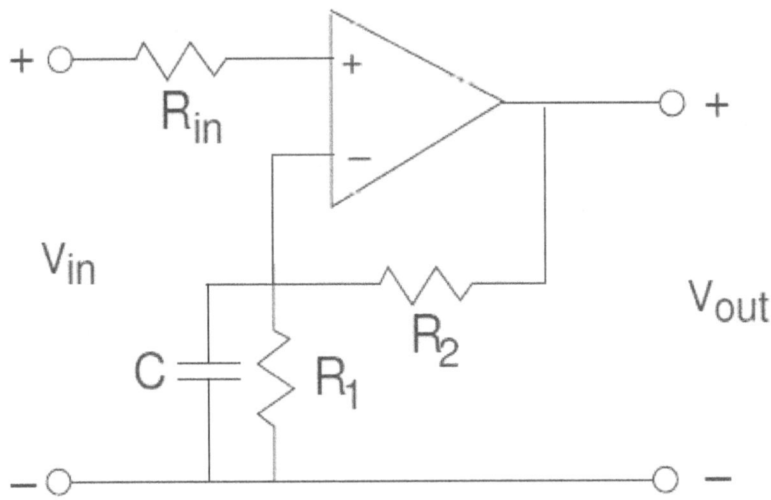


Figure 3.90

- What is this circuit's transfer function? Plot the magnitude and phase.
- If the input signal is the sinusoid sin(2πf0t), what will the output be when f0 is larger than the filter's "cutoff frequency"?

## Problem 3.50: Optical Receivers

In your optical telephone, the receiver circuit had the form shown (Figure 3.91).

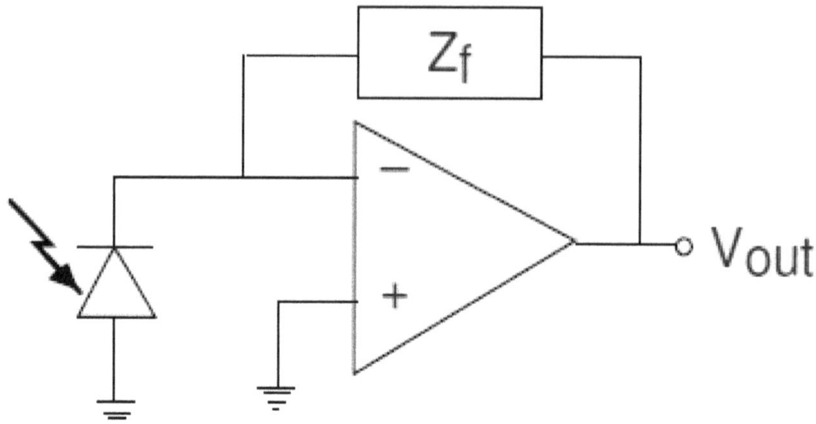


**Figure 3.91**

This circuit served as a transducer, converting light energy into a voltage vout. The photodiode acts as a current source, producing a current proportional to the light intesity falling upon it. As is often the case in this crucial stage, the signals are small and noise can be a problem. Thus, the op-amp stage serves to boost the signal and to filter out-of-band noise.

- Find the transfer function relating light intensity to vout.
- What should the circuit realizing the feedback impedance Zf be so that the transducer acts as a 5 kHz lowpass filter?
- A clever engineer suggests an alternative circuit (Figure 3.92) to accomplish the same task. Determine whether the idea works or not. If it does, find the impedance Zin that accomplishes the lowpass filtering task. If not, show why it does not work.

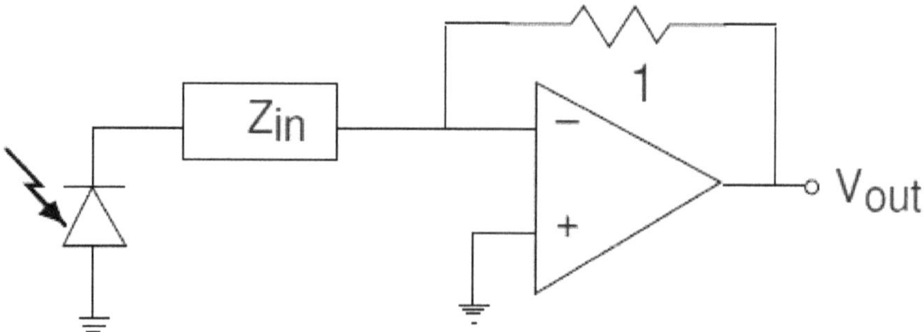


**Figure 3.92**

## Problem 3.51: Reverse Engineering

The depicted circuit (Figure 3.93) has been developed by the TBBG Electronics design group. They are trying to keep its use secret; we, representing RU Electronics, have discovered the schematic and want to figure out the intended application. Assume the diode is ideal.

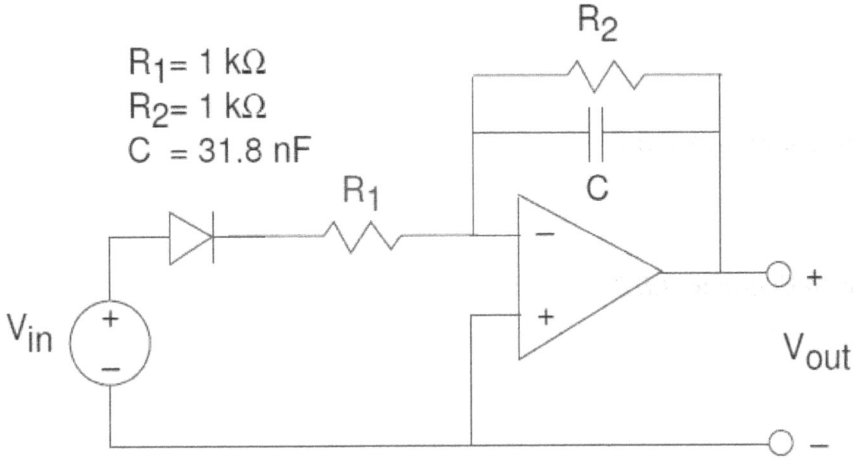


**Figure 3.93**

- Assuming the diode is a short-circuit (it has been removed from the circuit), what is the circuit's transfer function?
- With the diode in place, what is the circuit's output when the input voltage is sin(2πf0t)?
- What function might this circuit have?

## Solutions to Exercises in Chapter 3

### Solution to Exercise 3.1.1

One kilowatt-hour equals 3,600,000 watt-seconds, which indeed directly corresponds to 3,600,000 joules.

### Solution to Exercise 3.4.1

KCL says that the sum of currents entering or leaving a node must be zero. If we consider two nodes together as a "supernode", KCL applies as well to currents entering the combination. Since no currents enter an entire circuit, the sum of currents must be zero. If we had a two-node circuit, the KCL equation of one must be the negative of the other, We can combine all but one node in a circuit into a supernode; KCL for the supernode must be the negative of the remaining node's KCL equation. Consequently, specifying $n^{-1}$ KCL equations always specifies the remaining one.

### Solution to Exercise 3.4.2

The circuit serves as an amplifier having a gain of $\frac{R_2}{R_1+R_2}$.

### Solution to Exercise 3.5.1

The power consumed by the resistor R1 can be expressed as

$$(v_{\text{in}} - v_{\text{out}})\, i_{\text{out}} = \frac{R_1}{(R_1+R_2)^2} {v_{\text{in}}}^2$$

### Solution to Exercise 3.5.2

$$\frac{1}{R_1+R_2} {v_{\text{in}}}^2 = \frac{R_1}{(R_1+R_2)^2} {v_{\text{in}}}^2 + \frac{R_2}{(R_1+R_2)^2} {v_{\text{in}}}^2$$

### Solution to Exercise 3.6.1

Replacing the current source by a voltage source does not change the fact that the voltages are identical. Consequently, $v_{\text{in}} = R_2 i_{\text{out}}$ or $i_{\text{out}} = \frac{v_{\text{in}}}{R_2}$. This result does not depend on the resistor $R_1$, which means that we simply have a resistor (R2) across a voltage source. The two-resistor circuit has no apparent use.

### Solution to Exercise 3.6.2

$R_{\text{eq}} = \frac{R_2}{1+\frac{R_2}{R_L}}$. Thus, a 10% change means that the ratio $\frac{R_2}{R_L}$ must be less than 0.1. A 1% change means that $\frac{R_2}{R_L} < 0.01$.

### Solution to Exercise 3.6.3

In a series combination of resistors, the current is the same in each; in a parallel combination, the voltage is the same. For a series combination, the equivalent resistance is the sum of the resistances, which will be larger than any component resistor's value; for a parallel combination, the equivalent conductance is the sum of the component conductances, which is larger than any component conductance. The equivalent resistance is therefore smaller than any component resistance.

### Solution to Exercise 3.7.1

$v_{\text{oc}} = \frac{R_2}{R_1+R_2} v_{\text{in}}$ and $i_{\text{sc}} = -\frac{v_{\text{in}}}{R_1}$ (resistor R2 is shorted out in this case). Thus, $v_{\text{eq}} = \frac{R_2}{R_1+R_2} v_{\text{in}}$ and $R_{\text{eq}} = \frac{R_1 R_2}{R_1+R_2}$.

### Solution to Exercise 3.7.2

$i_{\text{eq}} = \frac{R_1}{R_1+R_2} i_{\text{in}}$ and $R_{\text{eq}} = R_3 \parallel R_1 + R_2$.

### Solution to Exercise 3.10.1

Division by j2πf arises from integrating a complex exponential. Consequently,

$$\frac{1}{j2\pi f} V \Leftrightarrow \int V e^{j2\pi f t}\, dt$$

### Solution to Exercise 3.11.1

For maximum power dissipation, the imaginary part of complex power should be zero. As the complex power is given by V I* = |V|| I|ej (Φ−Φ) , zero imaginary part occurs when the phases of the voltage and currents agree.

### Solution to Exercise 3.11.2

$P_{\text{ave}} = V_{\text{rms}} I_{\text{rms}} \cos(\phi - \theta)$. The cosine term is known as the power factor.

### Solution to Exercise 3.13.1

The key notion is writing the imaginary part as the difference between a complex exponential and its complex conjugate:

$$\text{Im}\left(Ve^{j2\pi ft}\right) = \frac{Ve^{j2\pi ft} - V^* e^{-(j2\pi ft)}}{2j} \tag{3.35}$$

The response to $Ve^{j2\pi ft}$ is $VH(f)e^{j2\pi ft}$, which means the response to $V^* e^{-(j2\pi ft)}$ is $V^* H(-f)e^{-(j2\pi ft)}$. As H (−f)=H(f)* , the Superposition Principle says that the output to the imaginary part is $\text{Im}\left(VH(f)e^{j2\pi ft}\right)$. The same argument holds for the real part: $Re\left(Ve^{j2\pi ft}\right) \to Re\left(VH(f)e^{j2\pi ft}\right)$.

### Solution to Exercise 3.15.1

To find the equivalent resistance, we need to find the current flowing through the voltage source. This current equals the current we have just found plus the current flowing through the other vertical 1 Ω resistor. This current equals $\frac{e_1}{1} = \frac{6}{13} v_{\text{in}}$, making the total current through the voltage source (flowing out of it) $\frac{11}{13} v_{\text{in}}$. Thus, the equivalent resistance is $\frac{13}{11}\Omega$.

### Solution to Exercise 3.15.2

Not necessarily, especially if we desire individual knobs for adjusting the gain and the cutoff frequency.

### Solution to Exercise 3.19.1

The ratio between adjacent values is about $\sqrt{2}$.

# 4 Chapter
# INTRODUCTION TO FREQUENCY DOMAIN

## 4.1 FREQUENCY DOMAIN

The impedance approach was devised as a means of facilitating the analysis of linear circuits, as it simplified the process of solving them. During the course of our progress, we formulated the concept of a circuit's frequency response or transfer function. This concept, which is also applicable to all linear, time-invariant systems, explains the behavior of the circuit when subjected to a sinusoidal input, expressed in terms of a complex exponential. In addition, we acquired knowledge of the Superposition Principle as it applies to linear systems: The output of the system when given an input that is the sum of two signals is equal to the sum of the outputs of the system for each individual component.

The study of the frequency domain integrates the concepts of a system's sinusoidal response and a linear system's output to a total of inputs being the sum of the individual outputs. This integration leads to the development of the fundamental concept of a signal's spectrum. The number of signals that can be expressed as a combination of sinusoids is extensive. Indeed, any signal may be represented as a combination of sinusoidal waves.

Throughout the course of this narrative, it will become evident that information systems strongly depend on spectral concepts. For instance, radio, television, and cellular telephones transmit using distinct frequency ranges within the electromagnetic spectrum. The spectrum is of utmost importance, to the extent that the Federal Communications Commission in the United States and the International Treaty govern the specific portions of the spectrum that can be utilized by communication systems (see to Frequency Allocations, Section 7.3). Determining the spectrum is a straightforward task: The Fourier transform provides a method for determining the spectrum of a signal.

## 4.2 COMPLEX FOURIER SERIES

In a previous lesson, we demonstrated that a square wave can be represented as the combination of many pulses. Although this decomposition proved helpful in this particular situation, it does not have broad applicability to other periodic signals. How can the combination of many pulses result in a coherent signal resembling a sinusoidal waveform? Due to the significance of sinusoids in linear systems, one would question the possibility of combining them to represent a wide range of periodic signals. You would be correct, and you would also be in esteemed company. Euler and Gauss expressed concern on this difficulty, but Jean Baptiste Fourier was credited for it, despite the fact that the resolution of complex mathematical issues occurred at a later time. Their focus was on the development of the Fourier series, which involves expressing any periodic signal as a combination of sinusoidal functions.

However, the Fourier series extends much beyond being merely another technique for decomposing signals. Instead, the Fourier series initiates our exploration of how a signal can be accurately characterized in either the time-domain or the frequency-domain without any trade-offs. Consider a periodic signal denoted as s(t) with a period of T. Our objective is to demonstrate that periodic signals, including those with segments of constant values such as a square wave, may be represented as the combination of sine waves that are harmonically connected. These sine waves have frequencies that are whole number multiples of the fundamental frequency.

Because the signal has period T, the fundamental frequency is $\frac{1}{T}$. The complex Fourier series expresses the signal as a superposition of complex exponentials having frequencies: $\frac{k}{T}$, k={....,-1,0,1,....}

$$s(t) = \sum_{k=-\infty}^{\infty} c_k e^{j\frac{2\pi kt}{T}} \tag{4.1}$$

With $c_k = \frac{1}{2}(a_k - jb_k)$. The real and imaginary parts of the Fourier coefficients ck are written in this unusual way for convenience in defining the classic Fourier series. The zeroth coefficient equals the signal's average value and is real- valued for real-valued signals: c0=a0. The family of functions $\{e^{j\frac{2\pi kt}{T}}\}$are called basis functions and form the foundation of the Fourier series. No matter what the periodic signal might be, these functions are always present and form the representation›s building blocks. They depend on the signal period T, and are indexed by k.

**Key Point:** Assuming we know the period, knowing the Fourier coefficients is equivalent to knowing the signal. Thus, it makes no difference if we have a time-domain or a frequency- domain characterization of the signal.

### Exercise 4.2.1

What is the complex Fourier series for a sinusoid?

To find the Fourier coefficients, we note the orthogonality property

$$\int_0^T e^{j\frac{2\pi kt}{T}} e^{(-j)\frac{2\pi lt}{T}} dt = \begin{cases} T \text{ if } k = l \\ 0 \text{ if } k \neq l \end{cases} \quad (4.2)$$

Assuming for the moment that the complex Fourier series "works," we can find a signal's complex Fourier coefficients, its spectrum, by exploiting the orthogonality properties of harmonically related complex exponentials. Simply multiply each side of (4.1) by e–(j2πlt) and integrate over the interval [0,T].

$$\begin{aligned} c_k &= \tfrac{1}{T} \int_0^T s(t)\, e^{-\left(j\frac{2\pi kt}{T}\right)} dt \\ c_0 &= \tfrac{1}{T} \int_0^T s(t)\, dt \end{aligned} \quad (4.3)$$

## Example 4.1

Finding the Fourier series coefficients for the square wave sqT(t) is very simple. Mathematically, this signal can be expressed as

$$\mathrm{sq}_T(t) = \begin{cases} 1 \text{ if } 0 < t < \frac{T}{2} \\ -1 \text{ if } \frac{T}{2} < t < T \end{cases}$$

The expression for the Fourier coefficients has the form

$$c_k = \frac{1}{T} \int_0^{\frac{T}{2}} e^{-\left(j\frac{2\pi kt}{T}\right)} dt - \frac{1}{T} \int_{\frac{T}{2}}^{T} e^{-\left(j\frac{2\pi kt}{T}\right)} dt \quad (4.4)$$

**Note:** When integrating an expression containing i, treat it just like any other constant.

The two integrals are very similar, one equaling the negative of the other. The final expression becomes

$$\begin{aligned} c_k &= \frac{-2}{j2\pi k}\left((-1)^k - 1\right) \\ &= \begin{cases} \frac{2}{j\pi k} \text{ if } k \text{ odd} \\ 0 \text{ if } k \text{ even} \end{cases} \end{aligned} \quad (4.5)$$

$$\mathrm{sq}(t) = \sum_{k \in \{\dots,-3,-1,1,3,\dots\}} \frac{2}{j\pi k} e^{(j)\frac{2\pi kt}{T}} \quad (4.6)$$

Consequently, the square wave equals a sum of complex exponentials, but only those having frequencies equal to odd multiples of the fundamental frequency $\frac{1}{T}$. The coefficients decay slowly as the frequency index k increases. This index corresponds to the k-th harmonic of the signal's period.

A signal's Fourier series spectrum ck has interesting properties.

**Property 4.1:** If s(t) is real, ck=c–k* (real-valued periodic signals have conjugate-symmetric spectra).

This result follows from the integral that calculates the ck from the signal. Furthermore, this result means that: R(ck)=R(c-k): The real part of the Fourier coefficients for real-valued signals is even. Similarly, I(ck)=I(c-k) :The imaginary parts of the Fourier coefficients have odd symmetry. Consequently, if you are given the Fourier coefficients for positive indices and zero and are told the signal is real-valued, you can find the negative-indexed coefficients, hence the entire spectrum. This kind of symmetry, ck=c-k *,,is known as conjugate symmetry.

**Property 4.2:** If s(-t) = s(t), which says the signal has even symmetry about the origin, c-k=ck

Given the previous property for real-valued signals, the Fourier coefficients of even signals are real-valued. A real-valued Fourier expansion amounts to an expansion in terms of only cosines, which is the simplest example of an even signal.

**Property 4.3:** If s(-t) = - s(t), which says the signal has odd symmetry, c-k=-ck

Therefore, the Fourier coefficients are purely imaginary. The square wave is a great example of an odd-symmetric signal.

**Property 4.4:** The spectral coefficients for a periodic signal delayed by τ, s(t-τ), are $c_k e^{-\frac{j2\pi k\tau}{T}}$, where ck denotes the spectrum of s(t). Delaying a signal by τ seconds results in a spectrum having a linear phase shift of $-\frac{2\pi k\tau}{T}$ in comparison to the spectrum of the undelayed signal. Note that the spectral magnitude is unaffected. Showing this property is easy.

**Proof:**

$$\begin{aligned} \frac{1}{T}\int_0^T s(t-\tau)\, e^{(-j)\frac{2\pi kt}{T}}\,dt &= \frac{1}{T}\int_{-\tau}^{T-\tau} s(t)\, e^{(-j)\frac{2\pi k(t+\tau)}{T}}\,dt \\ &= \frac{1}{T} e^{(-j)\frac{2\pi k\tau}{T}} \int_{-\tau}^{T-\tau} s(t)\, e^{(-j)\frac{2\pi kt}{T}}\,dt \end{aligned} \tag{4.7}$$

Note that the range of integration extends over a period of the integrand. Consequently, it should not matter how we integrate over a period, which means that $\int_{-\tau}^{T-\tau} (\cdot)\,dt = \int_0^T (\cdot)\,dt$, and we have our result.

The complex Fourier series obeys Parseval's Theorem, one of the most important results in signal analysis. This general mathematical result says you can calculate a signal's power in either the time domain or the frequency domain.

**Theorem 4.1:** Parseval's Theorem

Average power calculated in the time domain equals the power calculated in the frequency domain.

$$\frac{1}{T}\int_{0}^{T} s^2(t)\,dt = \sum_{k=-\infty}^{\infty}\left(|c_k|\right)^2 \qquad (4.8)$$

This result is a (simpler) re-expression of how to calculate a signal's power than with the real-valued Fourier series expression for power.

Let's calculate the Fourier coefficients of the periodic pulse signal shown here (Figure 4.1)

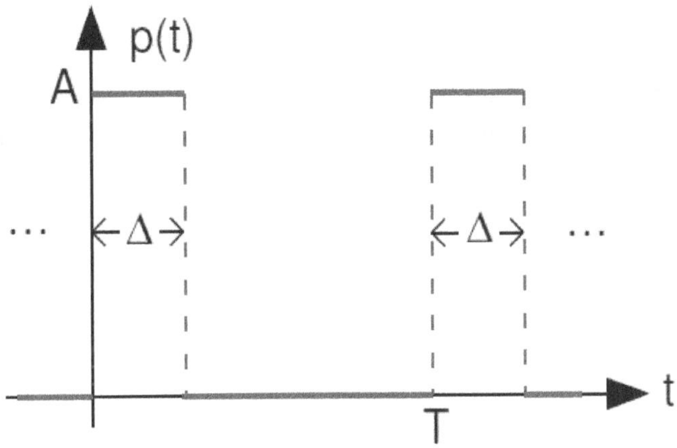


**Figure 4.1: Periodic Pulse Signal**

The pulse width is Δ, the period T, and the amplitude A. The complex Fourier spectrum of this signal is given by:

$$c_k = \frac{1}{T}\int_{0}^{\Delta} Ae^{-\frac{j2\pi kt}{T}}\,dt = -\left(\frac{A}{j2\pi k}\left(e^{-\frac{j2\pi k\Delta}{T}}-1\right)\right)$$

At this point, simplifying this expression requires knowing an interesting property.

$$1-e^{-(j\theta)} = e^{-\frac{j\theta}{2}}\left(e^{\frac{j\theta}{2}}-e^{-\frac{j\theta}{2}}\right) = e^{-\frac{j\theta}{2}}2j\sin\left(\frac{\theta}{2}\right)$$

Armed with this result, we can simply express the Fourier series coefficients for our pulse sequence.

$$c_k = Ae^{-\frac{j\pi k\Delta}{T}}\frac{\sin\left(\frac{\pi k\Delta}{T}\right)}{\pi k} \qquad (4.9)$$

Because this signal is real-valued, we find that the coefficients do indeed have conjugate symmetry: ck=c-k* .The periodic pulse signal has neither even nor odd symmetry; consequently, no additional symmetry exists in the spectrum. Because the spectrum is complex valued, to plot it we need to calculate its magnitude and phase.

$$|c_k| = A\left|\frac{\sin\left(\frac{\pi k \Delta}{T}\right)}{\pi k}\right| \tag{4.10}$$

$$\angle(c_k) = -\frac{\pi k \Delta}{T} + \pi \text{neg}\left(\frac{\sin\left(\frac{\pi k \Delta}{T}\right)}{\pi k}\right)\text{sign}(k)$$

The function neg(•) equals -1 if its argument is negative and zero otherwise. The somewhat complicated expression for the phase results because the sine term can be negative; magnitudes must be positive, leaving the occasional negative values to be accounted for as a phase shift of π.

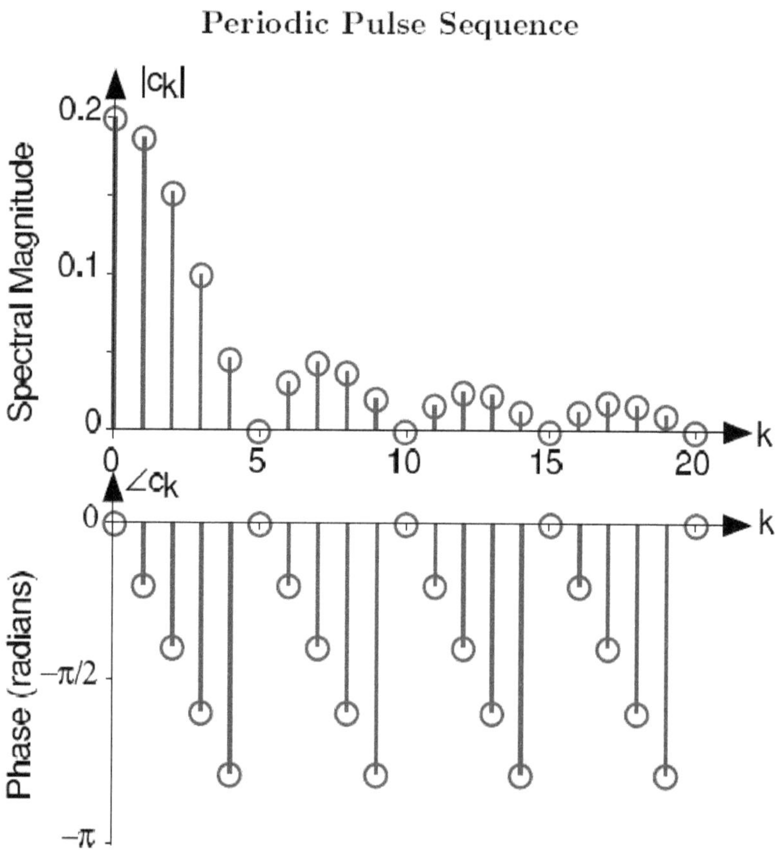


**Figure 4.2The magnitude and phase of the periodic pulse sequence's spectrum is shown for positive frequency indices. Here $\frac{\Delta}{T} = 0.2$ and A=1.**

Also note the presence of a linear phase term (the first term in ∠ck is proportional to frequency $\frac{k}{T}$). Comparing this term with that predicted from delaying a signal, a delay of $\frac{\Delta}{2}$ is present in our signal. Advancing the signal by this amount centers the pulse about the origin, leaving an even signal, which in turn means that its spectrum is real-valued. Thus, our calculated spectrum is consistent with the properties of the Fourier spectrum.

### Exercise 4.2.2

What is the value of c0? Recalling that this spectral coefficient corresponds to the signal's average value, does your answer make sense?

The phase plot shown in Figure 4.2 (Periodic Pulse Sequence) requires some explanation as it does not seem to agree with what (4.10) suggests. There, the phase has a linear component, with a jump of π every time the sinusoidal term changes sign. We must realize that any integer multiple of 2 π can be added to a phase at each frequency without affecting the value of the complex spectrum. We see that at frequency index 4 the phase is nearly – π. The phase at index 5 is undefined because the magnitude is zero in this example. At index 6, the formula suggests that the phase of the linear term should be less than – π (more negative). In addition, we expect a shift of – π in the phase between indices 4 and 6. Thus, the phase value predicted by the formula is a little less than –(2 π). Because we can add 2 π without affecting the value of the spectrum at index 6, the result is a slightly negative number as shown. Thus, the formula and the plot do agree. In phase calculations like those made in MATLAB, values are usually confined to the range [–π; π) by adding some (possibly negative) multiple of 2 π to each phase value

## 4.3 CLASSIC FOURIER SERIES

The classic Fourier series as derived originally expressed a periodic signal (period T) in terms of harmonically related sines and cosines.

$$s(t) = a_0 + \sum_{k=1}^{\infty} a_k \cos\left(\frac{2\pi kt}{T}\right) + \sum_{k=1}^{\infty} b_k \sin\left(\frac{2\pi kt}{T}\right) \quad (4.11)$$

The complex Fourier series and the sine-cosine series are identical, each representing a signal's spectrum. The Fourier coefficients, ak and bk, express the real and imaginary parts respectively of the spectrum while the coefficients ck of the complex Fourier series express the spectrum as a magnitude and phase. Equating the classic Fourier series (4.11) to the complex Fourier series, an extra factor of two and complex conjugate become necessary to relate the Fourier coefficients in each.

$$c_k = \frac{1}{2}(a_k - jb_k)$$

### Exercise 4.3.1

Derive this relationship between the coefficients of the two Fourier series.

Just as with the complex Fourier series, we can find the Fourier coeffcients using the orthogonality properties of sinusoids. Note that the cosine and sine of harmonically related frequencies, even the same frequency, are orthogonal.

$$\int_0^T \sin\left(\frac{2\pi kt}{T}\right)\cos\left(\frac{2\pi lt}{T}\right)dt = 0\ ,\ \ k \in \mathbb{Z}\ \ l \in \mathbb{Z} \tag{4.12}$$

$$\int_0^T \sin\left(\frac{2\pi kt}{T}\right)\sin\left(\frac{2\pi lt}{T}\right)dt = \begin{cases} \frac{T}{2} \text{ if } (k = l)\ \ and\ \ (k \neq 0)\ \ and\ \ (l \neq 0) \\ 0 \text{ if } (k \neq l)\ \ or\ \ (k = 0 = l) \end{cases}$$

$$\int_0^T \cos\left(\frac{2\pi kt}{T}\right)\cos\left(\frac{2\pi lt}{T}\right)dt = \begin{cases} \frac{T}{2} \text{ if } (k = l)\ \ and\ \ (k \neq 0)\ \ and\ \ (l \neq 0) \\ T \text{ if } k = 0 = l \\ 0 \text{ if } k \neq l \end{cases}$$

These orthogonality relations follow from the following important trigonometric identities.

$$\begin{aligned} \sin(\alpha)\sin(\beta) &= \frac{1}{2}\left(\cos(\alpha-\beta) - \cos(\alpha+\beta)\right) \\ \cos(\alpha)\cos(\beta) &= \frac{1}{2}\left(\cos(\alpha+\beta) + \cos(\alpha-\beta)\right) \\ \sin(\alpha)\cos(\beta) &= \frac{1}{2}\left(\sin(\alpha+\beta) + \sin(\alpha-\beta)\right) \end{aligned} \tag{4.13}$$

These identities allow you to substitute a sum of sines and/or cosines for a product of them. Each term in the sum can be integrated by noticing one of two important properties of sinusoids.

- The integral of a sinusoid over an integer number of periods equals zero.
- The integral of the square of a unit-amplitude sinusoid over a period T equals $\frac{T}{2}$.

To use these, let's, for example, multiply the Fourier series for a signal by the cosine of the lth harmonic $\cos\left(\frac{2\pi lt}{T}\right)$ and integrate. The idea is that, because integration is linear, the integration will sift out all but the term involving al.

$$\int_0^T s(t)\cos\left(\frac{2\pi lt}{T}\right)dt = \int_0^T a_0\cos\left(\frac{2\pi lt}{T}\right)dt + \sum_{k=1}^{\infty} a_k \int_0^T \cos\left(\frac{2\pi kt}{T}\right)\cos\left(\frac{2\pi lt}{T}\right)dt + \sum_{k=1}^{\infty} b_k \int_0^T \sin\left(\frac{2\pi kt}{T}\right)\cos\left(\frac{2\pi lt}{T}\right)dt \tag{4.14}$$

The first and third terms are zero; in the second, the only non-zero term in the sum results when the indices k and l are equal (but not zero), in which case we obtain: $\frac{a_l T}{2}$. If k=0=l, we obtain a0T.

Consequently,

$$a_l = \frac{2}{T}\int_0^T s(t)\cos\left(\frac{2\pi lt}{T}\right)dt\ ,\ \ l \neq 0$$

All of the Fourier coefficients can be found similarly.

$$\begin{aligned} a_0 &= \tfrac{1}{T}\int_0^T s(t)\,dt \\ a_k &= \tfrac{2}{T}\int_0^T s(t)\cos\left(\tfrac{2\pi kt}{T}\right)dt\ ,\ \ k\neq 0 \\ b_k &= \tfrac{2}{T}\int_0^T s(t)\sin\left(\tfrac{2\pi kt}{T}\right)dt \end{aligned} \quad (4.15)$$

**Exercise 4.3.2**

The expression for a0 is referred to as the average value of s(t). Why?

**Exercise 4.3.3**

What is the Fourier series for a unit-amplitude square wave?

**Exercise 4.2**

Let's find the Fourier series representation for the half-wave rectified sinusoid.

$$s(t) = \begin{cases} \sin\left(\frac{2\pi t}{T}\right) & \text{if } 0 \leq t < \frac{T}{2} \\ 0 \text{ if } \frac{T}{2} \leq t < T \end{cases} \quad (4.16)$$

Begin with the sine terms in the series; to find bk we must calculate the integral

$$b_k = \frac{2}{T}\int_0^{\frac{T}{2}} \sin\left(\frac{2\pi t}{T}\right)\sin\left(\frac{2\pi kt}{T}\right)dt \quad (4.17)$$

Using our trigonometric identities turns our integral of a product of sinusoids into a sum of integrals of individual sinusoids, which are much easier to evaluate.

$$\begin{aligned} \int_0^{\frac{T}{2}} \sin\left(\tfrac{2\pi t}{T}\right)\sin\left(\tfrac{2\pi kt}{T}\right)dt &= \tfrac{1}{2}\int_0^{\frac{T}{2}} \cos\left(\tfrac{2\pi(k-1)t}{T}\right) - \cos\left(\tfrac{2\pi(k+1)t}{T}\right)dt \\ &= \begin{cases} \frac{1}{2} \text{ if } k=1 \\ 0 \text{ otherwise} \end{cases} \end{aligned} \quad (4.18)$$

Thus,

$$b_1 = \frac{1}{2}$$

$$b_2 = b_3 = \cdots = 0$$

On to the cosine terms. The average value, which corresponds to a0, equals $\frac{1}{\pi}$. The remainder of the cosine coefficients are easy to find, but yield the complicated result:

$$a_k = \begin{cases} -\left(\frac{2}{\pi}\frac{1}{k^2-1}\right) & \text{if } k \in \{2,4,\dots\} \\ 0 \text{ if k odd} \end{cases} \quad (4.19)$$

Thus, the Fourier series for the half-wave rectified sinusoid has non-zero terms for the average, the fundamental, and the even harmonics.

## 4.4 A SIGNAL'S SPECTRUM

A periodic signal, such as the half-wave rectified sinusoid, consists of a sum of elemental sinusoids. A plot of the Fourier coefficients as a function of the frequency index, such as shown in Figure 4.3, displays the signal's spectrum. The word "spectrum" implies that the independent variable, here k, corresponds somehow to frequency. Each coefficient is directly related to a sinusoid having a frequency of $\frac{k}{T}$. Thus, if we half-wave rectified a 1 kHz sinusoid, k=1 corresponds to 1 kHz, k=2 to 2 kHz, etc.

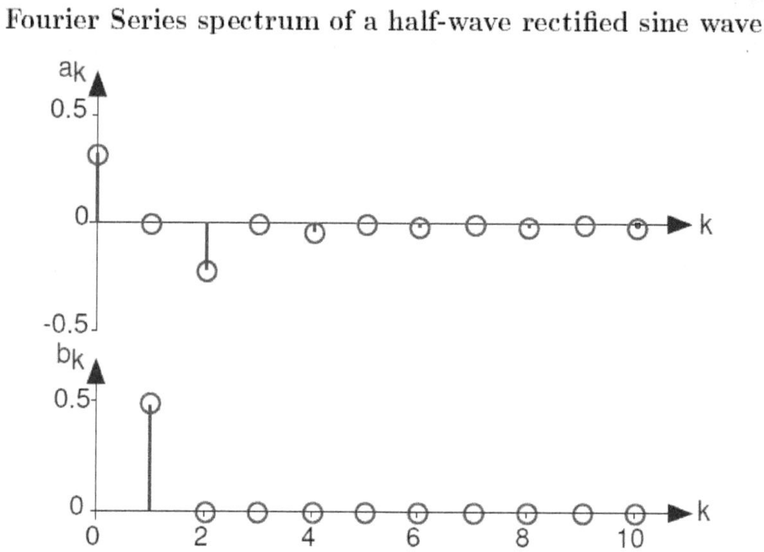


**Figure 4.3: Fourier Series spectrum of a half-wave rectified sine wave The index indicates the multiple of the fundamental frequency at which the signal has energy.**

A subtle, but very important, aspect of the Fourier spectrum is its uniqueness: You can unambiguously find the spectrum from the signal (decomposition) and the signal from the spectrum (composition). Thus, any aspect of the signal can be found from the spectrum and vice versa. A signal's frequency domain expression is its spectrum. A periodic signal can be defined either in the time domain (as a function) or in the frequency domain (as a spectrum).

A fundamental aspect of solving electrical engineering problems is whether the time or frequency domain provides the most understanding of a signal's properties and the simplest way of manipulating it. The uniqueness property says that either domain can provide the right answer. As a simple example, suppose we want to know the (periodic) signal's maximum value. Clearly the time domain provides the answer directly. To use a frequency domain approach would require us to find the spectrum, form the signal from the spectrum and calculate the maximum; we're back in the time domain!

Another feature of a signal is its average power. A signal's instantaneous power is defined to be its square. The average power is the average of the instantaneous power over some time interval. For a periodic signal, the natural time interval is clearly its period; for nonperiodic signals, a better choice would be entire time or time from onset. For a periodic signal, the average power is the square of its root-mean-squared (rms) value. We define the rms value of a periodic signal to be:

$$\mathrm{rms}(s) = \sqrt{\frac{1}{T}\int_0^T s^2(t)\,dt} \tag{4.20}$$

and thus its average power is

$$\begin{aligned}\mathrm{power}(s) &= \mathrm{rms}^2(s)\\ &= \tfrac{1}{T}\int_0^T s^2(t)\,dt\end{aligned} \tag{4.21}$$

### Exercise 4.4.1

What is the rms value of the half-wave rectified sinusoid?

To find the average power in the frequency domain, we need to substitute the spectral representation of the signal into this expression.

$$\mathrm{power}(s) = \frac{1}{T}\int_0^T \left(a_0 + \sum_{k=1}^{\infty} a_k \cos\left(\frac{2\pi kt}{T}\right) + \sum_{k=1}^{\infty} b_k \sin\left(\frac{2\pi kt}{T}\right)\right)^2 dt$$

The square inside the integral will contain all possible pairwise products. However, the orthogonality properties say that most of these crossterms integrate to zero. The survivors leave a rather simple expression for the power we seek.

$$\mathrm{power}(s) = {a_0}^2 + \frac{1}{2}\sum_{k=1}^{\infty} {a_k}^2 + {b_k}^2 \tag{4.22}$$

It could well be that computing this sum is easier than integrating the signal's square. Furthermore, the contribution of each term in the Fourier series toward representing the signal can be measured by its contribution to the signal's average power. Thus, the power contained in a signal at its kth harmonic is: $\frac{{a_k}^2+{b_k}^2}{2}$. The power spectrum, Ps(k) such as shown in Figure 4.4 (Power Spectrum of a Half-Wave Rectified Sinusoid), plots each harmonic's contribution to the total power.

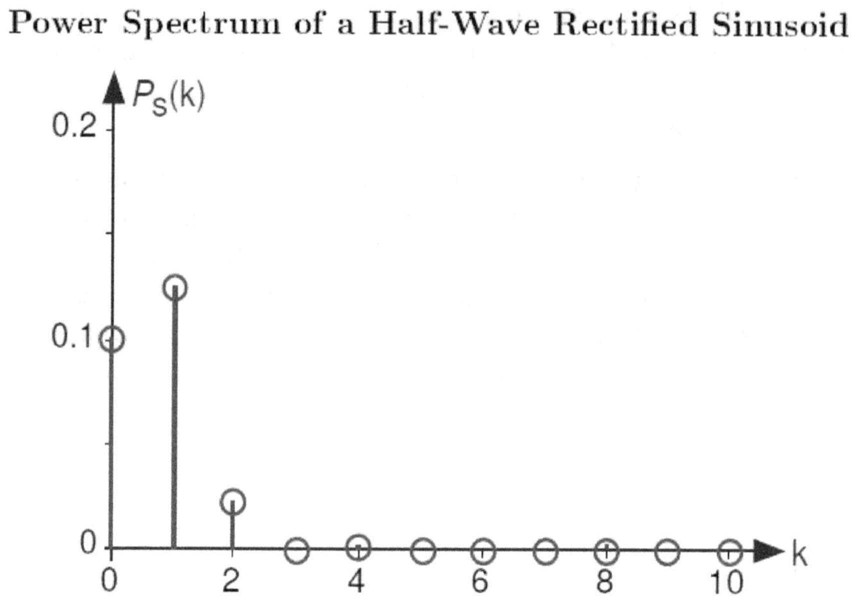


**Figure 4.4: Power spectrum of a half-wave rectified sinusoid**

### Exercise 4.4.2

In high-end audio, deviation of a sine wave from the ideal is measured by the total harmonic distortion, which equals the total power in the harmonics higher than the first compared to power in the fundamental. Find an expression for the total harmonic distortion for any periodic signal. Is this calculation most easily performed in the time or frequency domain?

## 4.5 FOURIER SERIES APPROXIMATION OF SIGNALS

It is interesting to consider the sequence of signals that we obtain as we incorporate more terms into the Fourier series approximation of the half-wave rectified sine wave. Define sK(t) to be the signal containing K+1 Fourier terms.

$$s_K(t) = a_0 + \sum_{k=1}^{K} a_k \cos\left(\frac{2\pi kt}{T}\right) + \sum_{k=1}^{K} b_k \sin\left(\frac{2\pi kt}{T}\right) \tag{4.23}$$

Figure 4.5 (Fourier Series spectrum of a half-wave rectified sine wave) shows how this sequence of signals portrays the signal more accurately as more terms are added.

We need to assess quantitatively the accuracy of the Fourier series approximation so that we can judge how rapidly the series approaches the signal. When we use a K+1-term series, the error–the difference between the signal and the K+1-term series–corresponds to the unused terms from the series.

Fourier Series spectrum of a half-wave rectified sine wave

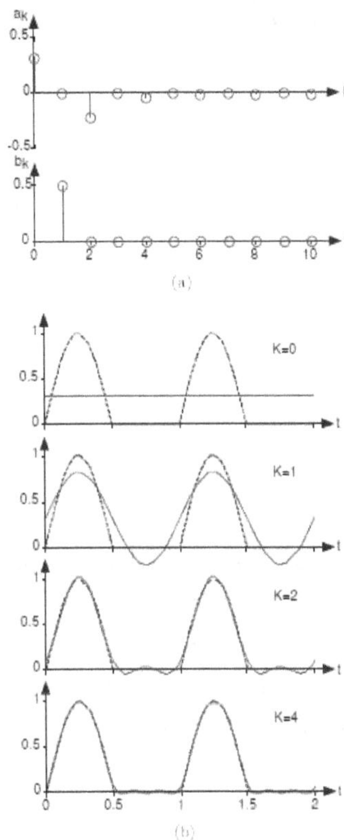


**Figure 4.5: The Fourier series spectrum of a half-wave rectified sinusoid is shown in the upper portion. The index indicates the multiple of the fundamental frequency at which the signal has energy. The cumulative effect of adding terms to the Fourier series for the half wave rectified sine wave is shown in the bottom portion. The dashed line is the actual signal, with the solid line showing the finite series approximation to the indicated number of terms, K+1.**

$$\epsilon_K(t) = \sum_{k=K+1}^{\infty} a_k \cos\left(\frac{2\pi kt}{T}\right) + \sum_{k=K+1}^{\infty} b_k \sin\left(\frac{2\pi kt}{T}\right) \tag{4.24}$$

To find the rms error, we must square this expression and integrate it over a period. Again, the integral of most cross-terms is zero, leaving

$$\text{rms}(\epsilon_K) = \sqrt{\frac{1}{2}\sum_{k=K+1}^{\infty} a_k^{\,2} + b_k^{\,2}} \tag{4.25}$$

Figure 4.6 (Approximation error for a half-wave rectified sinusoid) shows how the error in the Fourier series for the half-wave rectified sinusoid decreases as more terms are incorporated. In particular, the use of four terms, as shown in the bottom plot of Figure 4.5 (Fourier Series spectrum of a half-wave rectified sine wave), has a rms error (relative to the rms value of the signal) of about 3%. The

Fourier series in this case converges quickly to the signal.

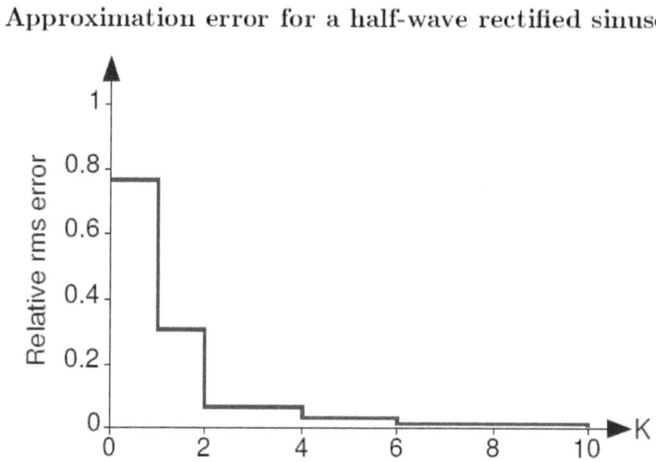


**Figure 4.6 The rms error calculated according to the equation above is shown as a function of the number of terms in the series for the half-wave rectified sinusoid. The error has been normalized by the rms value of the signal.**

We can look at Figure 4.7 (Power spectrum and approximation error for a square wave) to see the power spectrum and the rms approximation error for the square wave.

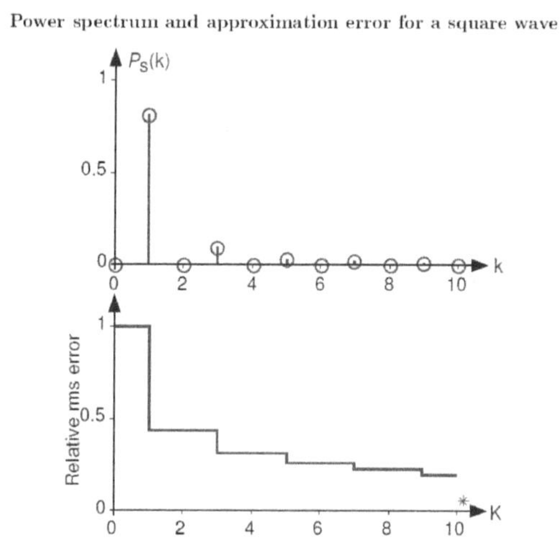


**Figure 4.7 The upper plot shows the power spectrum of the square wave, and the lower plot the rms error of the finite-length Fourier series approximation to the square wave. The asterisk denotes the rms error when the number of terms K in the Fourier series equals 99.**

Because the Fourier coefficients decay more slowly here than for the half-wave rectified sinusoid, the rms error is not decreasing quickly. Said another way, the square-wave's spectrum contains more power at higher frequencies than does the half-wave-rectified sinusoid. This difference between the two Fourier series results because the half-wave rectified sinusoid's Fourier coefficients are proportional to $\frac{1}{k^2}$while those of the square wave are proportional to $\frac{1}{k}$. If fact, after 99 terms of the square wave's approximation, the error is bigger than 10 terms of the approximation for the half-wave rectified sinusoid. Mathematicians have shown that no signal has an rms approximation error that decays more slowly than it does for the square wave.

## Exercise 4.5.1

Calculate the harmonic distortion for the square wave.

More than just decaying slowly, Fourier series approximation shown in Figure 4.8 (Fourier series approximation of a square wave) exhibits interesting behavior.

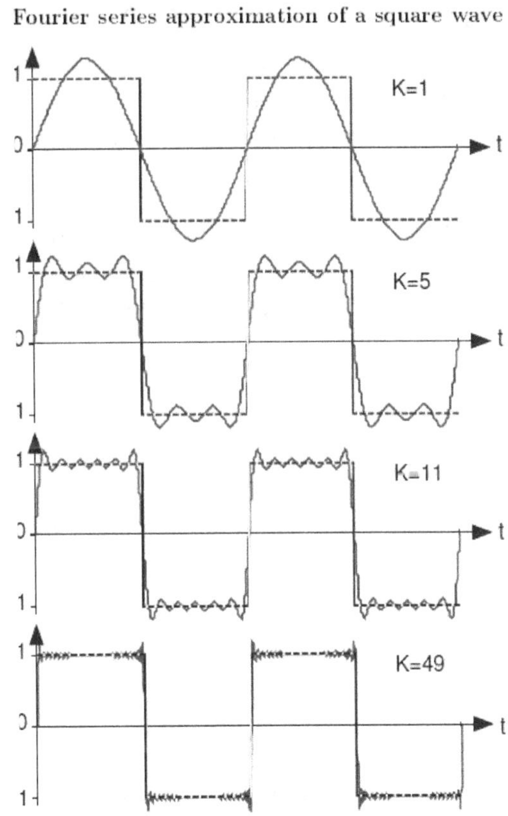


**Figure 4.8 Fourier series approximation to sq(t). The number of terms in the Fourier sum is indicated in each plot, and the square wave is shown as a dashed line over two periods.**

Although the square wave's Fourier series requires more terms for a given representation accuracy, when comparing plots it is not clear that the two are equal. Does the Fourier series really equal the square wave at all values of t? In particular, at each step-change in the square wave, the Fourier series exhibits a peak followed by rapid oscillations. As more terms are added to the series, the oscillations seem to become more rapid and smaller, but the peaks are not decreasing. For the Fourier series approximation for the half-wave rectified sinusoid (Figure 4.5: Fourier Series spectrum of a half-wave rectified sine wave), no such behaviour occurs. What is happening?

Consider this mathematical question intuitively: Can a discontinuous function, like the square wave, be expressed as a sum, even an infinite one, of continuous signals? One should at least be suspicious, and in fact, it can't be thus expressed. This issue brought Fourier much criticism from the French Academy of Science (Laplace, Lagrange, Monge and LaCroix comprised the review committee) for several years after its presentation on 1807. It was not resolved for almost a century, and its resolution is interesting and important to understand from a practical viewpoint.

The extraneous peaks in the square wave's Fourier series never disappear; they are termed Gibb's phenomenon after the American physicist Josiah Willard Gibbs. They occur whenever the signal is discontinuous, and will always be present whenever the signal has jumps.

Let's return to the question of equality; how can the equal sign in the definition of the Fourier series be justified? The partial answer is that pointwise—each and every value of t—equality is not guaranteed. However, mathematicians later in the nineteenth century showed that the rms error of the Fourier series was always zero.

$$\operatorname*{limit}_{K\to\infty} \mathrm{rms}\left(\epsilon_K\right) = 0$$

What this means is that the error between a signal and its Fourier series approximation may not be zero, but that its rms value will be zero! It is through the eyes of the rms value that we redefine equality: The usual definition of equality is called pointwise equality: Two signals s1(t), s2(t) are said to be equal pointwise if s1(t)=s2(t) for all values of t. A new definition of equality is mean-square equality: Two signals are said to be equal in the mean square if rms(s1-s2)=0. For Fourier series, Gibb's phenomenon peaks have finite height and zero width. The error differs from zero only at isolated points—whenever the periodic signal contains discontinuities—and equals about 9% of the size of the discontinuity. The value of a function at a finite set of points does not affect its integral. This effect underlies the reason why defining the value of a discontinuous function, like we refrained from doing in defining the step function (Section 2.2.4: Unit Step), at its discontinuity is meaningless. Whatever you pick for a value has no practical relevance for either the signal's spectrum or for how a system responds to the signal. The Fourier series value "at" the discontinuity is the average of the values on either side of the jump.

## 4.6 ENCODING INFORMATION IN THE FREQUENCY DOMAIN

To emphasize the fact that every periodic signal has both a time and frequency domain representation, we can exploit both to encode information into a signal. Refer to the Fundamental Model of Communication. We have an information source, and want to construct a transmitter that produces a signal x(t). For the source, let's assume we have information to encode every T seconds. For example, we want to represent typed letters produced by an extremely good typist (a key is struck every T seconds). Let's consider the complex Fourier series formula in the light of trying to encode information.

$$x(t) = \sum_{k=-K}^{K} c_k e^{j\frac{2\pi kt}{T}} \tag{4.26}$$

We use a finite sum here merely for simplicity (fewer parameters to determine). An important aspect of the spectrum is that each frequency component ck can be manipulated separately: Instead of finding the Fourier spectrum from a time-domain specification, let›s construct it in the frequency domain by selecting the ck according to some rule that relates coefficient values to the alphabet. In defining this rule, we want to always create a real-valued signal x(t). Because of the Fourier spectrum's properties, the spectrum must have conjugate symmetry. This requirement means that we can only assign positive-indexed coefficients (positive frequencies), with negative-indexed ones equaling the complex conjugate of the corresponding positive-indexed ones.

Assume we have N letters to encode: {a1,...,aN}. One simple encoding rule could be to make a single Fourier coefficient be non-zero and all others zero for each letter. For example, if an occurs, we make: cn=1 and ck=0, k≠n. In this way, the nth harmonic of the frequency 1/T is used to represent a letter. Note that the bandwidth—the range of frequencies required for the encoding—equals N/T. Another possibility is to consider the binary representation of the letter's index. For example, if the letter a13 occurs, converting 13 to its base 2 representation, we have 13=11012. We can use the pattern of zeros and ones to represent directly which Fourier coefficients we "turn on" (set equal to one) and which we "turn off".

### Exercise 4.6.1

Compare the bandwidth required for the direct encoding scheme (one nonzero Fourier coefficient for each letter) to the binary number scheme. Compare the bandwidths for a 128-letter alphabet. Since both schemes represent information without loss -- we can determine the typed letter uniquely from the signal's spectrum - both are viable. Which makes more efficient use of bandwidth and thus might be preferred?

## Exercise 4.6.2

Can you think of an information-encoding scheme that makes even more efficient use of the spectrum? In particular, can we use only one Fourier coefficient to represent N letters uniquely?

We can create an encoding scheme in the frequency domain to represent an alphabet of letters. But, as this information-encoding scheme stands, we can represent one letter for all time. However, we note that the Fourier coefficients depend only on the signal›s characteristics over a single period. We could change the signal›s spectrum every T as each letter is typed. In this way, we turn spectral coefficients on and off as letters are typed, thereby encoding the entire typed document. For the receiver (see the Fundamental Model of Communication) to retrieve the typed letter, it would simply use the Fourier formula for the complex Fourier spectrum for each T-second interval to determine what each typed letter was. Figure 4.9 (Encoding Signals) shows such a signal in the time-domain.

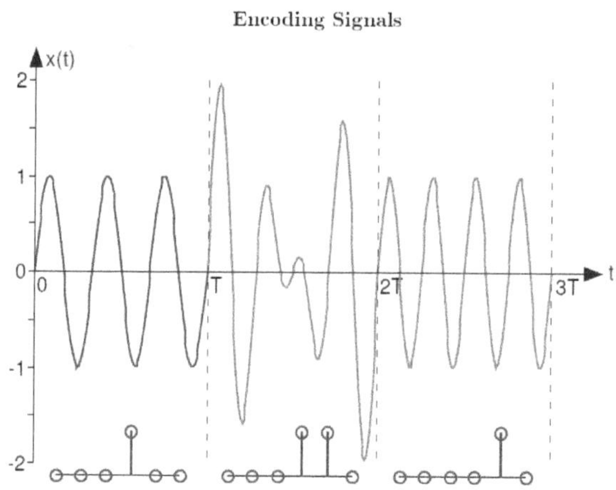


**Figure 4.9: The encoding of signals via the Fourier spectrum is shown over three "periods." In this example, only the third and fourth harmonics are used, as shown by the spectral magnitudes corresponding to each T-second interval plotted below the waveforms. Can you determine the phase of the harmonics from the waveform?**

In this Fourier-series encoding scheme, we have used the fact that spectral coefficients can be independently specified and that they can be uniquely recovered from the time-domain signal over one "period." Do note that the signal representing the entire document is no longer periodic. By understanding the Fourier series' properties (in particular that coefficients are determined only over a T-second interval, we can construct a communications system. This approach represents a simplification of how modern modems represent text that they transmit over telephone lines.

## 4.7 FILTERING PERIODIC SIGNALS

The Fourier series representation of a periodic signal makes it easy to determine how a linear, time-invariant filter reshapes such signals in general. The fundamental property of a linear system is that its input-output relation obeys superposition: L(a1s1(t)+a2s2(t))=a1L(s1(t))+a2L(s2(t)). Because the Fourier series represents a periodic signal as a linear combination of complex exponentials, we can exploit the superposition property. Furthermore, we found for linear circuits that their output to a complex exponential input is just the frequency response evaluated at the signal's frequency times the complex exponential. Said mathematically, if $x(t) = e^{j\frac{2\pi kt}{T}}$, then the output $y(t) = H\left(\frac{k}{T}\right)e^{j\frac{2\pi kt}{T}}$ because $f = \frac{k}{T}$. Thus, if x(t) is periodic thereby having a Fourier series, a linear circuit's output to this signal will be the superposition of the output to each component.

$$y(t) = \sum_{k=-\infty}^{\infty} c_k H\left(\frac{k}{T}\right) e^{j\frac{2\pi kt}{T}} \tag{4.27}$$

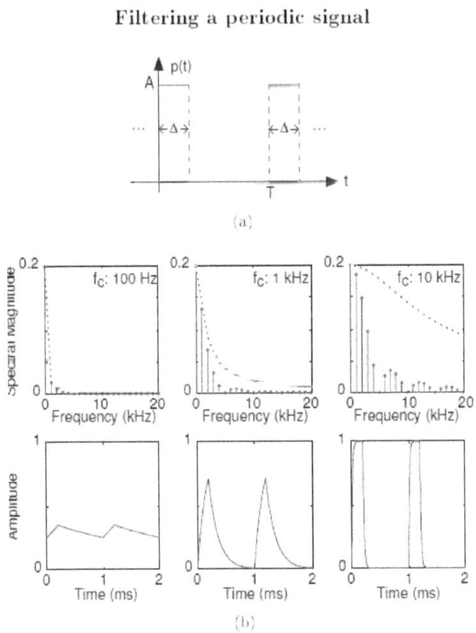


**Figure 4.10 A periodic pulse signal, such as shown on the left part ($\frac{\Delta}{T} = 0.2$), serves as the input to an RC lowpass filter. The input's period was 1 ms (millisecond). The filter's cutoff frequency was set to the various values indicated in the top row, which display the output signal's spectrum and the filter's transfer function. The bottom row shows the output signal derived from the Fourier series coefficients shown in the top row. (a) Periodic pulse signal (b) Top plots show the pulse signal's spectrum for various cutoff frequencies. Bottom plots show the filter's output signals.**

Thus, the output has a Fourier series, which means that it too is periodic. Its Fourier coefficients equal $c_k H\left(\frac{k}{T}\right)$. To obtain the spectrum of the output, we simply multiply the input spectrum by the frequency response. The circuit modifies the magnitude and phase of each Fourier coefficient. Note especially that while the Fourier coefficients do not depend on the signal's period, the circuit's transfer function does depend on frequency, which means that the circuit's output will differ as the period varies.

### Example 4.3

The periodic pulse signal shown on the left above serves as the input to a RC-circuit that has the transfer function (calculated elsewhere (Figure 3.30: Magnitude and phase of the transfer function))

$$H(f) = \frac{1}{1 + j2\pi fRC} \tag{4.28}$$

Figure 4.10 (Filtering a periodic signal ) shows the output changes as we vary the filter's cutoff frequency. Note how the signal's spectrum extends well above its fundamental frequency. Having a cutoff frequency ten times higher than the fundamental does perceptibly change the output waveform, rounding the leading and trailing edges. As the cutoff frequency decreases (center, then left), the rounding becomes more prominent, with the leftmost waveform showing a small ripple.

### Example 4.7.1

What is the average value of each output waveform? The correct answer may surprise you.

This example also illustrates the impact a lowpass filter can have on a waveform. The simple RC filter used here has a rather gradual frequency response, which means that higher harmonics are smoothly suppressed. Later, we will describe filters that have much more rapidly varying frequency responses, allowing a much more dramatic selection of the input's Fourier coefficients.

More importantly, we have calculated the output of a circuit to a periodic input without writing, much less solving, the differential equation governing the circuit›s behavior. Furthermore, we made these calculations entirely in the frequency domain. Using Fourier series, we can calculate how any linear circuit will respond to a periodic input.

## 4.8 DERIVATION OF THE FOURIER TRANSFORM

Fourier series clearly open the frequency domain as an interesting and useful way of determining how circuits and systems respond to periodic input signals.

Can we use similar techniques for nonperiodic signals? What is the response of the filter to a single pulse? Addressing these issues requires us to find the Fourier spectrum of all signals, both periodic and nonperiodic ones. We need a definition for the Fourier spectrum of a signal, periodic or not. This spectrum is calculated by what is known as the Fourier transform.

Let sT(t) be a periodic signal having period T. We want to consider what happens to this signal's spectrum as we let the period become longer and longer. We denote the spectrum for any assumed value of the period by ck(T). We calculate the spectrum according to the familiar formula:

$$c_k(T) = \frac{1}{T}\int_{-\frac{T}{2}}^{\frac{T}{2}} s_T(t)\, e^{-\frac{j2\pi kt}{T}}\, dt \tag{4.29}$$

where we have used a symmetric placement of the integration interval about the origin for subsequent derivational convenience. Let f be a fixed frequency equaling $\frac{k}{T}$; we vary the frequency index k proportionally as we increase the period. Define

$$S_T(f) \equiv T c_k(T) = \int_{-\frac{T}{2}}^{\frac{T}{2}} s_T(t)\, e^{-(j2\pi ft)}\, dt \tag{4.30}$$

making the corresponding Fourier series

$$s_T(t) = \sum_{k=-\infty}^{\infty} S_T(f)\, e^{j2\pi ft} \frac{1}{T} \tag{4.31}$$

As the period increases, the spectral lines become closer together, becoming a continuum. Therefore,

$$\lim_{T\to\infty} s_T(t) \equiv s(t) = \int_{-\infty}^{\infty} S(f)\, e^{j2\pi ft}\, df \tag{4.32}$$

with

$$S(f) = \int_{-\infty}^{\infty} s(t)\, e^{-(j2\pi ft)}\, dt \tag{4.33}$$

S(f) is the Fourier transform of s(t) (the Fourier transform is symbolically denoted by the uppercase version of the signal's symbol) and is defined for any signal for which the integral ((4.33)) converges.

**Example 4.4**

Let's calculate the Fourier transform of the pulse signal (Section 2.2.5: Pulse), p(t).

$$P(f) = \int_{-\infty}^{\infty} p(t)\, e^{-(j2\pi f t)} dt = \int_{0}^{\Delta} e^{-(j2\pi f t)} dt = \frac{1}{-(j2\pi f)} \left(e^{-(j2\pi f \Delta)} - 1\right)$$

$$P(f) = e^{-(j\pi f \Delta)} \frac{\sin(\pi f \Delta)}{\pi f}$$

Note how closely this result resembles the expression for Fourier series coefficients of the periodic pulse signal.

Figure 4.11 (Spectrum) shows how increasing the period does indeed lead to a continuum of coefficients, and that the Fourier transform does correspond to what the continuum becomes. The quantity $\frac{\sin(t)}{t}$ has a special name, the sinc (pronounced «sink») function, and is denoted by sinc(t). Thus, the magnitude of the pulse's Fourier transform equals $|\Delta \text{sinc}(\pi f \Delta)|$.

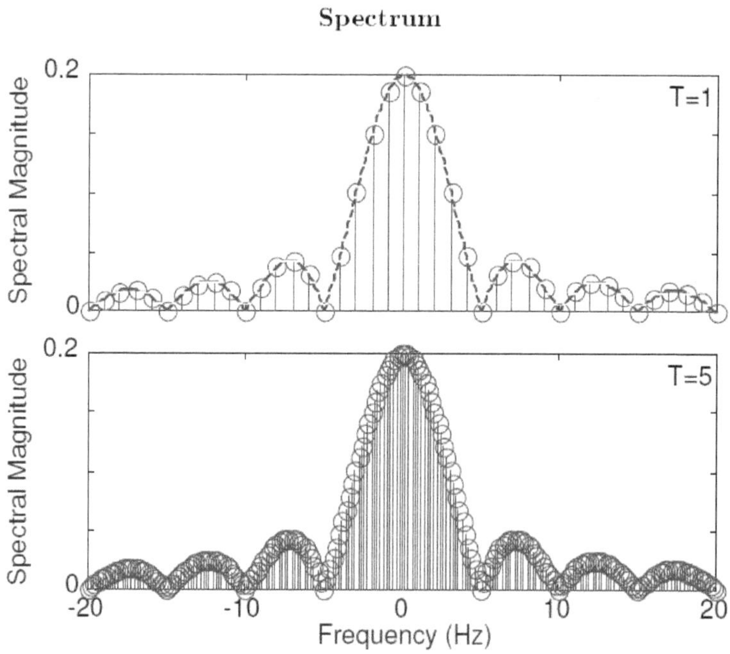


**Figure 4.11: The upper plot shows the magnitude of the Fourier series spectrum for the case of T=1 with the Fourier transform of p(t) shown as a dashed line. For the bottom panel, we expanded the period to T=5, keeping the pulse's duration fixed at 0.2, and computed its Fourier series coefficients.**

The Fourier transform relates a signal's time and frequency domain representations to each other. The direct Fourier transform (or simply the Fourier transform) calculates a signal's frequency domain representation from its time-domain variant. See equation below. The inverse Fourier transform (Equation) finds the time-domain representation from the frequency domain. Rather than explicitly writing the required integral, we often symbolically express these transform calculations as F(s) and F-1(S), respectively.

$$\begin{aligned} \mathcal{F}(s) &= S(f) \\ &= \int_{-\infty}^{\infty} s(t)\, e^{-(j2\pi ft)} dt \end{aligned} \tag{4.34}$$

$$\begin{aligned} \mathcal{F}^{-1}(S) &= s(t) \\ &= \int_{-\infty}^{\infty} S(f)\, e^{j2\pi ft} df \end{aligned} \tag{4.35}$$

We must have s(t)=F-1(F(s(t))) and S(f)=F(F-1(S(f))), and these results are indeed valid with minor exceptions.

Note: Recall that the Fourier series for a square wave gives a value for the signal at the discontinuities equal to the average value of the jump. This value may differ from how the signal is defined in the time domain, but being unequal at a point is indeed minor.

Showing that you "get back to where you started" is difficult from an analytic viewpoint, and we won't try here. Note that the direct and inverse transforms differ only in the sign of the exponent.

### Exercise 4.8.1

The differing exponent signs means that some curious results occur when we use the wrong sign. What is F(S(f))? In other words, use the wrong exponent sign in evaluating the inverse Fourier transform.

Properties of the Fourier transform and some useful transform pairs are provided in the accompanying tables (Table 4.1: Short Table of Fourier Transform Pairs and Table 4.2: Fourier Transform Properties). Especially important among these properties is Parseval's Theorem, which states that power computed in either domain equals the power in the other.

$$\int_{-\infty}^{\infty} s^2(t)\, dt = \int_{-\infty}^{\infty} (|S(f)|)^2 df \tag{4.36}$$

Of practical importance is the conjugate symmetry property: When s(t) is real-valued, the spectrum at negative frequencies equals the complex conjugate of the spectrum at the corresponding positive frequencies. Consequently, we need only plot the positive frequency portion of the spectrum (we can easily determine the remainder of the spectrum).

**Table 4.1 Short Table of Fourier Transform Pairs**

| s(t) | s(f) |
|---|---|
| e–(at) u(t) | $\frac{1}{j2\pi f+a}$ |
| e-(a\|t\|) | $\frac{2a}{4\pi^2 f^2+a^2}$ |
| $p(t)=\begin{cases}1 \text{ if } \|t\|<\frac{\Delta}{2}\\0 \text{ if } \|t\|>\frac{\Delta}{2}\end{cases}$ | $\frac{\sin(\pi f\Delta)}{\pi f}$ |
| $\frac{\sin(2\pi Wt)}{\pi t}$ | $S(f)=\begin{cases}1 \text{ if } \|f\|<W\\0 \text{ if } \|f\|>W\end{cases}$ |

## Exercise 4.8.2

How many Fourier transform operations need to be applied to get the original signal back: F(...(F(s)))=s(t)

Note that the mathematical relationships between the time domain and frequency domain versions of the same signal are termed transforms. We are transforming (in the nontechnical meaning of the word) a signal from one representation to another. We express Fourier transform pairs as s(t)↔S(f). A signal's time and frequency domain representations are uniquely related to each other. A signal thus "exists" in both the time and frequency domains, with the Fourier transform bridging between the two. We can define an information carrying signal in either the time or frequency domains; it behooves the wise engineer to use the simpler of the two. There is a prevalent misconception that a signal may only be expressed in either the time domain or the frequency domain, but not both simultaneously. Both cannot coexist simultaneously. This scenario reflects the behavior of complex amplitudes in circuits: By elucidating the workings and architecture of communication systems, we shall define signals solely in the frequency domain, without explicitly determining their time domain counterparts. This concept is demonstrated in a different module, where we establish Fourier series coefficients based on the letter that is being communicated. Therefore, a signal, although commonly described in the time-domain, can also be defined just as effectively (and perhaps more conveniently) in the frequency domain. Impedances are contingent upon frequency, and the time variable is not applicable.

It is most convenient to compute the output of a linear, time-invariant system in the time domain by first identifying the spectrum of the input signal, conducting a straightforward computation in the frequency domain, and then using the inverse transform to obtain the result. Moreover, comprehending communications and information processing systems necessitates a comprehensive grasp of signal structure and the functioning of systems in both the temporal and spectral domains.

The sole challenge in computing the Fourier transform of any signal arises when dealing with periodic signals in any domain. Recognizing that the Fourier series is a specific instance of the Fourier transform, we merely compute the Fourier series coefficients and display them alongside the spectra of nonperiodic signals on the identical frequency axis.

**Table 4.2 Fourier Transform Properties**

| | Time-Domain | Frequency Domain |
|---|---|---|
| Linearity | a1s1(t) + a2s2 (t) | a1s1(f) + a2s2 (f) |
| Conjugate Symmetry | s(t) Î R | S(f) = S (–f)* |
| Even Symmetry | s(t) = s (–t) | s(f) = s (–f) |
| Old Symmetry | s(t) = –s (–t) | s(f) = –S (–f) |
| Scale change | s(at) | $\frac{1}{\lvert a\rvert}S\left(\frac{f}{a}\right)$ |
| Time Delay | S (t–r) | e–(j2πfr) S (f) |
| Complex Modulation | ej2πf0t s (t) | S(f–f0) |
| Amplitude Modulation by Cosine | s(t) cos (2πf0t) | $\frac{S(f-f_0)+S(f+f_0)}{2}$ |
| Amplitude Modulation by Sine | s(t) sin (2πf0t) | $\frac{S(f-f_0)-S(f+f_0)}{2j}$ |
| Differentiation | $\frac{d}{dt}s\left(t\right)$ | J2πfS (f) |
| Integration | $\int_{-\infty}^{t} s\left(\alpha\right) d\alpha$ | $\frac{1}{j2\pi f}S\left(f\right)\text{ if }S\left(0\right)=0$ |
| Multiplication by t | ts (t) | $\frac{1}{-(j2\pi)}\frac{dS(f)}{df}$ |
| Area | $\int_{-\infty}^{\infty} s\left(t\right) dt$ | S (0) |
| Value at Origin | s (0) | $\int_{-\infty}^{\infty} S\left(f\right) df$ |
| Parseval's Theorem | $\int_{-\infty}^{\infty} \left(\lvert s\left(t\right)\rvert\right)^2 dt$ | $\int_{-\infty}^{\infty} \left(\lvert S\left(f\right)\rvert\right)^2 df$ |

### Exercise 4.5

In communications, a very important operation on a signal s(t) is to amplitude modulate it. Using this operation more as an example rather than elaborating the communications aspects here, we want to compute the Fourier transform — the spectrum — of

$$\left(1+s\left(t\right)\right)\cos\left(2\pi f_c t\right)$$

Thus,

$$\left(1+s\left(t\right)\right)\cos\left(2\pi f_c t\right)=\cos\left(2\pi f_c t\right)+s\left(t\right)\cos\left(2\pi f_c t\right)$$

For the spectrum of cos(2πfct), we use the Fourier series. Its period is $\frac{1}{f_c}$, and its only nonzero Fourier coefficients are $c_{\pm 1}=\frac{1}{2}$. The second term is not periodic unless s(t) has the same period as the sinusoid. Using Euler's reason, the spectrum of the second term can be derived as

$$s(t)\cos(2\pi f_c t) = \int_{-\infty}^{\infty} S(f)\, e^{j2\pi ft} df \cos(2\pi f_c t)$$

Using Euler's relation for the cosine,

$$(s(t)\cos(2\pi f_c t)) = \frac{1}{2}\int_{-\infty}^{\infty} S(f)\, e^{j2\pi(f+f_c)t} df + \frac{1}{2}\int_{-\infty}^{\infty} S(f)\, e^{j2\pi(f-f_c)t} df$$

$$(s(t)\cos(2\pi f_c t)) = \frac{1}{2}\int_{-\infty}^{\infty} S(f-f_c)\, e^{j2\pi ft} df + \frac{1}{2}\int_{-\infty}^{\infty} S(f+f_c)\, e^{j2\pi ft} df$$

$$(s(t)\cos(2\pi f_c t)) = \int_{-\infty}^{\infty} \frac{S(f-f_c)+S(f+f_c)}{2} e^{j2\pi ft} df$$

Exploiting the uniqueness property of the Fourier transform, we have

$$\mathcal{F}(s(t)\cos(2\pi f_c t)) = \frac{S(f-f_c)+S(f+f_c)}{2} \tag{4.37}$$

This component of the spectrum consists of the original signal's spectrum delayed and advanced in frequency. The spectrum of the amplitude modulated signal is shown in Figure 4.12.

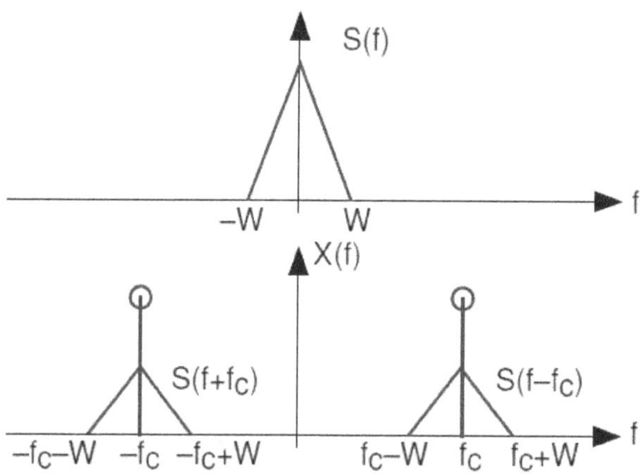


**Figure 4.12: A signal which has a triangular shaped spectrum is shown in the top plot. Its highest frequency — the largest frequency containing power — is W Hz. Once amplitude modulated, the resulting spectrum has "lines" corresponding to the Fourier series components at +(fc) and the original triangular spectrum shifted to components at +(fc) and scaled by $\frac{1}{2}$.**

Note how in this figure the signal s(t) is defined in the frequency domain. To find its time domain representation, we simply use the inverse Fourier transform.

### Exercise 4.8.3

What is the signal s(t) that corresponds to the spectrum shown in the upper panel of Figure 4.12?

### Exercise 4.8.4

What is the power in x(t), the amplitude-modulated signal? Try the calculation in both the time and frequency domains.

In this example, we call the signal s(t) a baseband signal because its power is contained at low frequencies. Signals such as speech and the Dow Jones averages are baseband signals. The baseband signal›s bandwidth equals W, the highest frequency at which it has power. Since x(t)'s spectrum is confined to a frequency band not close to the origin (we assume fc»W), we have a bandpass signal. The bandwidth of a bandpass signal is not its highest frequency, but the range of positive frequencies where the signal has power. Thus, in this example, the bandwidth is 2W Hz. Why a signal's bandwidth should depend on its spectral shape will become clear once we develop communications systems.

## 4.9 LINEAR TIME INVARIANT SYSTEMS

When we apply a periodic input to a linear, time-invariant system, the output is periodic and has Fourier series coefficients equal to the product of the system's frequency response and the input's Fourier coefficients (Filtering Periodic Signals). The way we derived the spectrum of non-periodic signal from periodic ones makes it clear that the same kind of result works when the input is not periodic. If x(t) serves as the input to a linear, time-invariant system having frequency response H(f), the spectrum of the output is X(f)H(f).

### Example 4.6

Let's use this frequency-domain input-output relationship for linear, time-invariant systems to find a formula for the RC-circuit's response to a pulse input. We have expressions for the input's spectrum and the system's frequency response.

$$P(f) = e^{-(j\pi f\Delta)}\frac{\sin(\pi f\Delta)}{\pi f} \tag{4.38}$$

$$H(f) = \frac{1}{1+j2\pi fRC} \tag{4.39}$$

Thus, the output's Fourier transform equals

$$Y(f) = e^{-(j\pi f\Delta)}\frac{\sin(\pi f\Delta)}{\pi f}\frac{1}{1+j2\pi fRC} \tag{4.40}$$

You won't find this Fourier transform in our table, and the required integral is difficult to evaluate as the expression stands. This situation requires cleverness and an understanding of the Fourier transform's properties. In particular, recall Euler's relation for the sinusoidal term and note the fact that multiplication by a complex exponential in the frequency domain amounts to a time delay. Let's momentarily make the expression for Y(f) more complicated.

$$\begin{aligned} e^{-(j\pi f\Delta)}\frac{\sin(\pi f\Delta)}{\pi f} &= e^{-(j\pi f\Delta)}\frac{e^{j\pi f\Delta}-e^{-(j\pi f\Delta)}}{j2\pi f} \\ &= \frac{1}{j2\pi f}\left(1-e^{-(j2\pi f\Delta)}\right) \end{aligned} \tag{4.41}$$

Consequently,

$$Y(f) = \frac{1}{j2\pi f}\left(1-e^{-(j\pi f\Delta)}\right)\frac{1}{1+j2\pi fRC} \tag{4.42}$$

The table of Fourier transform properties suggests thinking about this expression as a product of terms.

- Multiplication by $\frac{1}{j2\pi f}$ means integration.
- Multiplication by the complex exponential $e^{-(j2\pi f\Delta)}$ means delay by Δ seconds in the time domain.
- The term $1-e^{-(j2\pi f\Delta)}$ means, in the time domain, subtract the time-delayed signal from its original.
- The inverse transform of the frequency response is $\frac{1}{RC}e^{-\frac{t}{RC}}u(t)$.

We can translate each of these frequency-domain products into time-domain operations in any order we like because the order in which multiplications occur doesn›t affect the result. Let›s start with the product of $\frac{1}{j2\pi f}$ (integration in the time domain) and the transfer function:

$$\frac{1}{j2\pi f}\frac{1}{1+j2\pi fRC} \leftrightarrow \left(1-e^{-\frac{t}{RC}}\right)u(t) \tag{4.43}$$

The middle term in the expression for Y(f) consists of the difference of two terms: the constant 1 and the complex exponential e-(i2πfΔ) .Because of the Fourier transform's linearity, we simply subtract the results.

$$Y(f) \leftrightarrow \left(1-e^{-\frac{t}{RC}}\right)u(t) - \left(1-e^{-\frac{t-\Delta}{RC}}\right)u(t-\Delta) \tag{4.44}$$

Note that in delaying the signal how we carefully included the unit step. The second term in this result does not begin until t = Δ. Thus, the waveforms shown in the Filtering Periodic Signals example mentioned above are exponentials. We say that the time constant of an exponentially decaying signal equals the time it takes to decrease by 1/e of its original value. Thus, the time-constant of the rising and falling portions of the output equal the product of the circuit›s resistance and capacitance.

## Exercise 4.9.1

Derive the filter's output by considering the terms in (4.41) in the order given. Integrate last rather than first. You should get the same answer.

In this example, we used the table extensively to find the inverse Fourier transform, relying mostly on what multiplication by certain factors, like $\frac{1}{j2\pi f}$ and $e^{-(j2\pi f\Delta)}$, meant. We essentially treated multiplication by these factors as if they were transfer functions of some fictitious circuit. The transfer function $\frac{1}{j2\pi f}$ corresponded to a circuit that integrated, and $e^{-(j2\pi f\Delta)}$ to one that delayed. We even implicitly interpreted the circuit's transfer function as the input's spectrum! This approach to finding inverse transforms -- breaking down a complicated expression into products and sums of simple components -- is the engineer's way of breaking down the problem into several subproblems that are much easier to solve and then gluing the results together. Along the way we may make the system serve as the input, but in the rule Y(f)=X(f)H(f),which term is the input and which is the transfer function is merely a notational matter (we labeled one factor with an X and the other with an H).

## 4.9.1 Transfer Functions

The concept of a transfer function is applicable to a wide range of systems, not just limited to linear circuits. While we currently lack the necessary resources to provide a conclusive demonstration, it is universally true that all linear, time-invariant systems possess a frequency-domain input-output relationship that can be expressed as the multiplication of the Fourier transform of the input and the transfer function of the system. Linear circuits can be considered as a specific type of linear, time-invariant systems. As we address increasingly complex issues in the transmission, manipulation, and reception of information, we shall consider linear systems with specific qualities (transfer functions) without concerning themselves with the specific circuit that possesses these desirable properties. At this juncture, you might have reservations about the superficial nature of this method, and your concerns are justified. Later, we will demonstrate that by utilizing software, we may avoid the need to worry about generating a transfer function from circuit parts and op-amps.

## 4.9.2 Commutative Transfer Functions

Another interesting notion arises from the commutative property of multiplication (exploited in the above example). We can rather arbitrarily choose an order in which to apply each product. Consider a cascade of two linear, time-invariant systems. Because the Fourier transform of the first system's output is X(f) H1(f) and it serves as the second system's input, the cascade's output spectrum is X(f)H1(f)H2(f) Because this product also equals X(f)H2(f)H1(f), the cascade having the linear systems in the opposite order yields the same result. Furthermore, the cascade acts like a single linear system, having transfer function H1(f)H2(f). This result applies to other configurations of linear, time-invariant systems as well; see this Frequency Domain Problem. Engineers exploit this property by determining what transfer function they want, then breaking it down into components arranged according to standard configurations. Using the fact that op-amp circuits can be connected in cascade with the transfer function equaling the product of its component's transfer function (see this analog signal processing problem), we find a ready way of realizing designs. We now understand why op-amp implementations of transfer functions are so important.

## 4.10 MODELING THE SPEECH SIGNAL

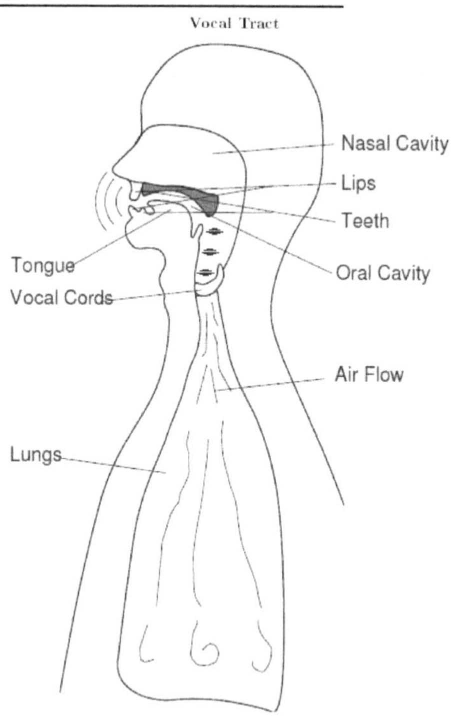


**Figure 4.13: The vocal tract is shown in cross-section. Air pressure produced by the lungs forces air through the vocal cords that, when under tension, produce puffs of air that excite resonances in the vocal and nasal cavities. What are not shown are the brain and the musculature that control the entire speech production process.**

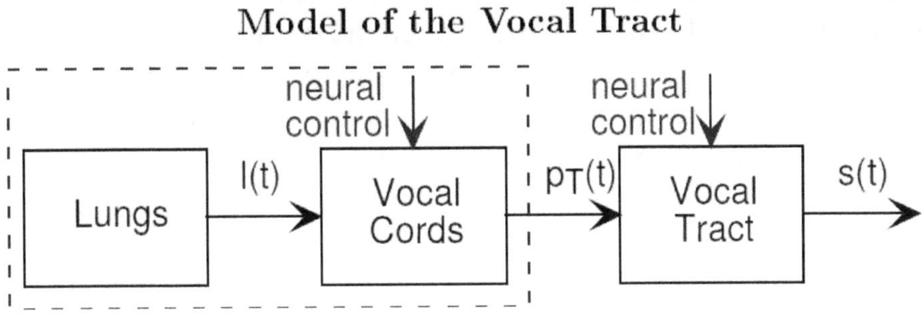


**Figure 4.14: The systems model for the vocal tract. The signals l(t), pT(t), and s(t) are the air pressure provided by the lungs, the periodic pulse output provided by the vocal cords, and the speech output respectively. Control signals from the brain are shown as entering the systems from the top. Clearly, these come from the same source, but for modeling purposes we describe them separately since they control different aspects of the speech signal.**

The spoken word transmits information through the voice signal. In order to examine various speech transmission and processing systems, it is necessary to comprehend the structure of the speech signal, including its unique characteristics, and develop methods to characterize and model speech generation. This modeling endeavor involves determining the system's depiction of how somewhat disorganized signals, originating from uncomplicated origins, acquire organization by being transmitted across a network of systems to produce speech. The selection of a system, whether it be for speech or any other context, is determined by the fundamental principles of physics that control the process of production. Due to the linear nature of the wave equation, which is the fundamental equation of acoustics, we may accurately utilize linear systems in our model. The applicability of linear system models for speech is limited to speech-related scenarios and does not extend to other contexts. Often, the mathematics underlying the problem in physics, biology, and/or chemistry are nonlinear, which means that linear systems models are only approximations. Nonlinear models are currently more challenging to comprehend compared to linear models, and information engineers often favor linear models due to the higher degree of comfort they offer. However, this does not necessarily guarantee a satisfactory level of accuracy.

Figure 4.13 (Vocal Tract) shows the actual speech production system and Figure 4.14 (Model of the Vocal Tract) shows the model speech production system. The properties of the model vary depending on whether you are articulating a vowel or a consonant. Our primary focus is on the mechanics of vowel creation. When the muscles surrounding the vocal cords tighten, the air pressure from the lungs causes the vocal cords to oscillate. To illustrate this phenomenon, grasp a rubber band and position it in front of your lips. When the object is kept open while air is blown through it, the air flows relatively unobstructed. This condition is referred to as the "breathing mode". When the rubber band is stretched tightly and positioned closely, blowing through the opening induces vibrations on its sides. This phenomenon is most effective when using a broad elastic band. One might envision the airflow characteristics on the opposing side of the rubber band or the vocal cords. The primary source mentioned previously is your lung capacity, which may be represented as a consistent provision of air pressure. The vocal cords react to this input by oscillating, resulting in a periodic function as the output of this system.

### Exercise 4.10.1

Note that the vocal cord system takes a constant input and produces a periodic airflow that corresponds to its output signal. Is this system linear or nonlinear? Justify your answer.

Singers modify vocal cord tension to change the pitch to produce the desired musical note. Vocal cord tension is governed by a control input to the musculature; in system's models we represent control inputs as signals coming into the top or bottom of the system. Certainly in the case of speech and in many other cases as well, it is the control input that carries information, impressing it on the system's output. The change of signal structure resulting from varying the control input enables information to be conveyed by the signal, a process generically known as modulation. In singing, musicality is largely conveyed by pitch; in western speech, pitch is much less important. A sentence can be read in a monotone fashion without completely destroying the information expressed by the sentence. However, the difference between a statement and a question is frequently expressed by pitch changes. For example, note the sound differences between "Let's go to the park." and "Let's go to the park?"

For some consonants, the vocal cords vibrate just as in vowels. For example, the so-called nasal sounds "n" and "m" have this property. For others, the vocal cords do not produce a periodic output. Going back to mechanism, when consonants such as "f" are produced, the vocal cords are placed under much less tension, which results in turbulent flow. The resulting output airflow is quite erratic, so much so that we describe it as being noise. We define noise carefully later when we delve into communication problems.

The vocal cords' periodic output can be well described by the periodic pulse train pT(t), as shown in the periodic pulse signal (Figure 4.1), with T denoting the pitch period. The spectrum of this signal (4.9) contains harmonics of the frequency $\frac{1}{T}$, what is known as the pitch frequency or the fundamental frequency F0. The primary difference between adult male and female/prepubescent speech is pitch. Before puberty, pitch frequency for normal speech ranges between 150-400 Hz for both males and females. After puberty, the vocal cords of males undergo a physical change, which has the effect of lowering their pitch frequency to the range 80-160 Hz. If we could examine the vocal cord output, we could probably discern whether the speaker was male or female. This difference is also readily apparent in the speech signal itself.

To simplify our speech modeling effort, we shall assume that the pitch period is constant. With this simplification, we collapse the vocal-cord-lung system as a simple source that produces the periodic pulse signal (Figure 4.14 (Model of the Vocal Tract)). The sound pressure signal thus produced enters the mouth behind the tongue, creates acoustic disturbances, and exits primarily through the lips and to some extent through the nose. Speech specialists tend to name the mouth, tongue, teeth, lips, and nasal cavity the vocal tract. The physics governing the sound disturbances produced in the vocal tract and those of an organ pipe are quite similar. Whereas the organ pipe has the simple physical structure of a straight tube, the cross-section of the vocal tract "tube" varies along its length because of the positions of the tongue, teeth, and lips. It is these positions that are controlled by the brain to produce the vowel sounds. Spreading the lips, bringing the teeth together, and bringing the tongue toward the front portion of the roof of the mouth produces the sound "ee." Rounding the lips, spreading the teeth, and positioning the tongue toward the back of the oral cavity produces the sound "oh." These variations result in a linear, time-invariant system that has a frequency response typified by several peaks, as shown in Figure 4.15 (Speech Spectrum).

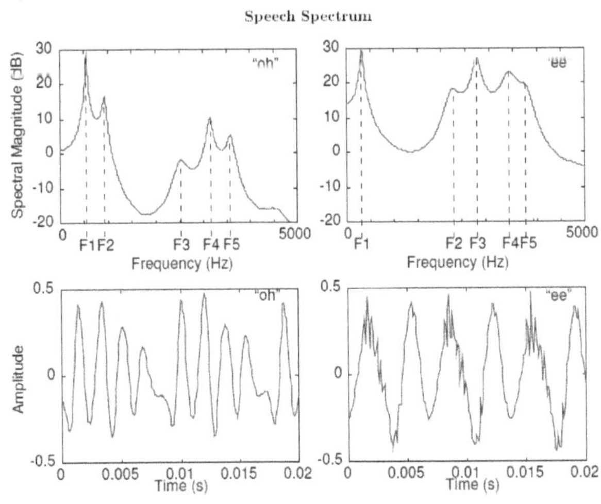


**Figure 4.15: The ideal frequency response of the vocal tract as it produces the sounds "oh" and "ee" are shown on the top left and top right, respectively. The spectral peaks are known as formants, and are numbered consecutively from low to high frequency. The bottom plots show speech waveforms corresponding to these sounds.**

These peaks are known as formants. Thus, speech signal processors would say that the sound "oh" has a higher first formant frequency than the sound "ee," with F2 being much higher during «ee." F2 and F3 (the second and third formants) have more energy in "ee" than in "oh." Rather than serving as a filter, rejecting high or low frequencies, the vocal tract serves to shape the spectrum of the vocal cords. In the time domain, we have a periodic signal, the pitch, serving as the input to a linear system. We know that the output—the speech signal we utter and that is heard by others and ourselves—will also be periodic. Example time-domain speech signals are shown in Figure 4.15 where the periodicity is quite apparent.

### Exercise 4.10.2

From the waveform plots shown in Figure 4.15, determine the pitch period and the pitch frequency.

Since speech signals are periodic, speech has a Fourier series representation given by a linear circuit's response to a periodic signal (4.27). Because the acoustics of the vocal tract are linear, we know that the spectrum of the output equals the product of the pitch signal's spectrum and the vocal tract's frequency response. We thus obtain the fundamental model of speech production.

$$S(f) = P_T(f)\, H_V(f) \tag{4.45}$$

Here, HV(f) is the transfer function of the vocal tract system. The Fourier series for the vocal cords' output, derived in this equation (p. 122), is

$$c_k = Ae^{-\frac{j\pi k\Delta}{T}} \frac{\sin\left(\frac{\pi k\Delta}{T}\right)}{\pi k} \tag{4.46}$$

and is plotted on the top in Figure 4.10.4a. If we had, for example, a male speaker with about a 110 Hz pitch (T ~ 9.1ms) saying the vowel "oh", the spectrum of his speech predicted by our model is shown in Figure 4.16(b) (voice spectrum).

The model spectrum idealizes the measured spectrum, and captures all the important features. The measured spectrum certainly demonstrates what are known as pitch lines, and we realize from our model that they are due to the vocal cord's periodic excitation of the vocal tract. The vocal tract's shaping of the line spectrum is clearly evident, but difficult to discern exactly, especially at the higher frequencies. The model transfer function for the vocal tract makes the formants much more readily evident.

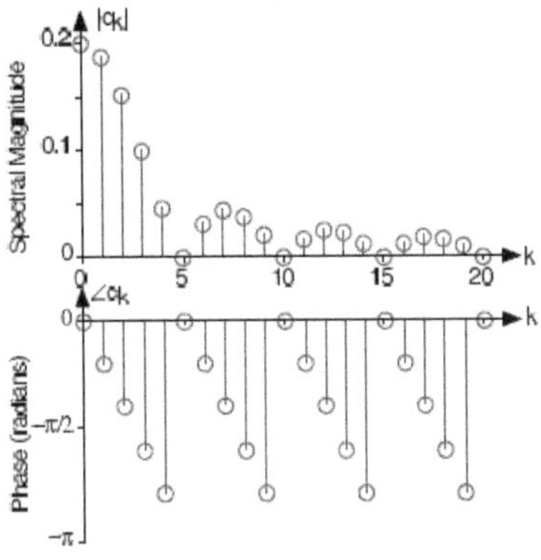


(a) pulse

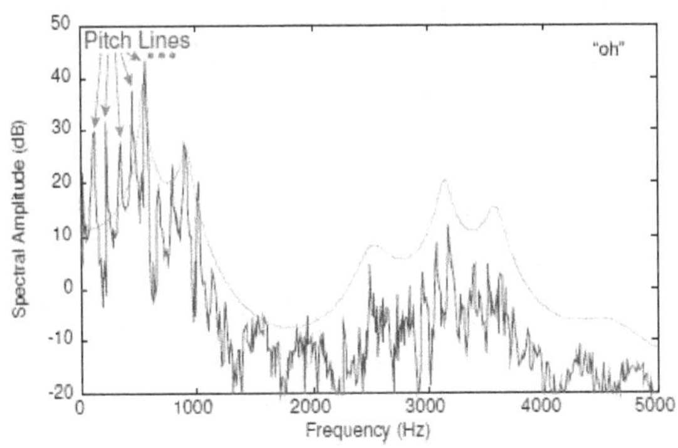


(b) voice spectrum

**Figure 4.16: The vocal tract's transfer function, shown as the thin, smooth line, is superimposed on the spectrum of actual male speech corresponding to the sound "oh." The pitch lines corresponding to harmonics of the pitch frequency are indicated. (a) The vocal cords' output spectrum PT (f). (b) The vocal tract's transfer function, HV (f) and the speech spectrum.**

## Exercise 4.10.3

The Fourier series coefficients for speech are related to the vocal tract's transfer function only at the frequencies $\frac{k}{T}$, $k \in \{1, 2, \ldots\}$;see previous result. Would male or female speech tend to have a more clearly identifiable formant structure when its spectrum is computed? Consider, for example, how the spectrum shown on the right in Figure 4.16 would change if the pitch were twice as high (≈ 300 Hz)?

When we speak, pitch and the vocal tract's transfer function are not static; they change according to their control signals to produce speech. Engineers typically display how the speech spectrum changes over time with what is known as a spectrogram. See Figure 16. Note how the line spectrum, which indicates how the pitch changes, is visible during the vowels, but not during the consonants (like the ce in «Rice»).

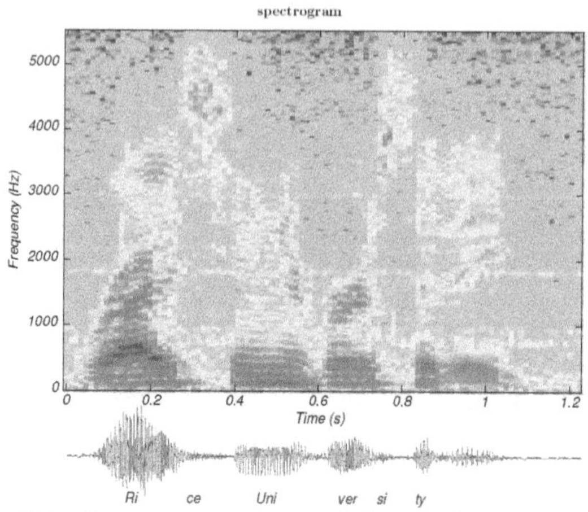


**Figure 4.17: Displayed is the spectrogram of the author saying "Rice University." Blue indicates low energy portion of the spectrum, with red indicating the most energetic portions. Below the spectrogram is the time-domain speech signal, where the periodicities can be seen.**

The fundamental model for speech indicates how engineers use the physics underlying the signal generation process and exploit its structure to produce a systems model that suppresses the physics while emphasizing how the signal is "constructed." From everyday life, we know that speech contains a wealth of information. We want to determine how to transmit and receive it. Efficient and effective speech transmission requires us to know the signal's properties and its structure (as expressed by the fundamental model of speech production). We see from Figure 4.17 (spectrogram), for example, that speech contains significant energy from zero frequency up to around 5 kHz.

Effective speech transmission systems must be able to cope with signals having this bandwidth. It is interesting that one system that does not support this 5 kHz bandwidth is the telephone: Telephone systems act like a bandpass filter passing energy between about 200 Hz and 3.2 kHz. The most important consequence of this filtering is the removal of high frequency energy. In our sample utterance, the «ce" sound in "Rice"" contains most of its energy above 3.2 kHz; this filtering effect is why it is extremely difficult to distinguish the sounds "s" and "f" over the telephone. Try this yourself: Call a friend and determine if they can distinguish between the words "six" and "fix". If you say these words in isolation so that no context provides a hint about which word you are saying, your friend will not be able to tell them apart. Radio does support this bandwidth (see more about AM and FM radio systems (Section 6.11)).

Efficient speech transmission systems exploit the speech signal›s special structure: What makes speech speech? You can conjure many signals that span the same frequencies as speech—car engine sounds, violin music, dog barks—but don't sound at all like speech. We shall learn later that transmission of any 5 kHz bandwidth signal requires about 80 kbps (thousands of bits per second) to transmit digitally. Speech signals can be transmitted using less than 1 kbps because of its special structure. To reduce the «digital bandwidth» so drastically means that engineers spent many years to develop signal processing and coding methods that could capture the special characteristics of speech without destroying how it sounds. If you used a speech transmission system to send a violin sound, it would arrive horribly distorted; speech transmitted the same way would sound fine.

Exploiting the special structure of speech requires going beyond the capabilities of analog signal processing systems. Many speech transmission systems work by finding the speaker's pitch and the formant frequencies. Fundamentally, we need to do more than filtering to determine the speech signal's structure; we need to manipulate signals in more ways than are possible with analog systems. Such flexibility is achievable (but not without some loss) with programmable digital systems.

## 4.11 FREQUENCY DOMAIN PROBLEMS

### Problem 4.1: Simple Fourier Series

Find the complex Fourier series representations of the following signals without explicitly calculating Fourier integrals. What is the signal's period in each case?

- s(t)=sin(t)
- s(t)=sin2(t)
- s(t)=cos(t)+2cos(2t)
- s(t)=cos(2t)cos(t)
- $s(t) = \cos\left(10\pi t + \frac{\pi}{6}\right)(1 + \cos(2\pi t))$
- s(t) given by the depicted waveform (Figure 4.18).

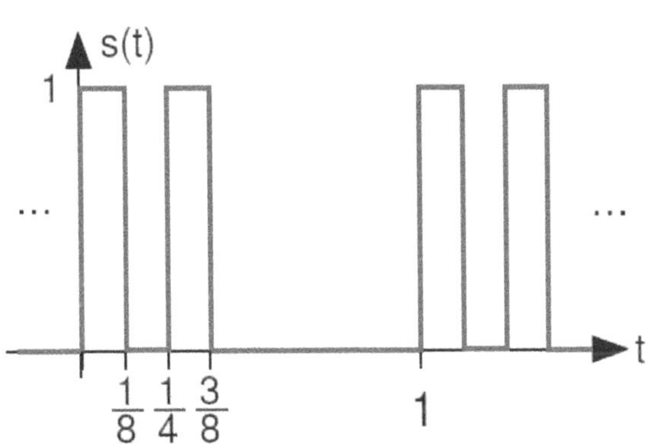


**Figure 4.18**

## Problem 4.2: Fourier Series

Find the Fourier series representation for the following periodic signals (Figure 4.19). For the third signal, find the complex Fourier series for the triangle wave without performing the usual Fourier integrals. Hint: How is this signal related to one for which you already have the series?

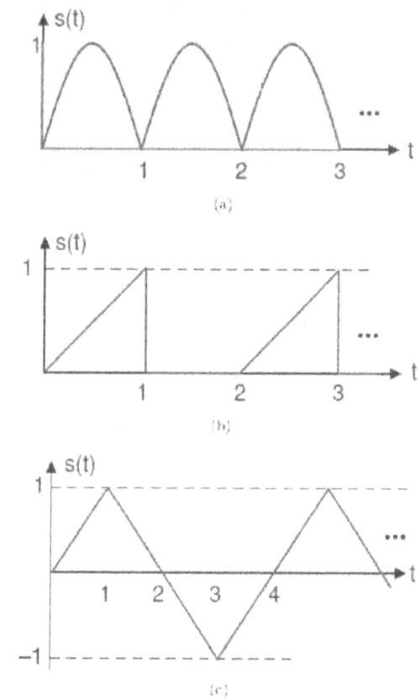

**Figure 4.19**

## Problem 4.3: Phase Distortion

We can learn about phase distortion by returning to circuits and investigate the following circuit (Figure 4.20).

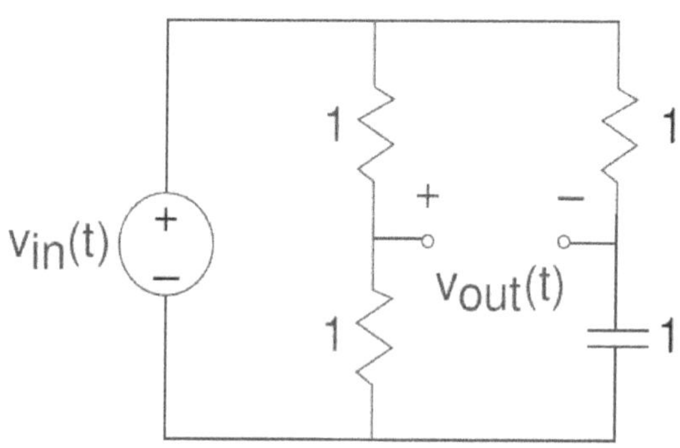


**Figure 4.20**

- Find this filter's transfer function.
- Find the magnitude and phase of this transfer function. How would you characterize this circuit?
- Let vin(t) be a square-wave of period T. What is the Fourier series for the output voltage?

- Use Matlab to find the output's waveform for the cases T=0.01 and T=2. What value of T delineates the two kinds of results you found? The software in fourier2.m might be useful.
- Instead of the depicted circuit, the square wave is passed through a system that delays its input, which applies a linear phase shift to the signal's spectrum. Let the delay τ be $\frac{T}{4}$. Use the transfer function of a delay to compute using Matlab the Fourier series of the output. Show that the square wave is indeed delayed.

## Problem 4.4: Approximating Periodic Signals

Often, we want to approximate a reference signal by a somewhat simpler signal. To assess the quality of an approximation, the most frequently used error measure is the mean-squared error. For a periodic signal s(t),

$$\epsilon^2 = \frac{1}{T}\int_0^T (s(t) - \tilde{s}(t))^2 dt$$

Where s(t) is the reference signal and $\tilde{s}(t)$ its approximation. One convenient way of finding approximations for periodic signals is to truncate their Fourier series.

$$\tilde{s}(t) = \sum_{k=-K}^{K} c_k e^{j\frac{2\pi k}{T}t}$$

The point of this problem is to analyze whether this approach is the best (i.e., always minimizes the mean-squared error).

- Find a frequency-domain expression for the approximation error when we use the truncated Fourier series as the approximation.
- Instead of truncating the series, let's generalize the nature of the approximation to including any set of 2K+1 terms: We'll always include the c0 and the negative indexed term corresponding to ck. What selection of terms minimizes the mean-squared error? Find an expression for the mean-squared error resulting from your choice.
- Find the Fourier series for the depicted signal (Figure 4.21). Use Matlab to find the truncated approximation and best approximation involving two terms. Plot the mean-squared error as a function of K for both approximations.

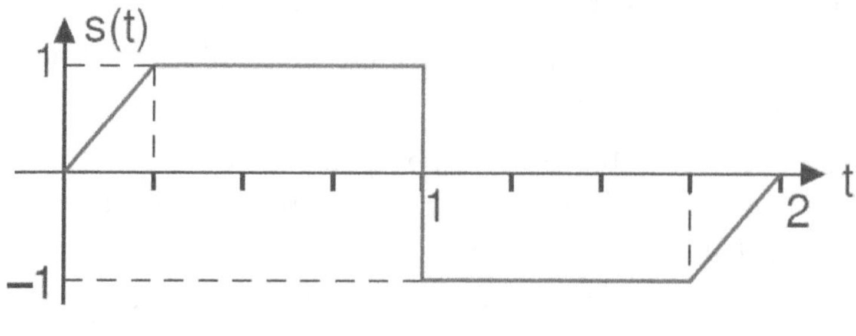


Figure 4.21

## Problem 4.5: Long, Hot Days

The daily temperature is a consequence of several effects, one of them being the sun's heating. If this were the dominant effect, then daily temperatures would be proportional to the number of daylight hours. The plot (Figure 4.22) shows that the average daily high temperature does not behave that way.

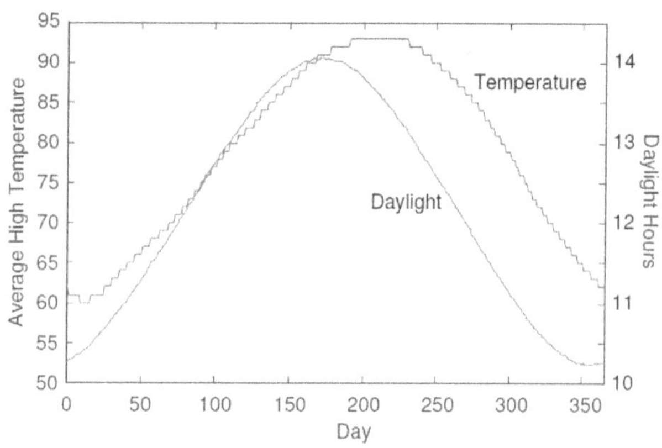


Figure 4.22

In this problem, we want to understand the temperature component of our environment using Fourier series and linear system theory. The file temperature. mat contains these data (daylight hours in the first row, corresponding average daily highs in the second) for Houston, Texas.

- Let the length of day serve as the sole input to a system having an output equal to the average daily temperature. Examining the plots of input and output, would you say that the system is linear or not? How did you reach you conclusion?
- Find the first five terms (c0,...,c4) of the complex Fourier series for each signal. Use the following formula that approximation the integral require to that find the Fourier coeffcients.

$$c_k = \frac{1}{366} \sum_{n=0}^{366} s(n)\, e^{-\left(j \frac{2\pi nk}{366}\right)}$$

- What is the harmonic distortion in the two signals? Exclude c0 from this calculation.
- Because the harmonic distortion is small, let's concentrate only on the first harmonic. What is the phase shift between input and output signals?
- Find the transfer function of the simplest possible linear model that would describe the data. Characterize and interpret the structure of this model. In particular, give a physical explanation for the phase shift.
- Predict what the output would be if the model had no phase shift. Would days be hotter? If so, by how much?

## Problem 4.6: Fourier Transform Pairs

Find the Fourier or inverse Fourier transform of the following:

- $x(t) = e^{-(a|t|)}$
- $x(t) = te^{-(at)} u(t)$
- $X(f) = \begin{cases} 1 & \text{if } |f| < W \\ 0 & \text{if } |f| > W \end{cases}$
- $x(t) = e^{-(at)} \cos(2\pi f_0 t)\, u(t)$

## Problem 4.7: Duality in Fourier Transforms

"Duality" means that the Fourier transform and the inverse Fourier transform are very similar. Consequently, the waveform s(t) in the time domain and the spectrum s(f) have a Fourier transform and an inverse Fourier transform, respectively, that are very similar.

- Calculate the Fourier transform of the signal shown below (Figure 4.23(a)).
- Calculate the inverse Fourier transform of the spectrum shown below (Figure 4.23(b)).
- How are these answers related? What is the general relationship between the Fourier transform of s(t) and the inverse transform of s(f)?

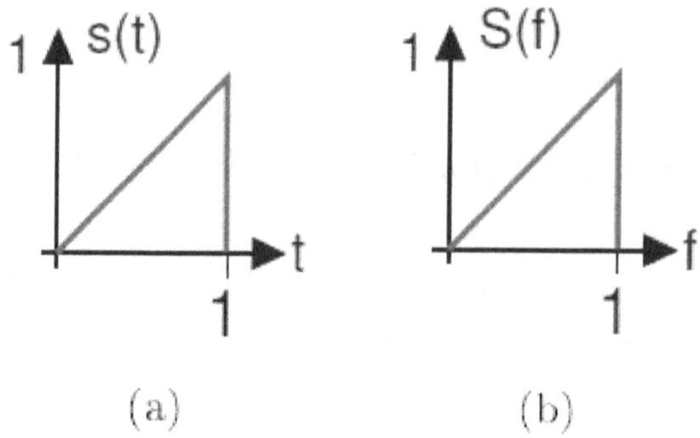

Figure 4.23

## Problem 4.8: Spectra of Pulse Sequences

Pulse sequences occur often in digital communication and in other fields as well. What are their spectral properties?

- Calculate the Fourier transform of the single pulse shown below (Figure 4.24(a)).
- Calculate the Fourier transform of the two-pulse sequence shown below (Figure 4.24(b)).
- Calculate the Fourier transform for the ten-pulse sequence shown in below (Figure 4.24 (c)). You should look for a general expression that holds for sequences of any length.
- Using Matlab, plot the magnitudes of the three spectra. Describe how the spectra change as the number of repeated pulses increases.

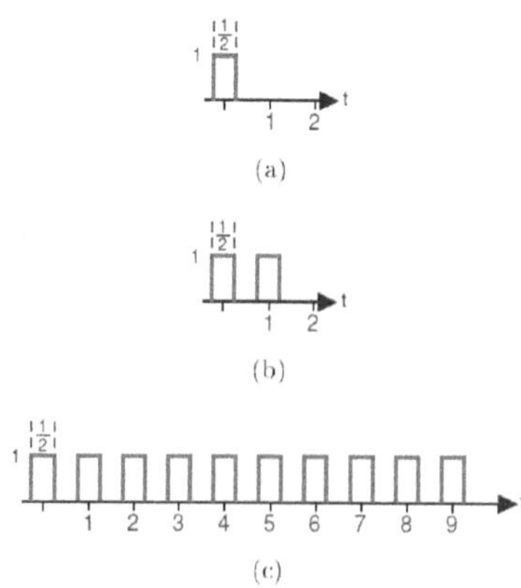

Figure 4.24

## Problem 4.9: Spectra of Digital Communication Signals

One way to represent bits with signals is shown in Figure 4.25. If the value of a bit is a "1", it is represented by a positive pulse of duration T. If it is a "0", it is represented by a negative pulse of the same duration. To represent a sequence of bits, the appropriately chosen pulses are placed one after the other.

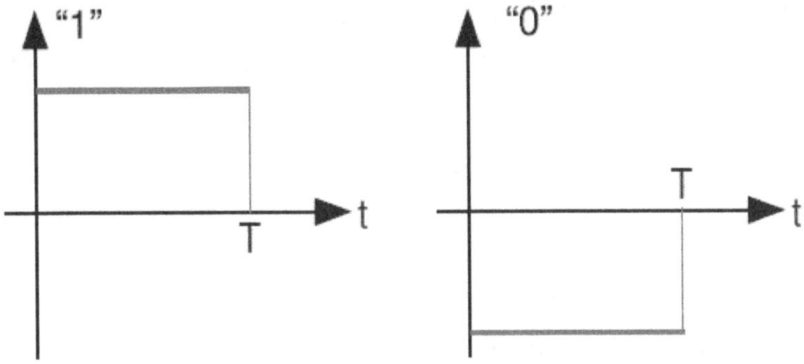


**Figure 4.25**

- What is the spectrum of the waveform that represents the alternating bit sequence "...01010101..."?
- This signal's bandwidth is defined to be the frequency range over which 90% of the power is contained. What is this signal's bandwidth?
- Suppose the bit sequence becomes "...00110011...". Now what is the bandwidth?

## Problem 4.10: Lowpass Filtering a Square Wave

Let a square wave (period T) serve as the input to a first-order lowpass system constructed as a RC filter. We want to derive an expression for the time-domain response of the filter to this input.

- First, consider the response of the filter to a simple pulse, having unit amplitude and width $\frac{T}{2}$. Derive an expression for the filter's output to this pulse.
- Noting that the square wave is a superposition of a sequence of these pulses, what is the filter's response to the square wave?
- The nature of this response should change as the relation between the square wave's period and the filter's cutoff frequency change. How long must the period be so that the response does not achieve a relatively constant value between transitions in the square wave? What is the relation of the filter's cutoff frequency to the square wave's spectrum in this case?

## Problem 4.11: Mathematics with Circuits

Simple circuits can implement simple mathematical operations, such as integration and differentiation. We want to develop an active circuit (it contains an op-amp) having an output that is proportional to the integral of its input. For example, you could use an integrator in a car to determine distance traveled from the speedometer.

- What is the transfer function of an integrator?
- Find an op-amp circuit so that its voltage output is proportional to the integral of its input for all signals.

## Problem 4.12: Where is that sound coming from?

We determine where sound is coming from because we have two ears and a brain. Sound travels at a relatively slow speed and our brain uses the fact that sound will arrive at one ear before the other. As shown here (Figure 4.26), a sound coming from the right arrives at the left ear τ seconds after it arrives at the right ear.

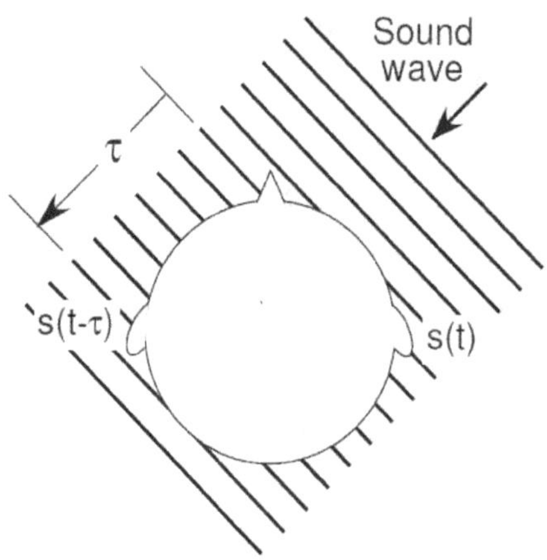


**Figure 4.26**

Once the brain finds this propagation delay, it can determine the sound direction. In an attempt to model what the brain might do, RU signal processors want to design an optimal system that delays each ear›s signal by some amount then adds them together. Δl and Δr are the delays applied to the left and right signals respectively. The idea is to determine the delay values according to some criterion that is based on what is measured by the two ears.

- What is the transfer function between the sound signal s(t) and the processor output y(t)? One way of determining the delay τ is to choose to maximize the power in y(t).
- How are these maximum-power processing delays related to τ?

## Problem 4.13: Arrangements of Systems

Architecting a system of modular components means arranging them in various configurations to achieve some overall input-output relation. For each of the following determine the overall transfer function between x(t) and y(t).

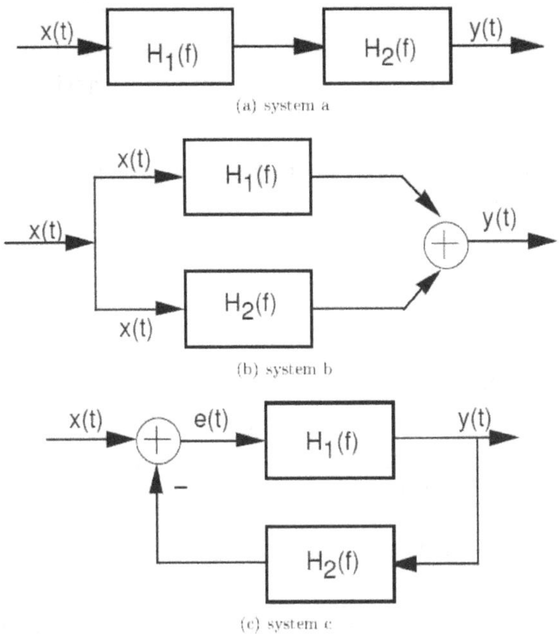


Figure 4.27

The overall transfer function for the cascade (first depicted system) is particularly interesting. What does it say about the effect of the ordering of linear, time-invariant systems in a cascade?

## Problem 4.14: Filtering

Let the signal $s(t) = \frac{\sin(\pi t)}{\pi t}$ be the input to a linear, time-invariant filter having the transfer function shown below (Figure 4.28). Find the expression for y(t), the filter's output.

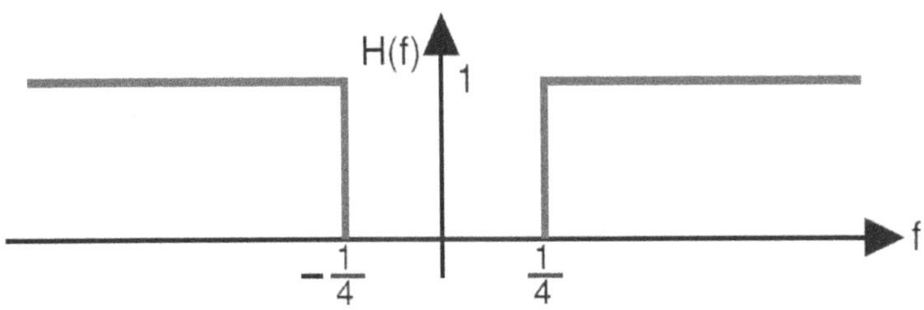


Figure 4.28

## Problem 4.15: Circuits Filter

A unit-amplitude pulse with duration of one second serves as the input to an RC-circuit having transfer function

$$H(f) = \frac{j2\pi f}{4 + j2\pi f}$$

- How would you categorize this transfer function: lowpass, highpass, bandpass, other?
- Find a circuit that corresponds to this transfer function.
- Find an expression for the filter's output.

## Problem 4.16: Reverberation

Reverberation corresponds to adding to a signal its delayed version.

- Assuming τ represents the delay, what is the input-output relation for a reverberation system? Is the system linear and time-invariant? If so, find the transfer function; if not, what linearity or time-invariance criterion does reverberation violate.
- A music group known as the ROwls is having trouble selling its recordings. The record company's engineer gets the idea of applying different delay to the low and high frequencies and adding the result to create a new musical effect. Thus, the ROwls' audio would be separated into two parts (one less than the frequency f0, the other greater than f0), these would be delayed by τl and τh respectively, and the resulting signals added. Draw a block diagram for this new audio processing system, showing its various components.
- How does the magnitude of the system's transfer function depend on the two delays?

## Problem 4.17: Echoes in Telephone Systems

A frequently encountered problem in telephones is echo. Here, because of acoustic coupling between the ear piece and microphone in the handset, what you hear is also sent to the person talking. That person thus not only hears you, but also hears her own speech delayed (because of propagation delay over the telephone network) and attenuated (the acoustic coupling gain is less than one). Furthermore, the same problem applies to you as well: The acoustic coupling occurs in her handset as well as yours.

- Develop a block diagram that describes this situation.
- Find the transfer function between your voice and what the listener hears.
- Each telephone contains a system for reducing echoes using electrical means. What simple system could null the echoes?

## Problem 4.18: Effective Drug Delivery

In most patients, it takes time for the concentration of an administered drug to achieve a constant level in the blood stream. Typically, if the drug concentration in the patient's intravenous line is $C_d u(t)$, the concentration in the patient's blood stream is $C_p(1 - e^{-(at)})u(t)$.

- Assuming the relationship between drug concentration in the patient's drug and the delivered concentration can be described as a linear, time-invariant system, what is the transfer function?
- Sometimes, the drug delivery system goes awry and delivers drugs with little control. What would the patient's drug concentration be if the delivered concentration were a ramp? More precisely, if it were $C_d t u(t)$?
- A clever doctor wants to have the flexibility to slow down or speed up the patient's drug concentration. In other words, the concentration is to be $C_p(1 - e^{-(bt)})u(t)$, with b bigger or smaller than a. How should the delivered drug concentration signal be changed to achieve this concentration profile?

## Problem 4.19: Catching Speeders with Radar

RU Electronics has been contracted to design a Doppler radar system. Radar transmitters emit a signal that bounces off any conducting object. Signal differences between what is sent and the radar return is processed and features of interest extracted. In Doppler systems, the object›s speed along the direction of the radar beam is the feature the design must extract. The transmitted signal is a sinsusoid: $x(t) = A\cos(2\pi f_c t)$. The measured return signal equals $B\cos(2\pi((f_c + \Delta f)t + \varphi))$, where the Doppler offset frequency Δf equals 10v where v is the car›s velocity coming toward the transmitter.

- Design a system that uses the transmitted and return signals as inputs and produces Δf.
- One problem with designs based on overly simplistic design goals is that they are sensitive to unmodeled assumptions. How would you change your design, if at all, so that whether the car is going away or toward the transmitter could be determined?
- Suppose two objects traveling different speeds provide returns. How would you change your design, if at all, to accomodate multiple returns?

## Problem 4.20: Demodulating an AM Signal

Let m(t) denote the signal that has been amplitude modulated.

$$x(t) = A(1 + m(t))\sin(2\pi f_c t)$$

Radio stations try to restrict the amplitude of the signal m(t) so that it is less than one in magnitude. The frequency fc is very large compared to the frequency content of the signal. What we are concerned about here is not transmission, but reception.

- The so-called coherent demodulator simply multiplies the signal x(t) by a sinusoid having the same frequency as the carrier and lowpass filters the result. Analyze this receiver and show that it works. Assume the lowpass filter is ideal.
- One issue in coherent reception is the phase of the sinusoid used by the receiver relative to that used by the transmitter. Assuming that the sinusoid of the receiver has a phase $\varphi$, how does the output depend on $\varphi$? What is the worst possible value for this phase?
- The incoherent receiver is more commonly used because of the phase sensitivity problem inherent in coherent reception. Here, the receiver full-wave rectifies the received signal and lowpass filters the result (again ideally). Analyze this receiver. Does its output differ from that of the coherent receiver in a significant way?

## Problem 4.21: Unusual Amplitude Modulation

We want to send a band-limited signal having the depicted spectrum (Figure 4.29 (a)) with amplitude modulation in the usual way. I.B. Different suggests using the square-wave carrier shown below (Figure 4.29 (b)). Well, it is different, but his friends wonder if any technique can demodulate it.

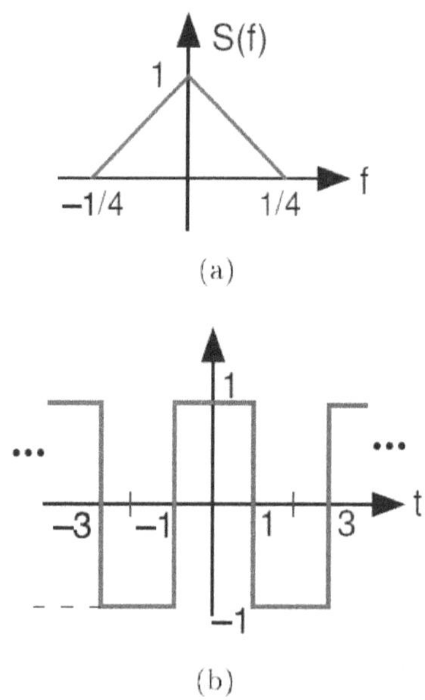


**Figure 4.29**

- Find an expression for X(f), the Fourier transform of the modulated signal.
- Sketch the magnitude of X(f), being careful to label important magnitudes and frequencies.

- What demodulation technique obviously works?
- I.B. challenges three of his friends to demodulate x(t) some other way. One friend suggests modulating x(t) with $\cos\left(\frac{\pi t}{2}\right)$, another wants to try modulating with cos(πt), and the third thinks $\cos\left(\frac{3\pi t}{2}\right)$ will work. Sketch the magnitude of the Fourier transform of the signal each student's approach produces. Which student comes closest to recovering the original signal? Why?

## Problem 4.22: Sammy Falls Asleep...

While sitting in ELEC 241 class, he falls asleep during a critical time when an AM receiver is being described. The received signal has the form $r(t) = A(1 + m(t))\cos(2\pi f_c t + \phi)$ where the phase φ is unknown. The message signal is m(t); it has a bandwidth of W Hz and a magnitude less than 1 ( |m(t)| <1). The phase φ is unknown. The instructor drew a diagram (Figure 4.30) for a receiver on the board; Sammy slept through the description of what the unknown systems where.

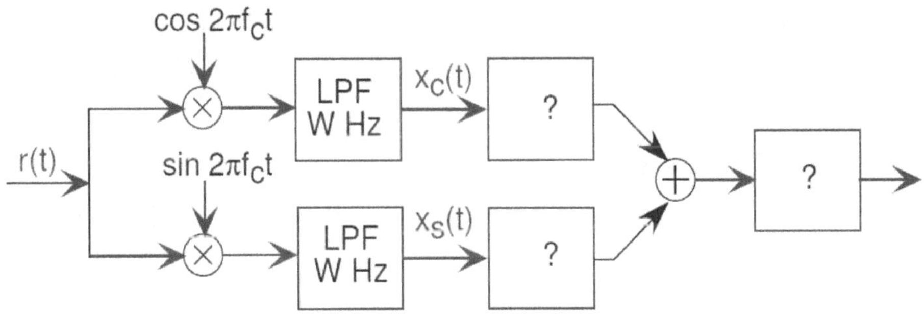


**Figure 4.30**

- What are the signals xc(t) and xs(t)?
- What would you put in for the unknown systems that would guarantee that the final output contained the message regardless of the phase?
- Hint: Think of a trigonometric identify that would prove useful.
- Sammy may have been asleep, but he can think of a far simpler receiver. What is it?

## Proble 4.23: Jamming

Sid Richardson college decides to set up its own AM radio station KSRR. The resident electrical engineer decides that she can choose any carrier frequency and message bandwidth for the station. A rival college decides to jam its transmissions by transmitting a high-power signal that interferes with radios that try to receive KSRR. The jamming signal jam(t) is what is known as a sawtooth wave (depicted in Figure 4.31) having a period known to KSRR's engineer.

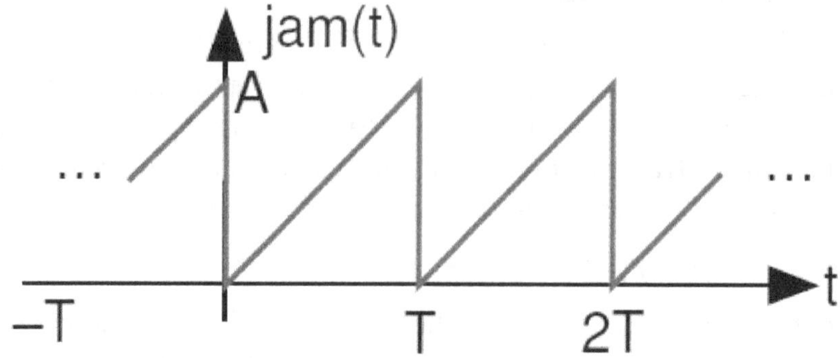


**Figure 4.31**

- Find the spectrum of the jamming signal.
- Can KSRR entirely circumvent the attempt to jam it by carefully choosing its carrier frequency and transmission bandwidth? If so, find the station's carrier frequency and transmission bandwidth in terms of T, the period of the jamming signal; if not, show why not.

### Problem 4.24: AM Stereo

A stereophonic signal consists of a "left" signal l(t) and a «right» signal r(t) that conveys sounds coming from an orchestra›s left and right sides, respectively. To transmit these two signals simultaneously, the transmitter first forms the sum signal $s_+(t)=l(t)+r(t)$ and the difference signal $s_-(t)=l(t)-r(t)$. Then, the transmitter amplitude-modulates the difference signal with a sinusoid having frequency 2W, where W is the bandwidth of the left and right signals. The sum signal and the modulated difference signal are added, the sum amplitude-modulated to the radio station's carrier frequency fc, and transmitted. Assume the spectra of the left and right signals are as shown (Figure 4.32).

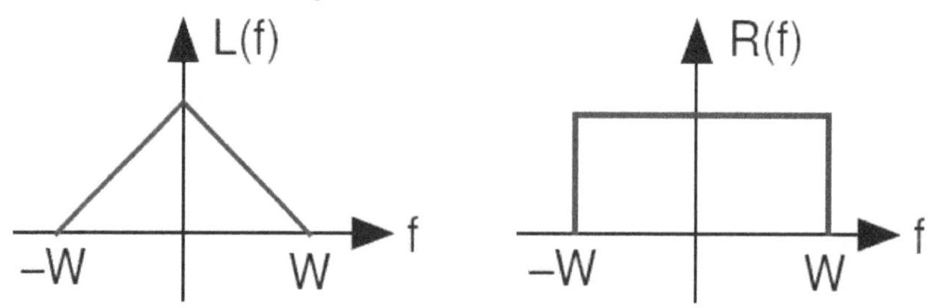


**Figure 4.32**

- What is the expression for the transmitted signal? Sketch its spectrum.
- Show the block diagram of a stereo AM receiver that can yield the left and right signals as separate outputs.

- What signal would be produced by a conventional coherent AM receiver that expects to receive a standard AM signal conveying a message signal having bandwidth W?

## Problem 4.25: Novel AM Stereo Method

A clever engineer has submitted a patent for a new method for transmitting two signals simultaneously in the same transmission bandwidth as commercial AM radio. As shown (Figure 4.33), her approach is to modulate the positive portion of the carrier with one signal and the negative portion with a second.

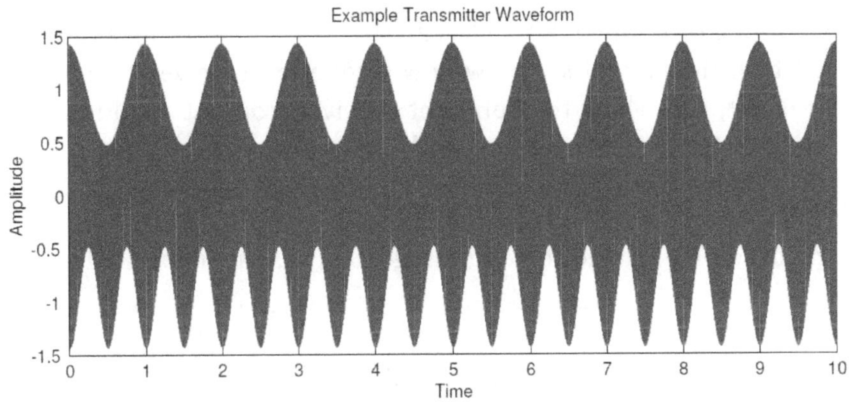


**Figure 4.33**

In detail the two message signals m1(t) and m2(t) are bandlimited to W Hz and have maximal amplitudes equal to 1. The carrier has a frequency fc much greater than W. The transmitted signal x(t) is given by:

$$x(t) = \begin{cases} A(1 + am_1(t))\sin(2\pi f_c t) & \text{if } \sin(2\pi f_c t) \geq 0 \\ A(1 + am_2(t))\sin(2\pi f_c t) & \text{if } \sin(2\pi f_c t) < 0 \end{cases}$$

In all cases, 0<a<1. The plot shows the transmitted signal when the messages are sinusoids: m1(t)=sin(2πfmt) and m2(t)=sin(2πfmt) where 2fm < W. You, as the patent examiner, must determine whether the scheme meets its claims and is useful.

- Provide a more concise expression for the transmitted signal x(t) than given above.
- What is the receiver for this scheme? It would yield both m1(t) and m2(t) from x(t).
- Find the spectrum of the positive portion of the transmitted signal.
- Determine whether this scheme satisfies the design criteria, allowing you to grant the patent. Explain your reasoning.

### Problem 4.26: A Radical Radio Idea

An ELEC 241 student has the bright idea of using a square wave instead of a sinusoid as an AM carrier. The transmitted signal would have the form:

$$x(t) = A(1 + m(t))\,\mathrm{sq}_T(t)$$

where the message signal m(t) would be amplitude-limited: |m(t)|<1

- Assuming the message signal is lowpass and has a bandwidth of W Hz, what values for the square wave's period T are feasible. In other words, do some combinations of W and T prevent reception?
- Assuming reception is possible, can standard radios receive this innovative AM transmission? If so, show how a coherent receiver could demodulate it; if not, show how the coherent receiver's output would be corrupted. Assume that the message bandwidth W=5 kHz.

### Problem 4.27: Secret Communication

An amplitude-modulated secret message m(t) has the following form:

$$r(t) = A(1 + m(t))\cos(2\pi(f_c + f_0)t)$$

The message signal has a bandwidth of W Hz and a magnitude less than 1 ( |m(t)| <1). The idea is to offset the carrier frequency by f0 Hz from standard radio carrier frequencies. Thus, «off-the-shelf» coherent demodulators would assume the carrier frequency has fc Hz. Here, f0 < W

- Sketch the spectrum of the demodulated signal produced by a coherent demodulator tuned to fc Hz.
- Will this demodulated signal be a "scrambled" version of the original? If so, how so; if not, why not?
- Can you develop a receiver that can demodulate the message without knowing the offset frequency fc?

### Problem 4.28: Signal Scrambling

An excited inventor announces the discovery of a way of using analog technology to render music unlistenable without knowing the secret recovery method. The idea is to modulate the bandlimited message m(t) by a special periodic signal s(t) that is zero during half of its period, which renders the message unlistenable and superficially, at least, unrecoverable (Figure 4.34).

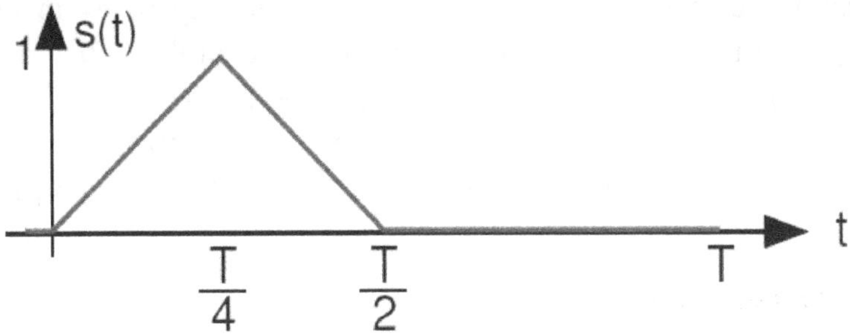


**Figure 4.34**

- What is the Fourier series for the periodic signal?
- What are the restrictions on the period T so that the message signal can be recovered from m(t)s(t)?
- ELEC 241 students think they have "broken" the inventor's scheme and are going to announce it to the world. How would they recover the original message without having detailed knowledge of the modulating signal?

## Solutions to Exercises in Chapter 4

### Solution to Exercise 4.2.1

Because of Euler's relation,

$$\sin(2\pi ft) = \frac{1}{2j}e^{j2\pi ft} - \frac{1}{2j}e^{(j2\pi ft)} \tag{4.47}$$

Thus, $c_1 = \frac{1}{2j}$, $c_{-1} = -\frac{1}{2j}$, and the other coefficients are zero.

### Solution to Exercise 4.2.2

$c_0 = \frac{A\Delta}{T}$. This quantity clearly corresponds to the periodic pulse signal's average value.

### Solution to Exercise 4.3.1

Write the coefficients of the complex Fourier series in Cartesian form as ck = Ak + jBk and substitute into the expression for the complex Fourier series.

$$\sum_{k=-\infty}^{\infty} c_k e^{j\frac{2\pi kt}{T}} = \sum_{k=-\infty}^{\infty} (A_k + jB_k)\, e^{j\frac{2\pi kt}{T}}$$

Simplifying each term in the sum using Euler's formula,

$$\begin{aligned}(A_k + jB_k)\, e^{j\frac{2\pi kt}{T}} &= (A_k + jB_k)\left(\cos\left(\tfrac{2\pi kt}{T}\right) + j\sin\left(\tfrac{2\pi kt}{T}\right)\right)\\ &= A_k\cos\left(\tfrac{2\pi kt}{T}\right) - B_k\sin\left(\tfrac{2\pi kt}{T}\right) + j\left(A_k\sin\left(\tfrac{2\pi kt}{T}\right) + B_k\cos\left(\tfrac{2\pi kt}{T}\right)\right)\end{aligned}$$

We now combine terms that have the same frequency index in magnitude. Because the signal is realvalued, the coefficients of the complex Fourier series have conjugate symmetry: c–k = ck* or A–k = Ak and B–k = –Bk. After we add the positive-indexed and negative-indexed terms, each term in the Fourier series becomes $2A_k\cos\left(\frac{2\pi kt}{T}\right) - 2B_k\sin\left(\frac{2\pi kt}{T}\right)$. To obtain the classic Fourier series (4.11), we must have 2Ak = ak and 2Bk = –bk.

### Solution to Exercise 4.3.2

The average of a set of numbers is the sum divided by the number of terms. Viewing signal integration as the limit of a Riemann sum, the integral corresponds to the average.

### Solution to Exercise 4.3.3

We found that the complex Fourier series coefficients are given by $c_k = \frac{2}{j\pi k}$. The coefficients are pure imaginary, which means ak = 0. The coefficients of the sine terms are given by bk=–(21m (ck)) so that

$$b_k = \begin{cases} \frac{4}{\pi k} \text{ if } k \text{ odd} \\ 0 \text{ if } k \text{ even} \end{cases}$$

Thus, the Fourier series for the square wave is

$$\mathrm{sq}(t) = \sum_{k\in\{1,3,\dots\}} \frac{4}{\pi k}\sin\left(\frac{2\pi kt}{T}\right) \tag{4.48}$$

### Solution to Exercise 4.4.1

The rms value of a sinusoid equals its amplitude divided by $\sqrt{22}$. As a half-wave rectified sine wave is zero during half of the period, its rms value is $\frac{A}{2}$ since the integral of the squared half-wave rectified sine wave equals half that of a squared sinusoid.

### Solution to Exercise 4.4.2

Total harmonic distortion equals $\frac{\sum_{k=2}^{\infty} {a_k}^2 + {b_k}^2}{{a_1}^2 + {b_1}^2}$. Clearly, this quantity is most easily computed in the frequency domain. However, the numerator equals the square of the signal's rms value minus the power in the average and the power in the first harmonic.

### Solution to Exercise 4.5.1

Total harmonic distortion in the square wave is $1 - \frac{1}{2}\left(\frac{4}{\pi}\right)^2 = 20\%$.

### Solution to Exercise 4.6.1

N signals directly encoded require a bandwidth of $\frac{N}{T}$. Using a binary representation, we need $\frac{\log_2 N}{T}$. For N = 128, the binary-encoding scheme has a factor of $\frac{7}{128} = 0.05$ smaller bandwidth. Clearly, binary encoding is superior.

### Solution to Exercise 4.6.2

We can use N different amplitude values at only one frequency to represent the various letters.

### Solution to Exercise 4.7.1

Because the filter's gain at zero frequency equals one, the average output values equal the respective average input values.

### Solution to Exercise 4.8.1

$$\mathcal{F}(S(f)) = \int_{-\infty}^{\infty} S(f)\, e^{-(j2\pi ft)} df = \int_{-\infty}^{\infty} S(f)\, e^{j2\pi f(-t)} df = s(-t)$$

### Solution to Exercise 4.8.2

$\mathcal{F}(\mathcal{F}(\mathcal{F}(\mathcal{F}(s(t))))) = s(t)$. We know that $\mathcal{F}(S(f)) = \int_{-\infty}^{\infty} S(f)\, e^{-(j2\pi ft)} df = \int_{-\infty}^{\infty} S(f)\, e^{j2\pi f(-t)} df = s(-t)$. Therefore, two Fourier transforms applied to s (t) yields s (–t). We need two more to get us back where we started.

### Solution to Exercise 4.8.3

The signal is the inverse Fourier transform of the triangularly shaped spectrum, and equals s (t) = $W\left(\frac{\sin(\pi Wt)}{\pi Wt}\right)^2$

### Solution to Exercise 4.8.4

The result is most easily found in the spectrum's formula: the power in the signal-related part of x (t) is half the power of the signal s (t).

### Solution to Exercise 4.9.1

The inverse transform of the frequency response is $\frac{1}{RC} e^{-\frac{t}{RC}} u(t)$. Multiplying the frequency response by $1 - e^{-(j2\pi f\Delta)}$ means subtract from the original signal its time-delayed version. Delaying the frequency response's time-domain version by Δ results in $\frac{1}{RC} e^{\frac{-(t-\Delta)}{RC}} u(t-\Delta)$. Subtracting from the undelayed signal yields $\frac{1}{RC} e^{\frac{-t}{RC}} u(t) - \frac{1}{RC} e^{\frac{-(t-\Delta)}{RC}} u(t-\Delta)$. Now we integrate this sum. Because the integral of a sum equals the sum of the component integrals (integration is linear), we can consider each separately. Because integration and signal-delay are linear, the integral of a delayed signal equals the delayed version of the integral. The integral is provided in the example (4.44).

### Solution to Exercise 4.10.1

If the glottis were linear, a constant input (a zero-frequency sinusoid) should yield a constant output. The periodic output indicates nonlinear behavior.

### *Solution to Exercise 4.10.2*

In the bottom-left panel, the period is about 0.009 s, which equals a frequency of 111 Hz. The bottom-right panel has a period of about 0.0065 s, a frequency of 154 Hz.

### *Solution to Exercise 4.10.3*

Because males have a lower pitch frequency, the spacing between spectral lines is smaller. This closer spacing more accurately reveals the formant structure. Doubling the pitch frequency to 300 Hz for Figure 4.16 (voice spectrum) would amount to removing every other spectral line.

# 5 Chapter INTRODUCTION TO DIGITAL SIGNAL PROCESSING

## 5.1 DIGITAL SIGNAL PROCESSING

In addition to analog signals, which are real- or complex-valued functions of a continuous variable like time or space, we can also define digital signals. Digital signals are discrete, mathematical representations that are defined solely for integer values. Therefore, we use the symbol s(n) to represent a discrete-time one-dimensional signal, such as a digital music recording. Similarly, we use the symbol s(m,n) to represent a discrete-time two-dimensional signal, such as a photo captured with a digital camera. Sequences are inherently distinct from continuous-time signals. Sequences do not possess any significance when it comes to continuity.

Although there are significant variations, the theory behind digital signal processing is similar to that of analog signals. Fourier transforms, linear filtering, and linear systems are analogous to the concepts discussed in previous chapters. These analogies facilitate comprehension of the concepts and their necessity, but it is important not to interpret them as imitations of analog systems. We will ascertain that digital signal processing is not a mere approximation of analog processing. It is essential to specifically concern ourselves with the accuracy of converting analog signals to digital signals. The presence of music stored on CDs, voice transmitted by digital cellular telephones, and video conveyed by digital television all demonstrate the successful conversion of analog signals into digital signals and vice versa.

The primary factor contributing to the technological superiority of digital signal processing systems in the present era is the computer. The ability to perform computations, such as the Fourier transform, at a rapid pace allows for real-time calculation as the signal is generated. Additionally, the programmability of these systems enables effortless modification of the signal processing methods. The inherent adaptability of this feature is clearly attractive and has gained significant traction in the industry. Programmability refers to the ability to carry out signal processing activities that are not achievable with analog systems or circuits. Furthermore, we shall uncover that digital systems possess an algorithmic superiority that enhances their ability to handle information quickly: Computation can be optimized by restructuring it in less apparent manners to enhance processing speed. However, this adaptability is accompanied by a cost, which is a result of the functioning of computers. What is the process by which computers carry out signal processing?

## 5.2 COMPUTER ORGANIZATION: AN INTRODUCTION

### 5.2.1 Computer Architecture

In order to comprehend digital signal processing systems, it is important to have a basic understanding of the computational processes employed by computers. A computer is an electronic device that carries out computations on data and presents the outcomes to people or other computers in many potentially beneficial manners.

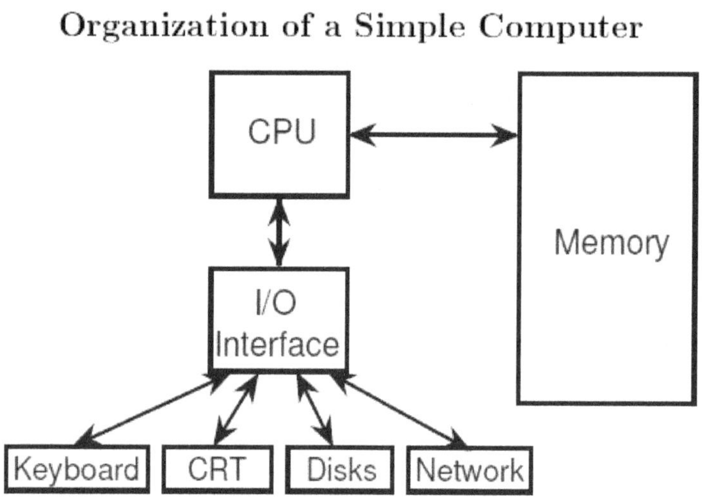


**Figure 5.1: Generic computer hardware organization**

The generic computer contains input devices (keyboard, mouse, A/D (analog-to-digital) converter, etc.), a computational unit, and output devices (monitors, printers, D/A converters). The computational unit is the computer's heart, and usually consists of a central processing unit (CPU), a memory, and an input/output (I/O) interface. What I/O devices might be present on a given computer vary greatly.

- A simple computer operates fundamentally in discrete time. Computers are clocked devices, in which computational steps occur periodically according to ticks of a clock. This description belies clock speed: When you say "I have a 1 GHz computer," you mean that your computer takes 1 nanosecond to perform each step. That is incredibly fast! A "step" does not, unfortunately, necessarily mean a computation like an addition; computers break such computations down into several stages, which means that the clock speed need not express the computational speed. Computational speed is expressed in units of millions of instructions/second (Mips). Your 1 GHz computer (clock speed) may have a computational speed of 200 Mips.

- Computers perform integer (discrete-valued) computations. Computer calculations can be numeric (obeying the laws of arithmetic), logical (obeying the laws of an algebra), or symbolic (obeying any law you like).1 Each computer instruction that performs an elementary numeric calculation --- an addition, a multiplication, or a division --- does so only for integers. The sum or product of two integers is also an integer, but the quotient of two integers is likely to not be an integer. How does a computer deal with numbers that have digits to the right of the decimal point? This problem is addressed by using the so-called floating-point representation of real numbers. At its heart, however, this representation relies on integer-valued computations.

## 5.2.2 Representing Numbers

Focusing on numbers, all numbers can represented by the positional notation system. 2 The b-ary positional representation system uses the position of digits ranging from 0 to b-1 to denote a number. The quantity b is known as the base of the number system. Mathematically, positional systems represent the positive integer n as:

$$n = \sum_{k=0}^{\infty} d_k b^k \quad , \quad d_k \in \{0, \ldots, b-1\} \tag{5.1}$$

and we succinctly express n in base-b as nb=dNdN-1...d0. The number 25 in base 10 equals 2×101+5×100, So that the digits representing this number are d0=5,d1=2, all other dk equal zero. This same number in binary (base 2) equals 11001(1×24+1×23+0×22+0×21+1×20) and 19 in hexadecimal (base 16). Fractions between zero and one are represented the same way.

$$f = \sum_{k=-\infty}^{-1} d_k b^k \quad , \quad d_k \in \{0, \ldots, b-1\} \tag{5.2}$$

All numbers can be represented by their sign, integer and fractional parts. Complex numbers can be thought of as two real numbers that obey special rules to manipulate them.

Humans use base 10, commonly assumed to be due to us having ten fingers. Digital computers use the base 2 or binary number representation, each digit of which is known as a bit (binary digit).

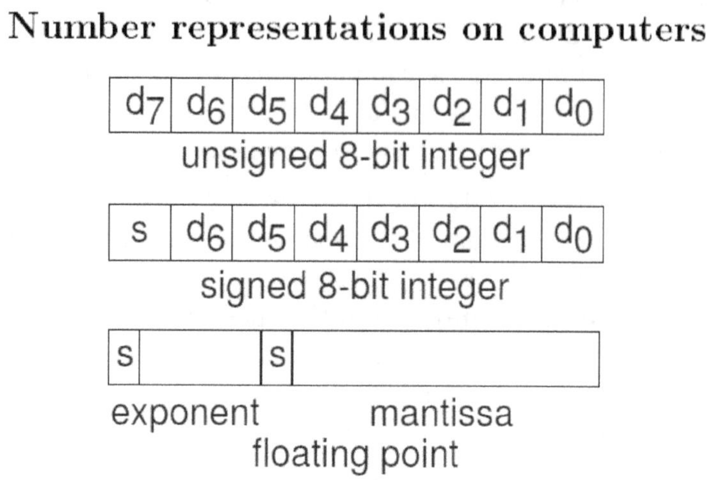


**Figure 5.2: The various ways numbers are represented in binary are illustrated. The number of bytes for the exponent and mantissa components of floating point numbers varies.**

Here, each bit is represented as a voltage that is either "high" or "low," thereby representing "1" or "0," respectively. To represent signed values, we tack on a special bit—the sign bit—to express the sign. The computer's memory consists of an ordered sequence of bytes, a collection of eight bits. A byte can therefore represent an unsigned number ranging from 0 to 255. If we take one of the bits and make it the sign bit, we can make the same byte to represent numbers ranging from -128 to 127. But a computer cannot represent all possible real numbers. The fault is not with the binary number system; rather having only a finite number of bytes is the problem. While a gigabyte of memory may seem to be a lot, it takes an infinite number of bits to represent $\pi$. Since we want to store many numbers in a computer's memory, we are restricted to those that have a finite binary representation. Large integers can be represented by an ordered sequence of bytes. Common lengths, usually expressed in terms of the number of bits, are 16, 32, and 64. Thus, an unsigned 32-bit number can represent integers ranging between 0 and 232-1(4,294,967,295), a number almost big enough to enumerate every human in the world!

### Exercise 5.2.1

For both 32-bit and 64-bit integer representations, what are the largest numbers that can be represented if a sign bit must also be included.

While this system represents integers well, how about numbers having nonzero digits to the right of the decimal point? In other words, how are numbers that have fractional parts represented? For such numbers, the binary representation system is used, but with a little more complexity. The floating-point system uses a number of bytes - typically 4 or 8 - to represent the number, but with one byte (sometimes two bytes) reserved to represent the exponent e of a power-of-two multiplier of the number-the mantissa m-expressed by the remaining bytes.

$$x = m2^e \tag{5.3}$$

The mantissa is usually taken to be a binary fraction having a magnitude in the range $[\frac{1}{2}, 1)$, which means that the binary representation is such that d-1=1. The number zero is an exception to this rule, and it is the only floating point number having a zero fraction. The sign of the mantissa represents the sign of the number and the exponent can be a signed integer.

A computer's representation of integers is either perfect or only approximate, the latter situation occurring when the integer exceeds the range of numbers that a limited set of bytes can represent. Floating point representations have similar representation problems: if the number x can be multiplied/divided by enough powers of two to yield a fraction lying between 1/2 and 1 that has a flinite binary-fraction representation, the number is represented exactly in floating point. Otherwise, we can only represent the number approximately, not catastrophically in error as with integers. For example, the number 2.5 equals 0:625 x 22, the fractional part of which has an exact binary representation. However, the number 2:6 does not have an exact binary representation, and only be represented approximately in floating point. In single precision floating point numbers, which require 32 bits (one byte for the exponent and the remaining 24 bits for the mantissa), the number 2.6 will be represented as 2:600000079:::. Note that this approximation has a much longer decimal expansion. This level of accuracy may not suffice in numerical calculations. Double precision floating point numbers consume 8 bytes, and quadruple precision 16 bytes. The more bits used in the mantissa, the greater the accuracy. This increasing accuracy means that more numbers can be represented exactly, but there are always some that cannot. Such inexact numbers have an infinite binary representation. Realizing that real numbers can be only represented approximately is quite important, and underlies the entire field of numerical analysis, which seeks to predict the numerical accuracy of any computation.

## Exercise 5.2.2

What are the largest and smallest numbers that can be represented in 32-bit floating point? in 64-bit floating point that has sixteen bits allocated to the exponent? Note that both exponent and mantissa require a sign bit.

So long as the integers aren't too large, they can be represented exactly in a computer using the binary positional notation. Electronic circuits that make up the physical computer can add and subtract integers without error. (This statement isn't quite true; when does addition cause problems?)

### 5.2.3 Computer Arithmetic and Logic

The binary addition and multiplication tables are

$$\begin{pmatrix} 0+0=0 \\ 0+1=1 \\ 1+1=10 \\ 1+0=1 \\ \\ 0\times 0=0 \\ 0\times 1=0 \\ 1\times 1=1 \\ 1\times 0=0 \end{pmatrix} \tag{5.4}$$

Note that if carries are ignored,7 subtraction of two single-digit binary numbers yields the same bit as addition. Computers use high and low voltage values to express a bit, and an array of such voltages express numbers akin to positional notation. Logic circuits perform arithmetic operations.

### Exercise 5.2.3

Add twenty-five and seven in base 2. Note the carries that might occur. Why is the result "nice"?

The variables of logic indicate truth or falsehood. A∩B, the AND of A and B, represents a statement that both A and B must be true for the statement to be true. You use this kind of statement to tell search engines that you want to restrict hits to cases where both of the events A and B occur. AUB, the OR of A and B, yields a value of truth if either is true. Note that if we represent truth by a "1" and falsehood by a "0," binary multiplication corresponds to AND and addition (ignoring carries) to XOR. XOR, the exclusive or operator, equals the union of AUB and A∩B. The Irish mathematician George Boole discovered this equivalence in the mid-nineteenth century. It laid the foundation for what we now call Boolean algebra, which expresses as equations logical statements. More importantly, any computer using base-2 representations and arithmetic can also easily evaluate logical statements. This fact makes an integer-based computational device much more powerful than might be apparent.

## 5.3 THE SAMPLING THEOREM

### 5.3.1 Analog-to-Digital Conversion

Because of the way computers are organized, signal must be represented by a finite number of bytes. This restriction means that both the time axis and the amplitude axis must be quantized: They must each be a multiple of the integers. 1 Quite surprisingly, the Sampling Theorem allows us to quantize the time axis without error for some signals. The signals that can be sampled without introducing error are interesting, and as described in the next section, we can make a signal «samplable" by filtering. In contrast, no one has found a way of performing the amplitude quantization step without introducing an unrecoverable error. Thus, a signal's value can no longer be any real number. Signals processed by digital computers must be discrete-valued: their values must be proportional to the integers. Consequently, analog-to-digital conversion introduces error.

### 5.3.2 The Sampling Theorem

Digital transmission of information and digital signal processing all require signals to first be "acquired" by a computer. One of the most amazing and useful results in electrical engineering is that signals can be converted from a function of time into a sequence of numbers without error: We can convert the numbers back into the signal with (theoretically) no error. Harold Nyquist, a Bell Laboratories engineer, first derived this result, known as the Sampling Theorem, in the 1920s. It found no real application back then. Claude Shannon, also at Bell Laboratories, revived the result once computers were made public after World War II.

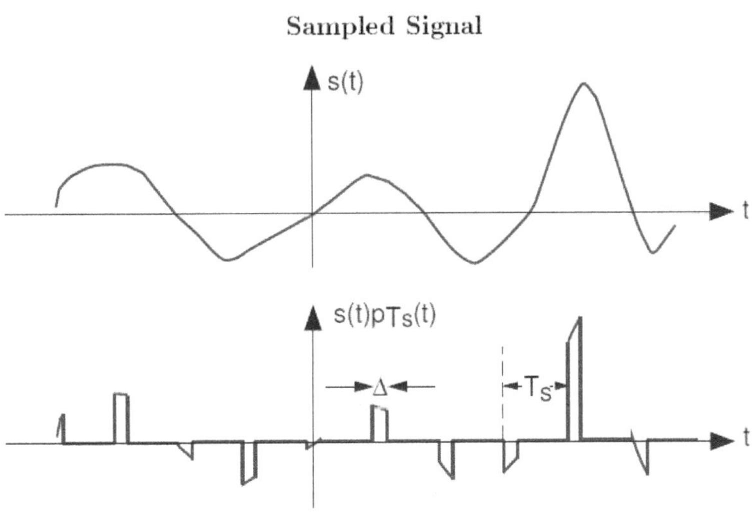


**Figure 5.3: The waveform of an example signal is shown in the top plot and its sampled version in the bottom.**

The sampled version of the analog signal s(t) is s(nTs), with Ts known as the sampling interval. Clearly, the value of the original signal at the sampling times is preserved; the issue is how the signal values between the samples can be reconstructed since they are lost in the sampling process. To characterize sampling, we approximate it as the product x(t)=s(t) PTs(t), with PTs being the periodic pulse signal. The resulting signal, as shown in Figure 5.3 (Sampled Signal), has nonzero values only during the time intervals $\left(nT_s - \frac{\Delta}{2}, nT_s + \frac{\Delta}{2}\right)$, $n \in \{\ldots, -1, 0, 1, \ldots\}$.

For our purposes here, we center the periodic pulse signal about the origin so that its Fourier series coefficients are real (the signal is even).

$$p_{T_s}(t) = \sum_{k=-\infty}^{\infty} c_k e^{\frac{j2\pi kt}{T_s}} \tag{5.5}$$

$$c_k = \frac{\sin\left(\frac{\pi k \Delta}{T_s}\right)}{\pi k} \tag{5.6}$$

If the properties of s(t) and the periodic pulse signal are chosen properly, we can recover s(t) from x(t) by filtering.

To understand how signal values between the samples can be "filled" in, we need to calculate the sampled signal's spectrum. Using the Fourier series representation of the periodic sampling signal,

$$x(t) = \sum_{k=-\infty}^{\infty} c_k e^{\frac{j2\pi kt}{T_s}} s(t) \tag{5.7}$$

Considering each term in the sum separately, we need to know the spectrum of the product of the complex exponential and the signal. Evaluating this transform directly is quite easy.

$$\int_{-\infty}^{\infty} s(t) e^{\frac{j2\pi kt}{T_s}} e^{-(j2\pi ft)} dt = \int_{-\infty}^{\infty} s(t) e^{-\left(j2\pi\left(f - \frac{k}{T_s}\right)t\right)} dt = S\left(f - \frac{k}{T_s}\right) \tag{5.8}$$

Thus, the spectrum of the sampled signal consists of weighted (by the coefficients ck) and delayed versions of the signal's spectrum (Figure 5.4 (aliasing)).

$$X(f) = \sum_{k=-\infty}^{\infty} c_k S\left(f - \frac{k}{T_s}\right) \tag{5.9}$$

In general, the terms in this sum overlap each other in the frequency domain, rendering recovery of the original signal impossible. This unpleasant phenomenon is known as aliasing.

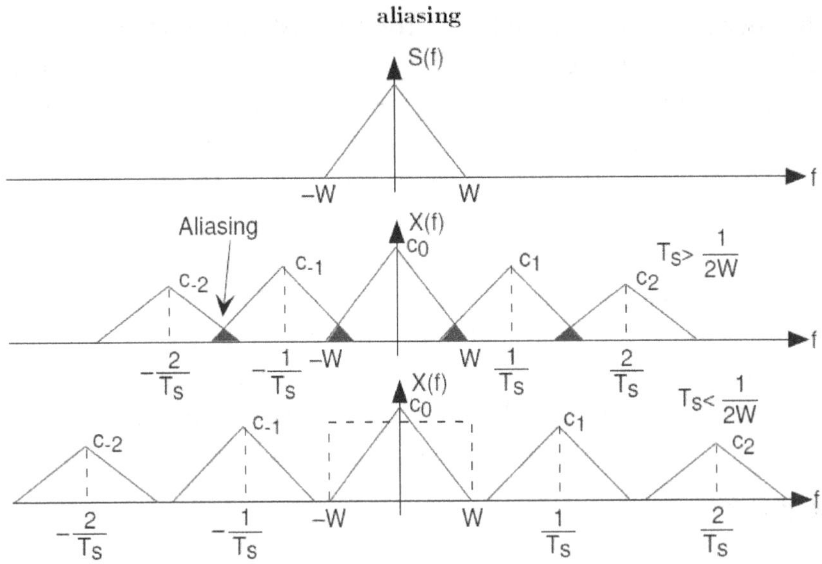


**Figure 5.4: The spectrum of some bandlimited (to W Hz) signal is shown in the top plot. If the sampling interval Ts is chosen too large relative to the bandwidth W, aliasing will occur. In the bottom plot, the sampling interval is chosen sufficiently small to avoid aliasing. Note that if the signal were not bandlimited, the component spectra would always overlap.**

If, however, we satisfy two conditions:

- The signal s(t) is bandlimited—has power in a restricted frequency range—to W Hz
- The sampling interval Ts is small enough so that the individual components in the sum do not overlap - Ts < 1/2 W,

Aliasing will not occur. In this delightful case, we can recover the original signal by lowpass filtering x(t) with a filter having a cutoff frequency equal to W Hz. These two conditions ensure the ability to recover a bandlimited signal from its sampled version: We thus have the Sampling Theorem.

## Exercise 5.3.1

The Sampling Theorem (as stated) does not mention the pulse width Δ. What is the effect of this parameter on our ability to recover a signal from its samples (assuming the Sampling Theorem's two conditions are met)?

## Solution

The only effect of pulse duration is to unequally weight the spectral repetitions. Because we are only concerned with the repetition centered about the origin, the pulse duration has no significant effect on recovering a signal from its samples.

The frequency $\frac{1}{2T_s}$, known today as the Nyquist frequency and the Shannon sampling frequency, corresponds to the highest frequency at which a signal can contain energy and remain compatible with the Sampling Theorem. High-quality sampling systems ensure that no aliasing occurs by unceremoniously lowpass filtering the signal (cutoff frequency being slightly lower than the Nyquist frequency) before sampling. Such systems therefore vary the anti-aliasing filter›s cutoff frequency as the sampling rate varies. Because such quality features cost money, many sound cards do not have anti-aliasing filters or, for that matter, post-sampling filters. They sample at high frequencies, 44.1 kHz for example, and hope the signal contains no frequencies above the Nyquist frequency (22.05 kHz in our example). If, however, the signal contains frequencies beyond the sound card›s Nyquist frequency, the resulting aliasing can be impossible to remove.

### Exercise 5.3.2

To gain a better appreciation of aliasing, sketch the spectrum of a sampled square wave. For simplicity consider only the spectral repetitions centered at $-\frac{1}{T_s}, 0, \frac{1}{T_s}$. Let the sampling interval Ts be 1; consider two values for the square wave's period: 3.5 and 4. Note in particular where the spectral lines go as the period decreases; some will move to the left and some to the right. What property characterizes the ones going the same direction?

If we satisfy the Sampling Theorem's conditions, the signal will change only slightly during each pulse. As we narrow the pulse, making Δ smaller and smaller, the nonzero values of the signal $s(t)\,p_{T_s}(t)$ will simply be $s(nT_s)$, the signal's samples. If indeed the Nyquist frequency equals the signal's highest frequency, at least two samples will occur within the period of the signal's highest frequency sinusoid. In these ways, the sampling signal captures the sampled signal's temporal variations in a way that leaves all the original signal's structure intact.

### Exercise 5.3.3

What is the simplest bandlimited signal? Using this signal, convince yourself that less than two samples/period will not suffice to specify it. If the sampling rate $\frac{1}{T_s}$ is not high enough, what signal would your resulting undersampled signal become?

## 5.4 AMPLITUDE QUANTIZATION

The Sampling Theorem says that if we sample a bandlimited signal s(t) fast enough, it can be recovered without error from its samples s(nTs),n∈{...,−1,0,1,...}. Sampling is only the first phase of acquiring data into a computer: Computational processing further requires that the samples be quantized: analog values are converted into digital (Section 1.2.2: Digital Signals) form. In short, we will have performed analog-to-digital (A/D) conversion.

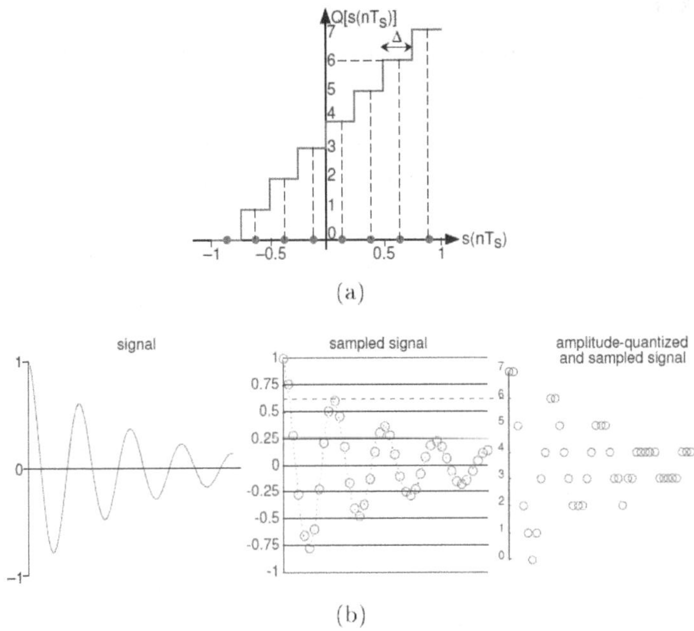


**Figure 5.5: A three-bit A/D converter assigns voltage in the range [-1,1] to one of eight integers between 0 and 7. For example, all inputs having values lying between 0.5 and 0.75 are assigned the integer value six and, upon conversion back to an analog value, they all become 0.625. The width of a single quantization interval Δ equals 2/2B. The bottom panel shows a signal going through the analog-to-digital converter, where B is the number of bits used in the A/D conversion process (3 in the case depicted here). First it is sampled, then amplitude-quantized to three bits. Note how the sampled signal waveform becomes distorted after amplitude quantization. For example the two signal values between 0.5 and 0.75 become 0.625. This distortion is irreversible; it can be reduced (but not eliminated) by using more bits in the A/D converter.**

A phenomenon reminiscent of the errors incurred in representing numbers on a computer prevents signal amplitudes from being converted with no error into a binary number representation. In analog-to-digital conversion, the signal is assumed to lie within a predefined range. Assuming we can scale the signal without affecting the information it expresses, we'll define this range to be [-1,1]. Furthermore, the A/D converter assigns amplitude values in this range to a set of integers. A B-bit converter produces one of the integers {0,1,...,2B−1} for each sampled input. Figure 5.5 shows how a three-bit A/D converter assigns input values to the integers. We define a quantization interval to be the range of values assigned to the same integer. Thus, for our example three-bit A/D converter, the quantization interval Δ is 0.25; in general, it is $\frac{2}{2^B}$.

## Exercise 5.4.1

Recalling the plot of average daily highs in this frequency domain problem, why is this plot so jagged? Interpret this effect in terms of analog-to-digital conversion.

Because values lying anywhere within a quantization interval are assigned the same value for computer processing, the original amplitude value cannot be recovered without error. Typically, the D/A converter, the device that converts integers to amplitudes, assigns an amplitude equal to the value lying halfway in the quantization interval. The integer 6 would be assigned to the amplitude 0.625 in this scheme. The error introduced by converting a signal from analog to digital form by sampling and amplitude quantization then back again would be half the quantization interval for each amplitude value. Thus, the so-called A/D error equals half the width of a quantization interval: $\frac{1}{2^B}$. As we have fixed the input-amplitude range, the more bits available in the A/D converter, the smaller the quantization error.

To analyze the amplitude quantization error more deeply, we need to compute the signal-to-noise ratio, which equals the ratio of the signal power and the quantization error power. Assuming the signal is a sinusoid, the signal power is the square of the rms amplitude: $\text{power}(s) = \left(\frac{1}{\sqrt{2}}\right)^2 = \frac{1}{2}$. The illustration (Figure 5.6) details a single quantization interval.

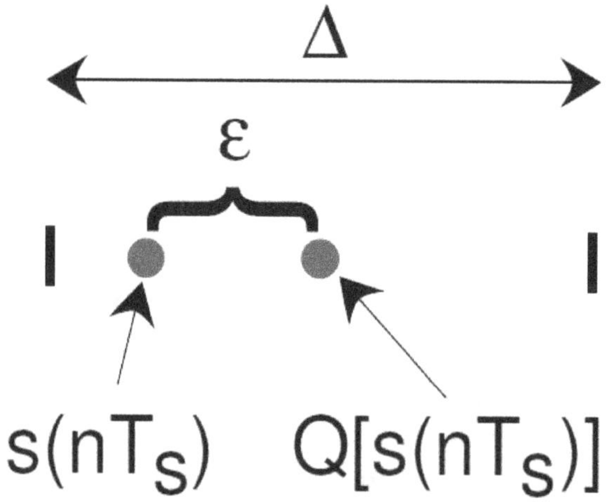


**Figure 5.6: A single quantization interval is shown, along with a typical signal's value before amplitude quantization s(nTs) and after Q(s(nTs). ε denotes the error thus incurred.**

Its width is Δ and the quantization error is denoted by ε. To find the power in the quantization error, we note that no matter into which quantization interval the signal's value falls, the error will have the same characteristics. To calculate the rms value, we must square the error and average it over the interval.

$$\begin{aligned} \text{rms}(\epsilon) &= \sqrt{\frac{1}{\Delta}\int_{-\frac{\Delta}{2}}^{\frac{\Delta}{2}} \epsilon^2 d\epsilon} \\ &= \left(\frac{\Delta^2}{12}\right)^{\frac{1}{2}} \end{aligned} \tag{5.10}$$

Since the quantization interval width for a B-bit converter equals $\frac{2}{2^B} = 2^{-(B-1)}$, we find that the signal-to-noise ratio for the analog-to-digital conversion process equals

$$\text{SNR} = \frac{\frac{1}{2}}{\frac{2^{-(2(B-1))}}{12}} = \frac{3}{2}2^{2B} = 6B + 10\log 1.5\text{dB} \tag{5.11}$$

Thus, every bit increase in the A/D converter yields a 6 dB increase in the signal-to-noise ratio. The constant term 10 log1.5 equals 1.76.

### Exercise 5.4.2

This derivation assumed the signal's amplitude lay in the range [-1,1]. What would the amplitude quantization signal-to-noise ratio be if it lay in the range [-A,A]?

### Exercise 5.4.3

How many bits would be required in the A/D converter to ensure that the maximum amplitude quantization error was less than 60 db smaller than the signal's peak value?

### Exercise 5.4.4

Music on a CD is stored to 16-bit accuracy. To what signal-to-noise ratio does this correspond?

Once we have acquired signals with an A/D converter, we can process them using digital hardware or software. It can be shown that if the computer processing is linear, the result of sampling, computer processing, and unsampling is equivalent to some analog linear system. Why go to all the bother if the same function can be accomplished using analog techniques? Knowing when digital processing excels and when it does not is an important issue.

## 5.5 DISCRETE-TIME SIGNALS AND SYSTEMS

Analog signals can be mathematically described as functions that depend on continuous variables, such as space and time. Discrete-time signals are mathematical functions that are defined specifically on the set of integers. These functions can be thought of as sequences of values. Similar to analog signals, we aim to break down discrete-time signals into more basic elements. By utilizing this method, we may enhance our comprehension of signal composition, enabling us to utilize this composition to encode information and subsequently retrieve the encoded information. When dealing with signals that have symbolic values, the technique is distinct: We create a standardized representation for all signals with symbolic values in order to consolidate and express the information they carry in a cohesive manner. From the standpoint of representing information, the primary concern is efficiency in both real-valued and symbolic-valued signals. This entails finding the most concise and compact method of representing information, allowing for easy extraction at a later time.

### 5.5.1 Real- and Complex-valued Signals

A discrete-time signal is represented symbolically as s(n) where n={...,-1,0,1,...}

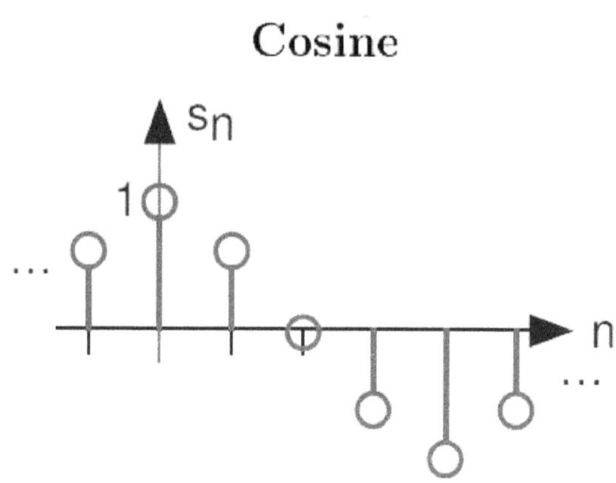


**Figure 5.7: The discrete-time cosine signal is plotted as a stem plot. Can you find the formula for this signal?**

We usually draw discrete-time signals as stem plots to emphasize the fact they are functions defined only on the integers. We can delay a discrete-time signal by an integer just as with analog ones. A signal delayed by m samples has the expression s(n-m).

### 5.5.2 Complex Exponentials

The most important signal is, of course, the complex exponential sequence.

$$s(n) = e^{j2\pi fn} \tag{5.12}$$

Note that the frequency variable f is dimensionless and that adding an integer to the frequency of the discrete-time complex exponential has no effect on the signal›s value.

$$\begin{aligned} e^{j2\pi(f+m)n} &= e^{j2\pi fn}e^{j2\pi mn} \\ &= e^{j2\pi fn} \end{aligned} \tag{5.13}$$

This derivation follows because the complex exponential evaluated at an integer multiple of 2π equals one. Thus, we need only consider frequency to have a value in some unit-length interval.

### 5.5.3 Sinusoids

Discrete-time sinusoids have the obvious form s(n)=Acos(2πfn+φ). As opposed to analog complex exponentials and sinusoids that can have their frequencies be any real value, frequencies of their discrete-time counterparts yield unique waveforms only when f lies in the interval $\left(-\frac{1}{2},\frac{1}{2}\right]$. This choice of frequency interval is arbitrary; we can also choose the frequency to lie in the interval [0,1). How to choose a unit-length interval for a sinusoid's frequency will become evident later.

### 5.5.4 Unit Sample

The second-most important discrete-time signal is the unit sample, which is defined to be

$$\delta(n)=\begin{cases}1 \text{ if } n=0\\ 0 \text{ otherwise}\end{cases} \quad (5.14)$$

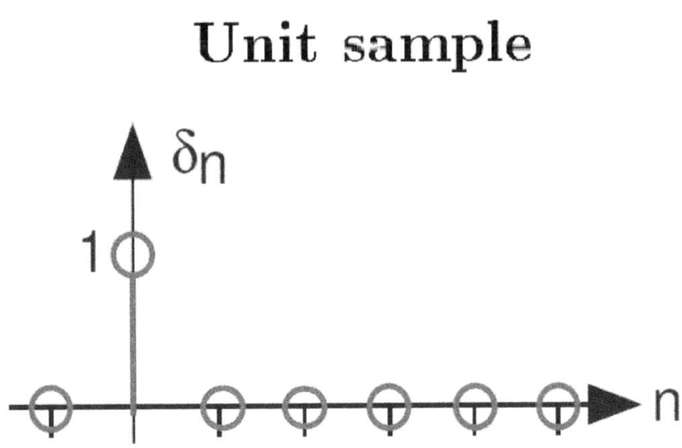


**Figure 5.8: The unit sample.**

Examination of a discrete-time signal's plot, like that of the cosine signal shown in Figure 5.8, reveals that all signals consist of a sequence of delayed and scaled unit samples. Because the value of a sequence at each integer m is denoted by s(m) and the unit sample delayed to occur at m is written δ(n-m), we can decompose any signal as a sum of unit samples delayed to the appropriate location and scaled by the signal value.

$$s(n) = \sum_{m=-\infty}^{\infty} s(m)\,\delta(n-m) \qquad (5.15)$$

This kind of decomposition is unique to discrete-time signals, and will prove useful subsequently.

### 5.5.5 Unit Step

The unit step in discrete-time is well-defined at the origin, as opposed to the situation with analog signals.

$$u(n) = \begin{cases} 1 & \text{if } n \geq 0 \\ 0 & \text{if } n < 0 \end{cases} \qquad (5.16)$$

### 5.5.6 Symbolic Signals

An interesting aspect of discrete-time signals is that their values do not need to be real numbers. We do have real-valued discrete-time signals like the sinusoid, but we also have signals that denote the sequence of characters typed on the keyboard. Such characters certainly aren't real numbers, and as a collection of possible signal values, they have little mathematical structure other than that they are members of a set. More formally, each element of the symbolic-valued signal s(n) takes on one of the values {a1,...,ak} which comprise the alphabet A. This technical terminology does not mean we restrict symbols to being members of the English or Greek alphabet. They could represent keyboard characters, bytes (8-bit quantities), integers that convey daily temperature. Whether controlled by software or not, discrete-time systems are ultimately constructed from digital circuits, which consist entirely of analog circuit elements. Furthermore, the transmission and reception of discrete-time signals, like e-mail, is accomplished with analog signals and systems. Understanding how discrete-time and analog signals and systems intertwine is perhaps the main goal of this course.

### 5.5.7 Discrete-Time Systems

Discrete-time systems can act on discrete-time signals in ways similar to those found in analog signals and systems. Because of the role of software in discrete-time systems, many more different systems can be envisioned and "constructed" with programs than can be with analog signals. In fact, a special class of analog signals can be converted into discrete-time signals, processed with software, and converted back into an analog signal, all without the incursion of error. For such signals, systems can be easily produced in software, with equivalent analog realizations difficult, if not impossible, to design.

## 5.6 DISCRETE-TIME FOURIER TRANSFORM (DTFT)

The Fourier transform of the discrete-time signal s(n) is defined to be

$$S\left(e^{j2\pi f}\right) = \sum_{n=-\infty}^{\infty} s(n)\, e^{-(j2\pi fn)} \tag{5.17}$$

Frequency here has no units. As should be expected, this definition is linear, with the transform of a sum of signals equaling the sum of their transforms. Real-valued signals have conjugate-symmetric spectra: $S\left(e^{-(j2\pi f)}\right) = S\left(e^{j2\pi f}\right)^{*}$.

### Exercise 5.6.1

A special property of the discrete-time Fourier transform is that it is periodic with period one: $S\left(e^{j2\pi(f+1)}\right) = S\left(e^{j2\pi f}\right)$. Derive this property from the definition of the DTFT. Because of this periodicity, we need only plot the spectrum over one period to understand completely the spectrum's structure; typically, we plot the spectrum over the frequency range $\left[-\frac{1}{2}, \frac{1}{2}\right]$. When the signal is real-valued, we can further simplify our plotting chores by showing the spectrum only over $\left[0, \frac{1}{2}\right]$; the spectrum at negative frequencies can be derived from positive-frequency spectral values.

When we obtain the discrete-time signal via sampling an analog signal, the Nyquist frequency corresponds to the discrete-time frequency $\frac{1}{2}$. To show this, note that a sinusoid having a frequency equal to the Nyquist frequency $\frac{1}{2T_s}$ has a sampled waveform that equals

$$\cos\left(2\pi \times \frac{1}{2T_s} nT_s\right) = \cos(\pi n) = (-1)^n$$

The exponential in the DTFT at frequency $\frac{1}{2}$ equals $e^{-\frac{j2\pi n}{2}} = e^{-(j\pi n)} = (-1)^n$, meaning that discrete-time frequency equals analog frequency multiplied by the sampling interval

$$f_D = f_A T_s \tag{5.18}$$

fD and fA represent discrete-time and analog frequency variables, respectively. The aliasing figure provides another way of deriving this result. As the duration of each pulse in the periodic sampling signal pTs(t) narrows, the amplitudes of the signal›s spectral repetitions, which are governed by the Fourier series coefficients (4.10) of pTs(t), become increasingly equal. Examination of the periodic pulse signal (Figure 4.1) reveals that as Δ decreases, the value of c0, the largest Fourier coefficient, decreases to zero: $|c_0| = \frac{A\Delta}{T_s}$. Thus, to maintain a mathematically viable Sampling Theorem, the amplitude A must increase as $\frac{1}{\Delta}$, becoming infinitely large as the pulse duration decreases. Practical systems use a small value of Δ, say 0.1.Ts and use amplifiers to rescale the signal. Thus, the sampled signal's spectrum becomes periodic with period $\frac{1}{T_s}$. Thus, the Nyquist frequency $\frac{1}{2T_s}$ corresponds to the frequency $\frac{1}{2}$.

### Example 5.1

Let's compute the discrete-time Fourier transform of the exponentially decaying sequence s(n)=anu(n), where u(n) is the unit-step sequence. Simply plugging the signal›s expression into the Fourier transform formula,

$$\begin{aligned} S\left(e^{j2\pi f}\right) &= \sum_{n=-\infty}^{\infty} a^n u\left(n\right) e^{-(j2\pi f n)} \\ &= \sum_{n=0}^{\infty} \left(a e^{-(j2\pi f)}\right)^n \end{aligned} \tag{5.19}$$

This sum is a special case of the geometric series.

$$\sum_{n=0}^{\infty} \alpha^n = \frac{1}{1-\alpha} \,, \quad |\alpha| < 1 \tag{5.20}$$

Thus, as long as |a|<1, we have our Fourier transform.

$$S\left(e^{j2\pi f}\right) = \frac{1}{1 - a e^{-(j2\pi f)}} \tag{5.21}$$

Using Euler's relation, we can express the magnitude and phase of this spectrum.

$$|S\left(e^{j2\pi f}\right)| = \frac{1}{\sqrt{\left(1 - a\cos\left(2\pi f\right)\right)^2 + a^2 \sin^2\left(2\pi f\right)}} \tag{5.22}$$

$$\angle\left(S\left(e^{j2\pi f}\right)\right) = -\tan^{-1}\left(\frac{a\sin\left(2\pi f\right)}{1 - a\cos\left(2\pi f\right)}\right) \tag{5.23}$$

No matter what value of a we choose, the above formulae clearly demonstrate the periodic nature of the spectra of discrete-time signals. Figure 5.9 (Spectrum of expontial signal) shows indeed that the spectrum is a periodic function. We need only consider the spectrum between $-\frac{1}{2}$ and $\frac{1}{2}$ to unambiguously define it. When a>0, we have a lowpass spectrum—the spectrum diminishes as frequency increases from 0 to $\frac{1}{2}$ —with increasing a leading to a greater low frequency content; for a<0, we have a highpass spectrum (Figure 5.10 (Spectra of exponential signals)).

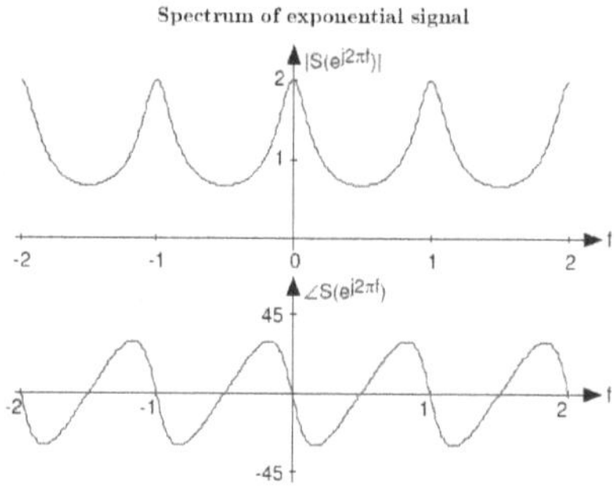


**Figure 5.9: The spectrum of the exponential signal (a = 0.5) is shown over the frequency range [-2,2], clearly demonstrating the periodicity of all discrete-time spectra. The angle has units of degrees.**

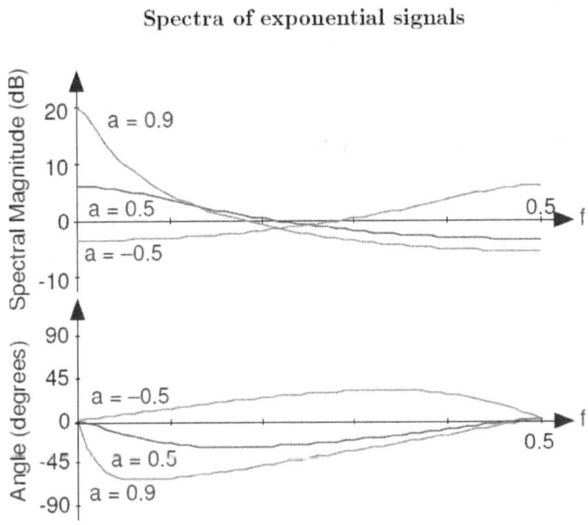


**Figure 5.10 The spectra of several exponential signals are shown. What is the apparent relationship between the spectra for a=0.5 and a=-0.5?**

## Example 5.2

Analogous to the analog pulse signal, let's find the spectrum of the length- N pulse sequence.

$$s(n) = \begin{cases} 1 & \text{if } 0 \leq n \leq N-1 \\ 0 & \text{otherwise} \end{cases} \quad (5.24)$$

The Fourier transform of this sequence has the form of a truncated geometric series.

$$S\left(e^{j2\pi f}\right) = \sum_{n=0}^{N-1} e^{-(j2\pi fn)} \quad (5.25)$$

For the so-called finite geometric series, we know that

$$\sum_{n=n_0}^{N+n_0-1} \alpha^n = \alpha^{n_0} \frac{1-\alpha^N}{1-\alpha} \quad (5.26)$$

for all values of α.

## Exercise 5.6.2

Derive this formula for the finite geometric series sum. The "trick" is to consider the difference between the series' sum and the sum of the series multiplied by α.

Applying this result yields (Figure 5.11 (Spectrum of length-ten pulse).)

$$\begin{aligned} S\left(e^{j2\pi f}\right) &= \frac{1-e^{-(j2\pi f N)}}{1-e^{-(j2\pi f)}} \\ &= e^{-(j\pi f(N-1))}\frac{\sin(\pi f N)}{\sin(\pi f)} \end{aligned} \tag{5.27}$$

The ratio of sine functions has the generic form of $\frac{\sin(Nx)}{\sin(x)}$, which is known as the discrete-time sinc function dsinc (x). Thus, our transform can be concisely expressed as $S\left(e^{j2\pi f}\right) = e^{-(j\pi f(N-1))}$ dsinc (πf). The discrete-time pulse's spectrum contains many ripples, the number of which increase with N, the pulse's duration.

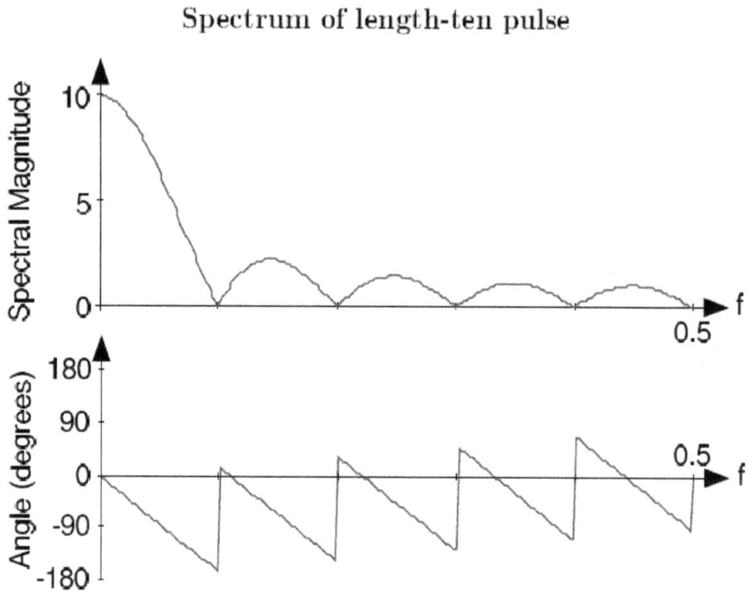


**Figure 5.11: The spectrum of a length-ten pulse is shown. Can you explain the rather complicated appearance of the phase?**

The inverse discrete-time Fourier transform is easily derived from the following relationship:

$$\begin{aligned} \int_{-\frac{1}{2}}^{\frac{1}{2}} e^{-(j2\pi f m)} e^{j2\pi f n}\, df &= \begin{cases} 1 \text{ if } m = n \\ 0 \text{ if } m \neq n \end{cases} \\ &= \delta(m-n) \end{aligned} \tag{5.28}$$

Therefore, we find that

$$\begin{aligned} \int_{-\frac{1}{2}}^{\frac{1}{2}} S\left(e^{j2\pi f}\right) e^{j2\pi f n}\, df &= \int_{-\frac{1}{2}}^{\frac{1}{2}} \sum_{mm} s(m)\, e^{-(j2\pi f m)} e^{j2\pi f n}\, df \\ &= \sum_{mm} s(m) \int_{-\frac{1}{2}}^{\frac{1}{2}} e^{(-(j2\pi f))(m-n)}\, df \\ &= s(n) \end{aligned} \tag{5.29}$$

The Fourier transform pairs in discrete-time are

$$\begin{aligned} S\left(e^{j2\pi f}\right) &= \sum_{n=-\infty}^{\infty} s\left(n\right) e^{-(j2\pi f n)} \\ s\left(n\right) &= \int_{-\frac{1}{2}}^{\frac{1}{2}} S\left(e^{j2\pi f}\right) e^{j2\pi f n} df \end{aligned} \tag{5.30}$$

The properties of the discrete-time Fourier transform mirror those of the analog Fourier transform. The DTFT properties table below shows similarities and differences. One important common property is Parseval's Theorem.

$$\sum_{n=-\infty}^{\infty} \left(\left|s\left(n\right)\right|\right)^2 = \int_{-\frac{1}{2}}^{\frac{1}{2}} \left(\left|S\left(e^{j2\pi f}\right)\right|\right)^2 df \tag{5.31}$$

To show this important property, we simply substitute the Fourier transform expression into the frequency-domain expression for power.

$$\begin{aligned} \int_{-\frac{1}{2}}^{\frac{1}{2}} \left(\left|S\left(e^{j2\pi f}\right)\right|\right)^2 df &= \int_{-\frac{1}{2}}^{\frac{1}{2}} \sum_{nn} s\left(n\right) e^{-(j2\pi f n)} \sum_{mm} s\left(n\right)^* e^{j2\pi f m} df \\ &= \sum_{n,mn,m} s\left(n\right) s\left(n\right)^* \int_{-\frac{1}{2}}^{\frac{1}{2}} e^{j2\pi f(m-n)} df \end{aligned} \tag{5.32}$$

Using the orthogonality relation (5.28), the integral equals δ(m-n), where δ(n) is the unit sample (Figure 5.8: Unit Sample). Thus, the double sum collapses into a single sum because nonzero values occur only when n=m, giving Parseval's Theorem as a result. We term $\sum_{nn} s^2(n)$ the energy in the discrete-time signal s(n) in spite of the fact that discrete-time signals don›t consume (or produce for that matter) energy. This terminology is a carry-over from the analog world.

### Exercise 5.6.3

Suppose we obtained our discrete-time signal from values of the product s(t)pTs(t), where the duration of the component pulses in pTs(t) is Δ. How is the discrete-time signal energy related to the total energy contained in s(t)? Assume the signal is bandlimited and that the sampling rate was chosen appropriate to the Sampling Theorem's conditions.

## 5.7 DISCRETE FOURIER TRANSFORMS (DFT)

The discrete-time Fourier transform (and the continuous-time transform as well) can be computed when we possess an analytical representation of the signal. Let's consider a signal, like the voice signal discussed in the previous chapter, that does not have a mathematical formula.

How then would you compute the spectrum? For example, how did we compute a spectrogram such as the one shown in the speech signal example? The Discrete Fourier Transform (DFT) allows the computation of spectra from discrete-time data. While in discrete-time we can exactly calculate spectra, for analog signals no similar exact spectrum computation exists. For analog-signal spectra, use must build special devices, which turn out in most cases to consist of A/D converters and discrete-time computations. Certainly discrete-time spectral analysis is more flexible than continuous-time spectral analysis.

The formula for the DTFT is a sum, which conceptually can be easily computed save for two issues.

- Signal duration. The sum extends over the signal's duration, which must be finite to compute the signal's spectrum. It is exceedingly difficult to store an infinite-length signal in any case, so we'll assume that the signal extends over [0,N-1]
- Continuous frequency. Subtler than the signal duration issue is the fact that the frequency variable is continuous: It may only need to span one period, like $\left[-\frac{1}{2}, \frac{1}{2}\right]$ or $[0, 1]$, but the DTFT formula as it stands requires evaluating the spectra at all frequencies within a period. Let's compute the spectrum at a few frequencies; the most obvious ones are the equally spaced ones $f = \frac{k}{K}, k \in \{0, \ldots, K-1\}$.

We thus define the discrete Fourier transform (DFT) to be

$$S(k) = \sum_{n=0}^{N-1} s(n) e^{-\frac{j2\pi nk}{K}}, \quad k \in \{0, \ldots, K-1\} \tag{5.33}$$

Here, S(k) is shorthand for $S\left(e^{j2\pi\frac{k}{K}}\right)$.

We can compute the spectrum at as many equally spaced frequencies as we like. Note that you can think about this computationally motivated choice as sampling the spectrum; more about this interpretation later. The issue now is how many frequencies are enough to capture how the spectrum changes with frequency. One way of answering this question is determining an inverse discrete Fourier transform formula: given S(k), k={0,...,K-1} how do we find s(n), n={0,...,N-1}? Presumably, the formula will be of the form $s(n) = \sum_{k=0}^{K-1} S(k) e^{\frac{j2\pi nk}{K}}$. Substituting the DFT formula in this prototype inverse transform yields

$$s(n) = \sum_{k=0}^{K-1} \sum_{m=0}^{N-1} s(m) e^{-\left(j\frac{2\pi mk}{K}\right)} e^{j\frac{2\pi nk}{K}} \tag{5.34}$$

Note that the orthogonality relation we use so often has a different character now.

$$\sum_{k=0}^{K-1} e^{-\left(j\frac{2\pi km}{K}\right)} e^{j\frac{2\pi kn}{K}} = \begin{cases} K \text{ if } \ (m = \{n, n \pm K, n \pm 2K, \ldots\}) \\ 0 \text{ otherwise} \end{cases} \tag{5.35}$$

We obtain nonzero value whenever the two indices differ by multiples of K. We can express this result as $K \sum_{l} \delta(m - n - lK)$. Thus, our formula becomes

$$s(n) = \sum_{m=0}^{N-1} s(m) K \sum_{l=-\infty}^{\infty} \delta(m - n - lK)$$

C

The integers n and m both range over {0,...,N-1}. To have an inverse transform, we need the sum to be a single unit sample for m,n in this range. If it did not, then s(n) would equal a sum of values, and we would not have a valid transform: Once going into the frequency domain, we could not get back unambiguously! Clearly, the term l=0 always provides a unit sample (we'll take care of the factor of K soon). If we evaluate the spectrum at fewer frequencies than the signal's duration, the term corresponding to m=n+K will also appear for some values of m,n={0,...,N-1}. This situation means that our prototype transform equals s(n)+s(n+K) for some values of n. The only way to eliminate this problem is to require K≥N: We must have at least as many frequency samples as the signal›s duration. In this way, we can return from the frequency domain we entered via the DFT.

### Exercise 5.7.1

When we have fewer frequency samples than the signal's duration, some discrete-time signal values equal the sum of the original signal values. Given the sampling interpretation of the spectrum, characterize this effect a different way.

Another way to understand this requirement is to use the theory of linear equations. If we write out the expression for the DFT as a set of linear equations,

$$s(0) + s(1) + \cdots + s(N-1) = S(0) \tag{5.37}$$

$$s(0) + s(1) e^{(-j)\frac{2\pi}{K}} + \cdots + s(N-1) e^{(-j)\frac{2\pi(N-1)}{K}} = S(1)$$

$$s(0) + s(1) e^{(-j)\frac{2\pi(K-1)}{K}} + \cdots + s(N-1) e^{(-j)\frac{2\pi(N-1)(K-1)}{K}} = S(K-1)$$

we have K equations in N unknowns if we want to find the signal from its sampled spectrum. This requirement is impossible to fulfill if K<N; we must have K≥N. Our orthogonality relation essentially says that if we have a sufficient number of equations (frequency samples), the resulting set of equations can indeed be solved.

By convention, the number of DFT frequency values K is chosen to equal the signal›s duration N. The discrete Fourier transform pair consists of

### Discrete Fourier Transform Pair

$$\begin{aligned} S(k) &= \textstyle\sum_{n=0}^{N-1} s(n) e^{-\left(j\frac{2\pi nk}{N}\right)} \\ s(n) &= \frac{1}{N} \textstyle\sum_{k=0}^{N-1} S(k) e^{j\frac{2\pi nk}{N}} \end{aligned} \tag{5.38}$$

### Exercise 5.3

Use this demonstration to perform DFT analysis of a signal.

This media object is a LabVIEW VI. Please view or download it at <DF Tanalysis. llb>

### Exercise 5.4

Use this demonstration to synthesize a signal from a DFT sequence.

This media object is a LabVIEW VI. Please view or download it at

<DFT_Component_Manipulation.llb>

## 5.8 DFT: COMPUTATIONAL COMPLEXITY

We now have a way of computing the spectrum for an arbitrary signal: The Discrete Fourier Transform (DFT) computes the spectrum at N equally spaced frequencies from a length- N sequence. An issue that never arises in analog "computation," like that performed by a circuit, is how much work it takes to perform the signal processing operation such as filtering. In computation, this consideration translates to the number of basic computational steps required to perform the needed processing. The number of steps, known as the complexity, becomes equivalent to how long the computation takes (how long must we wait for an answer). Complexity is not so much tied to specific computers or programming languages but to how many steps are required on any computer. Thus, a procedure's stated complexity says that the time taken will be proportional to some function of the amount of data used in the computation and the amount demanded.

For example, consider the formula for the discrete Fourier transform. For each frequency we choose, we must multiply each signal value by a complex number and add together the results. For a real-valued signal, each real-times-complex multiplication requires two real multiplications, meaning we have 2N multiplications to perform. To add the results together, we must keep the real and imaginary parts separate. Adding N numbers requires N-1 additions. Consequently, each frequency requires 2N+2(N−1)=4N−2 basic computational steps. As we have N frequencies, the total number of computations is N(4N−2).

In complexity calculations, we only worry about what happens as the data lengths increase, and take the dominant term—here the 4N2 term—as reflecting how much work is involved in making the computation. As multiplicative constants don›t matter since we are making a «proportional to» evaluation, we find the DFT is an O(N2) computational procedure. This notation is read "order N-squared". Thus, if we double the length of the data, we would expect that the computation time to approximately quadruple.

### Exercise 5.8.1

In making the complexity evaluation for the DFT, we assumed the data to be real. Three questions emerge. First of all, the spectra of such signals have conjugate symmetry, meaning that negative frequency components $(k = \{\frac{N}{2}+1, \ldots, N+1\}$ in the DFT (5.33)) can be computed from the corresponding positive frequency components. Does this symmetry change the DFT's complexity? Secondly, suppose the data are complex-valued; what is the DFT's complexity now? Finally, a less important but interesting question is suppose we want K frequency values instead of N; now what is the complexity?

## 5.9 FAST FOURIER TRANSFORM (FFT)

One wonders if the DFT can be computed faster: Does another computational procedure -- an algorithm - exist that can compute the same quantity, but more efficiently. We could seek methods that reduce the constant of proportionality, but do not change the DFT›s complexity O(N2). Here, we have something more dramatic in mind: Can the computations be restructured so that a smaller complexity results?

In 1965, IBM researcher Jim Cooley and Princeton faculty member John Tukey developed what is now known as the Fast Fourier Transform (FFT). It is an algorithm for computing that DFT that has order O(N log N) for certain length inputs. Now when the length of data doubles, the spectral computational time will not quadruple as with the DFT algorithm; instead, it approximately doubles. Later research showed that no algorithm for computing the DFT could have a smaller complexity than the FFT. Surprisingly, historical work has shown that Gauss in the early nineteenth century developed the same algorithm, but did not publish it! After the FFT›s rediscovery, not only was the computation of a signal›s spectrum greatly speeded, but also the added feature of algorithm meant that computations had flexibility not available to analog implementations.

### Exercise 5.9.1

Before developing the FFT, let's try to appreciate the algorithm's impact. Suppose a short-length transform takes 1 ms. We want to calculate a transform of a signal that is 10 times longer. Compare how much longer a straightforward implementation of the DFT would take in comparison to an FFT, both of which compute exactly the same quantity.

To derive the FFT, we assume that the signal's duration is a power of two: N=2L. Consider what happens to the even-numbered and odd-numbered elements of the sequence in the DFT calculation.

Each term in square brackets has the form of a $\frac{N}{2}$-length DFT. The first one is a DFT of the even-numbered elements, and the second of the odd-numbered elements. The first DFT is combined with the second multiplied by the complex exponential $e^{-\frac{j2\pi k}{N}}$. The half-length transforms are each evaluated at frequency indices k=0,...,N-1. Normally, the number of frequency indices in a DFT calculation range between zero and the transform length minus one. The computational advantage of the FFT comes from recognizing the periodic nature of the discrete Fourier transform. The FFT simply reuses the computations made in the half-length transforms and combines them through additions and the multiplication by $e^{-\frac{j2\pi k}{N}}$, which is not periodic over $\frac{N}{2}$. Figure 5.12 (Length-8 DFT decomposition) illustrates this decomposition. As it stands, we now compute two length- $\frac{N}{2}$ transforms of (complexity $2O\left(\frac{N^2}{4}\right)$), multiply one of them by the complex exponential of complexity O(N) and add the results (complexity O(N)). At this point, the total complexity is still dominated by the half-length DFT calculations, but the proportionality coefficient has been reduced.

$$S(k) = s(0) + s(2)e^{(-j)\frac{2\pi 2k}{N}} + \cdots + s(N-2)e^{(-j)\frac{2\pi(N-2)k}{N}} + \quad (5.39)$$
$$s(1)e^{(-j)\frac{2\pi k}{N}} + s(3)e^{(-j)\frac{2\pi\times(2+1)k}{N}} + \cdots + s(N-1)e^{(-j)\frac{2\pi(N-(2-1))k}{N}} =$$
$$\left[s(0) + s(2)e^{(-j)\frac{2\pi k}{\frac{N}{2}}} + \cdots + s(N-2)e^{(-j)\frac{2\pi\left(\frac{N}{2}-1\right)k}{\frac{N}{2}}}\right] + \left[s(1) + s(3)e^{(-j)\frac{2\pi k}{\frac{N}{2}}} + \cdots + s(N-1)e^{(-j)\frac{2\pi\left(\frac{N}{2}-1\right)}{\frac{N}{2}}}\right]$$

Now for the fun. Because N=2L,each of the half-length transforms can be reduced to two quarter-length transforms, each of these to two eighth-length ones, etc. This decomposition continues until we are left with length-2 transforms. This transform is quite simple, involving only additions. Thus, the first stage of the FFT has $\frac{N}{2}$ length-2 transforms (see the bottom part of Figure 5.12 (Length-8 DFT decomposition)). Pairs of these transforms are combined by adding one to the other multiplied by a complex exponential. Each pair requires 4 additions and 2 multiplications, giving a total number of computations equalling $6 \cdot \frac{N}{4} = \frac{3N}{2}$. This number of computations does not change from stage to stage. Because the number of stages, the number of times the length can be divided by two, equals $\frac{3N}{2}\log_2 N$, which makes the complexity of the FFT $O(N\log_2 N)$.]

Doing an example will make computational savings more obvious. Let's look at the details of a length-8 DFT. As shown on Figure 5.13 (Batterfly), we first decompose the DFT into two length-4 DFTs, with the outputs added and subtracted together in pairs. Considering Figure 5.13 (Butterfly) as the frequency index goes from 0 through 7, we recycle values from the length-4 DFT into the final calculation because of the periodicity of the DFT output. Examining how pairs of outputs are collected together, we create the basis computational element known as butterfly (Figure 5.13 (Butterfly)).

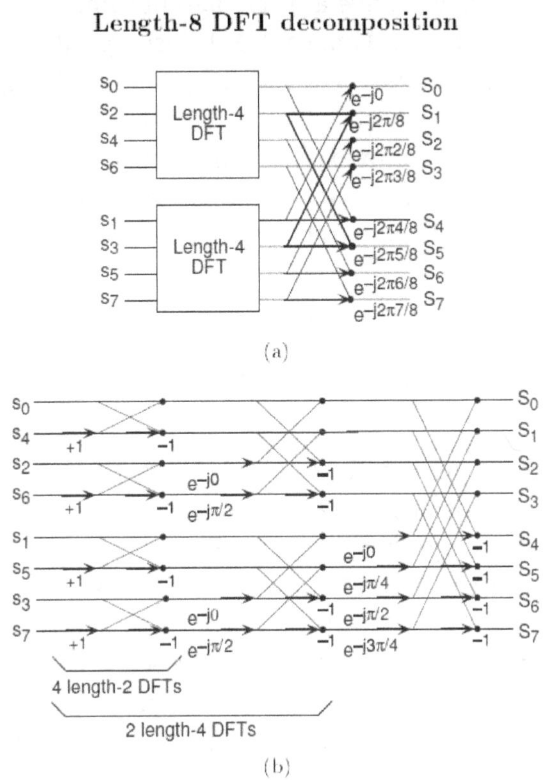


**Figure 5.12 The initial decomposition of a length-8 DFT into the terms using even- and odd-indexed inputs marks the first phase of developing the FFT algorithm. When these half-length transforms are successively decomposed, we are left with the diagram shown in the bottom panel that depicts the length-8 FFT computation.**

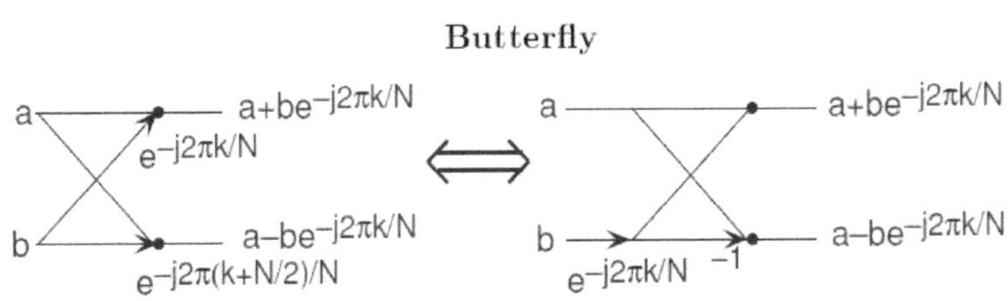


**Figure 5.13: The basic computational element of the fast Fourier transform is the butterfly. It takes two complex numbers, represented by a and b, and forms the quantities shown. Each butterfly requires one complex multiplication and two complex additions.**

By considering together the computations involving common output frequencies from the two half-length DFTs, we see that the two complex multiplies are related to each other, and we can reduce our computational work even further. By further decomposing the length-4 DFTs into two length-2 DFTs and combining their outputs, we arrive at the diagram summarizing the length-8 fast Fourier transform (Figure 5.12 (Length-8 DFT decomposition)). Although most of the complex multiplies are quite simple such as multiplying by $e^{-\left(j\frac{\pi}{2}\right)}$ means swapping real and imaginary parts and changing their signs. Let's count those for purposes of evaluating the complexity as full complex multiplies. We have $\frac{N}{2}=4$ complex multiplies and N = 8 complex additions for each stage and log2N=3 stages, making the number of basic computations $\frac{3N}{2}\log_2 N$ as predicted.

### Exercise 5.9.2

Note that the ordering of the input sequence in the two parts of Figure 5.12 (Length-8 DFT decomposition) aren't quite the same. Why not? How is the ordering determined?

Other "fast" algorithms were discovered, all of which make use of how many common factors the transform length N has. In number theory, the number of prime factors a given integer has measures how composite it is. The numbers 16 and 81 are highly composite (equaling 24 and 34 respectively), the number 18 is less so (21 . 32), and 17 not at all (it's prime). In over thirty years of Fourier transform algorithm development, the original Cooley-Tukey algorithm is far and away the most frequently used. It is so computationally efficient that power-of-two transform lengths are frequently used regardless of what the actual length of the data.

### Exercise 5.9.3

Suppose the length of the signal were 500? How would you compute the spectrum of this signal using the Cooley-Tukey algorithm? What would the length N of the transform be?

## 5.10 SPECTROGRAMS

We know how to acquire analog signals for digital processing (pre-filtering (Section 5.3), sampling (Section 5.3), and A/D conversion (Section 5.4)) and to compute spectra of discrete-time signals (using the FFT algorithm (Section 5.9)), let's put these various components together to learn how the spectrogram shown in Figure 5.14 (Speech Spectrogram), which is used to analyze speech, is calculated. The speech was sampled at a rate of 11.025 kHz and passed through a 16-bit A/D converter.

Point to Interest: Music compact discs (CDs) encode their signals at a sampling rate of 44.1 kHz. We'll learn the rationale for this number later. The 11.025 kHz sampling rate for the speech is 1/4 of the CD sampling rate, and was the lowest available sampling rate commensurate with speech signal bandwidths available on my computer.

## Exercise 5.10.1

Looking at Figure 5.14, the signal lasted a little over 1.2 seconds. How long was the sampled signal (in terms of samples)? What was the datarate during the sampling process in bps (bits per second)? Assuming the computer storage is organized in terms of bytes (8-bit quantities), how many bytes of computer memory does the speech consume?

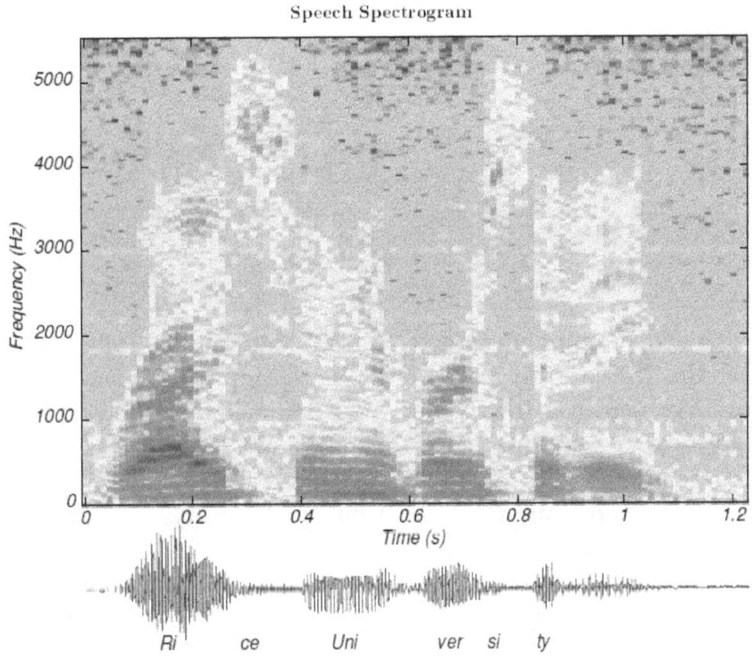


**Figure 5.14**

The resulting discrete-time signal, shown in the bottom of Figure 5.14 (Speech Spectrogram), clearly changes its character with time. To display these spectral changes, the long signal was sectioned into frames: comparatively short, contiguous groups of samples. Conceptually, a Fourier transform of each frame is calculated using the FFT. Each frame is not so long that significant signal variations are retained within a frame, but not so short that we lose the signal's spectral character. Roughly speaking, the speech signal's spectrum is evaluated over successive time segments and stacked side by side so that the x-axis corresponds to time and the y-axis frequency, with color indicating the spectral amplitude.

An important detail emerges when we examine each framed signal (Figure 5.15 (Spectrogram Hanning vs. Rectangular)).

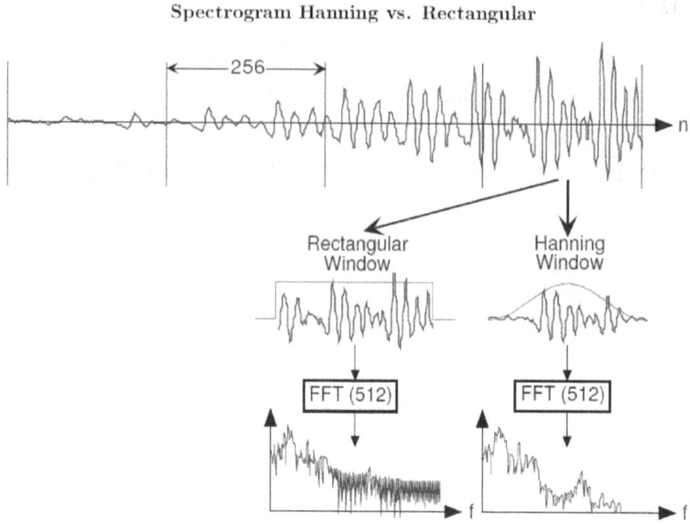


**Figure 5.15: The top waveform is a segment 1024 samples long taken from the beginning of the "Rice University" phrase. Computing Figure 5.14 (Speech Spectrogram) involved creating frames, here demarked by the vertical lines, that were 256 samples long and finding the spectrum of each. If a rectangular window is applied (corresponding to extracting a frame from the signal), oscillations appear in the spectrum (middle of bottom row). Applying a Hanning window gracefully tapers the signal toward frame edges, thereby yielding a more accurate computation of the signal's spectrum at that moment of time.**

At the frame's edges, the signal may change very abruptly, a feature not present in the original signal. A transform of such a segment reveals a curious oscillation in the spectrum, an artifact directly related to this sharp amplitude change. A better way to frame signals for spectrograms is to apply a window: Shape the signal values within a frame so that the signal decays gracefully as it nears the edges. This shaping is accomplished by multiplying the framed signal by the sequence w(n). In sectioning the signal, we essentially applied a rectangular window: $w(n) = 1, 0 \leq n \leq N-1$. A much more graceful window is the Hanning window; it has the cosine shape $w(n) = \frac{1}{2}\left(1 - \cos\left(\frac{2\pi n}{N}\right)\right)$. As shown in Figure 5.15 (Spectrogram Hanning vs. Rectangular), this shaping greatly reduces spurious oscillations in each frame's spectrum. Considering the spectrum of the Hanning windowed frame, we find that the oscillations resulting from applying the rectangular window obscured a formant (the one located at a little more than half the Nyquist frequency).

### Exercise 5.10.2

What might be the source of these oscillations? To gain some insight, what is the length-2N discrete Fourier transform of a length-N pulse? The pulse emulates the rectangular window, and certainly has edges. Compare your answer with the length-2N transform of a length-N Hanning window.

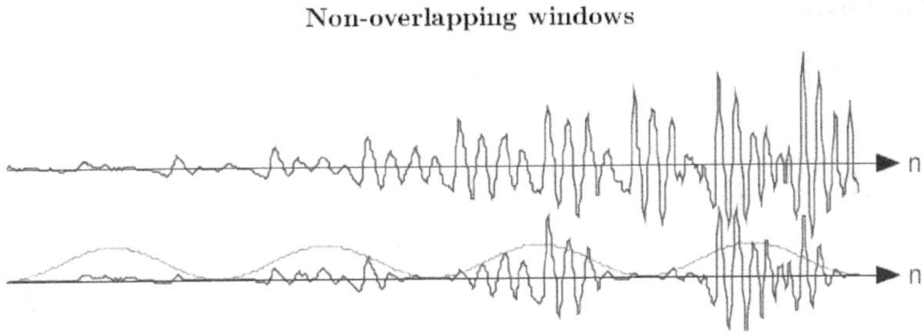


**Figure 5.16: In comparison with the original speech segment shown in the upper plot, the non-overlapped Hanning windowed version shown below it is very ragged. Clearly, spectral information extracted from the bottom plot could well miss important features present in the original.**

If you examine the windowed signal sections in sequence to examine windowing's effect on signal amplitude, we see that we have managed to amplitude-modulate the signal with the periodically repeated window (Figure 5.16 (Non-overlapping windows)). To alleviate this problem, frames are overlapped (typically by half a frame duration). This solution requires more Fourier transform calculations than needed by rectangular windowing, but the spectra are much better behaved and spectral changes are much better captured.

The speech signal, such as shown in the speech spectrogram (Figure 5.14: Speech Spectrogram), is sectioned into overlapping, equal-length frames, with a Hanning window applied to each frame. The spectra of each of these is calculated, and displayed in spectrograms with frequency extending vertically, window time location running horizontally, and spectral magnitude color-coded. Figure 5.17 (Overlapping windows for computing spectrograms) illustrates these computations.

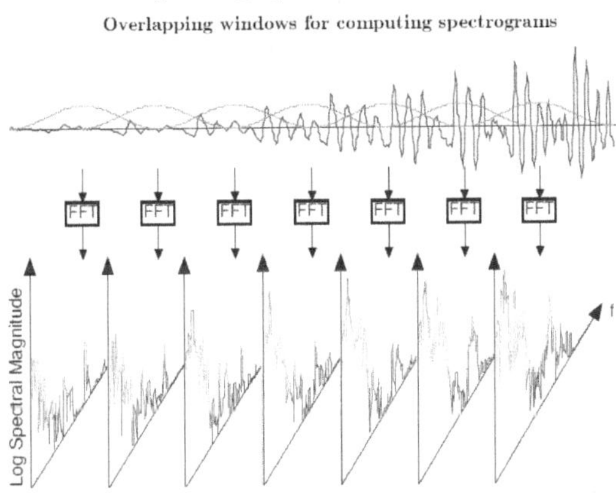


**Figure 5.17: The original speech segment and the sequence of overlapping Hanning windows applied to it are shown in the upper portion. Frames were 256 samples long and a Hanning window was applied with a half-frame overlap. A length-512 FFT of each frame was computed, with the magnitude of the first 257 FFT values displayed vertically, with spectral amplitude values color-coded.**

### Exercise 5.10.3

Why the specific values of 256 for N and 512 for K? Another issue is how was the length-512 transform of each length-256 windowed frame computed?

## 5.11 DISCRETE-TIME SYSTEMS

When we developed analog systems, interconnecting the circuit elements provided a natural starting place for constructing useful devices. In discrete-time signal processing, we are not limited by hardware considerations but by what can be constructed in software.

### Exercise 5.11.1

One of the first analog systems we described was the amplifier. We found that implementing an amplifier was difficult in analog systems, requiring an op-amp at least. What is the discrete-time implementation of an amplifier? Is this especially hard or easy?

In fact, we will discover that frequency-domain implementation of systems, wherein we multiply the input signal's Fourier transform by a frequency response, is not only a viable alternative, but also a computationally efficient one. We begin with discussing the underlying mathematical structure of linear, shift-invariant systems, and devise how software filters can be constructed.

## 5.12 DISCRETE-TIME SYSTEMS IN THE TIME-DOMAIN

A discrete-time signal s(n) is delayed by n0 samples when we write s(n-n0), with n0>0. Choosing n0 to be negative advances the signal along the integers. As opposed to analog delays (Section 2.6.3: Delay), discrete-time delays can only be integer valued. In the frequency domain, delaying a signal corresponds to a linear phase shift of the signal›s discrete-time Fourier transform: $s(n-n_0) \leftrightarrow e^{-(j2\pi f n_0)} S(e^{j2\pi f})$.

Linear discrete-time systems have the superposition property.

$$S(a_1 x_1(n) + a_2 x_2(n)) = a_1 S(x_1(n)) + a_2 S(x_2(n)) \tag{5.40}$$

A discrete-time system is called shift-invariant (analogous to time-invariant analog systems) if delaying the input delays the corresponding output. If S(x(n))=y(n), then a shift-invariant system has the property

$$S(x(n-n_0)) = y(n-n_0) \tag{5.41}$$

We use the term shift-invariant to emphasize that delays can only have integer values in discrete-time, while in analog signals, delays can be arbitrarily valued.

We want to concentrate on systems that are both linear and shift-invariant. It will be these that allow us the full power of frequency-domain analysis and implementations. Because we have no physical constraints in "constructing" such systems, we need only a mathematical specification. In analog systems, the differential equation specifies the input-output relationship in the time-domain. The corresponding discrete-time specification is the difference equation.

$$y(n) = a_1 y(n-1) + \cdots + a_p y(n-p) + b_0 x(n) + b_1 x(n-1) + \cdots + b_q x(n-q) \tag{5.42}$$

Here, the output signal y(n) is related to its past values y(n-1),l={1,...,p} and to the current and past values of the input signal x(n). The system's characteristics are determined by the choices for the number of coefficients p and q and the coefficients› values {a1,...,ap}and{b0,b1,...,bq}

Aside: There is an asymmetry in the coefficients: where is a0? This coefficient would multiply the y(n) term in (5.42). We have essentially divided the equation by it, which does not change the input-output relationship. We have thus created the convention that a0 is always one. As opposed to differential equations, which only provide an implicit description of a system (we must somehow solve the differential equation), difference equations provide an explicit way of computing the output for any input. We simply express the difference equation by a program that calculates each output from the previous output values, and the current and previous inputs. Difference equations are usually expressed in software with for loops. A MATLAB program that would compute the first 1000 values of the output has the form

```
for n=1:1000
    y(n) = sum(a.*y(n-1:-1:n-p)) + sum(b.*x(n:-1:n-q));
end
```

An important detail emerges when we consider making this program work; in fact, as written it has (at least) two bugs. What input and output values enter into the computation of y(1)? We need values for y(0), y(-1),..., values we have not yet computed. To compute them, we would need more previous values of the output, which we have not yet computed. To compute these values, we would need even earlier values, ad infinitum. The way out of this predicament is to specify the system›s initial conditions: we must provide the p output values that occurred before the input started. These values can be arbitrary, but the choice does impact how the system responds to a given input. One choice gives rise to a linear system: Make the initial conditions zero. The reason lies in the definition of a linear system: The only way that the output to a sum of signals can be the sum of the individual outputs occurs when the initial conditions in each case are zero.

## Exercise 5.12.1

The initial condition issue resolves making sense of the difference equation for inputs that start at some index. However, the program will not work because of a programming, not conceptual, error. What is it? How can it be "fixed?"

### Exercise 5.5

Let's consider the simple system having p=1 and q=0.

$$y(n) = ay(n-1) + bx(n) \tag{5.43}$$

To compute the output at some index, this difference equation says we need to know what the previous output y(n-1) and what the input signal is at that moment of time. In more detail, let›s compute this system›s output to a unit-sample input: x(n)=δ(n). Because the input is zero for negative indices, we start by trying to compute the output at n = 0.

$$y(0) = ay(-1) + b \tag{5.44}$$

What is the value of y(-1)? Because we have used an input that is zero for all negative indices, it is reasonable to assume that the output is also zero. Certainly, the difference equation would not describe a linear system if the input that is zero for all time did not produce a zero output. With this assumption, y(-1) = 0, leaving y(0) = b. For n > 0, the input unit-sample is zero, which leaves us with the difference equation y(n) = ay (n-1), n>0. We can envision how the filter responds to this input by making a table.

$$y(n) = ay(n-1) + b\delta(n) \tag{5.45}$$

**Table 5.1**

| n | x(n) | y(n) |
|---|---|---|
| -1 | 0 | 0 |
| 0 | 1 | B |
| 1 | 0 | ba |
| 2 | 0 | ba2 |
| : | 0 | : |
| N | 0 | ban |

Coefficient values determine how the output behaves. The parameter b can be any value, and serves as a gain. The effect of the parameter a is more complicated (Table 5.1). If it equals zero, the output simply equals the input times the gain b. For all non-zero values of a, the output lasts forever; such systems are said to be IIR (Infinite Impulse Response). The reason for this terminology is that the unit sample also known as the impulse (especially in analog situations), and the system's response to the "impulse" lasts forever. If a is positive and less than one, the output is a decaying exponential. When a = 1, the output is a unit step. If a is negative and greater than -1, the output oscillates while decaying exponentially. When a = -1, the output changes sign forever, alternating between b and -b. More dramatic effects when |a| > 1; whether positive or negative, the output signal becomes larger and larger, growing exponentially.

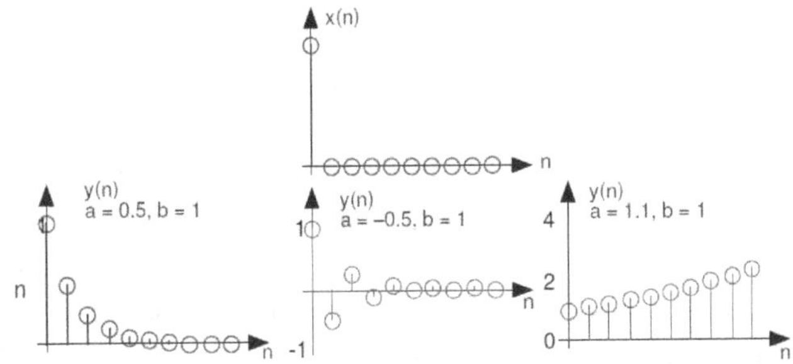


**Figure 5.18: The input to the simple example system, a unit sample, is shown at the top, with the outputs for several system parameter values shown below.**

Positive values of a are used in population models to describe how population size increases over time. Here, n might correspond to generation. The difference equation says that the number in the next generation is some multiple of the previous one. If this multiple is less than one, the population becomes extinct; if greater than one, the population flourishes. The same difference equation also describes the effect of compound interest on deposits. Here, n indexes the times at which compounding occurs (daily, monthly, etc.), a equals the compound interest rate plus one, and b = 1 (the bank provides no gain). In signal processing applications, we typically require that the output remain bounded for any input. For our example, that means that we restrict |a| < 1 and choose values for it and the gain according to the application.

## Exercise 5.12.2

Note that the difference equation (5.42),

$$y(n) = a_1 y(n-1) + \cdots + a_p y(n-p) + b_0 x(n) + b_1 x(n-1) + \cdots + b_q x(n-q)$$

does not involve terms like y(n+1) or x(n+1) on the equation›s right side. Can such terms also be included? Why or why not?

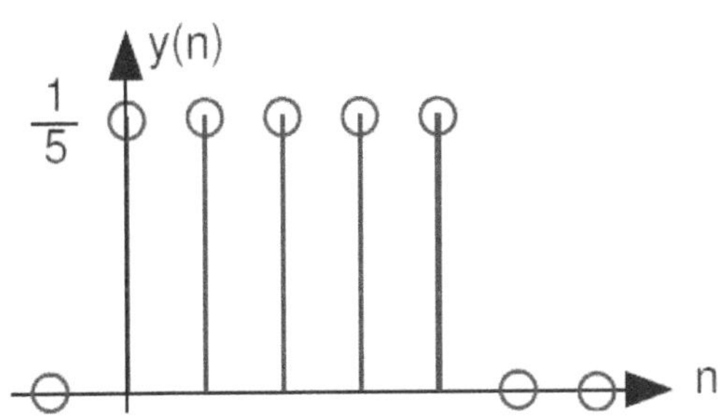


**Figure 5.19: The plot shows the unit-sample response of a length-5 boxcar filter.**

### Example 5.6

A somewhat different system has no "a" coefficients. Consider the difference equation:

$$y(n) = \frac{1}{q}\left(x(n) + \cdots + x(n-q+1)\right) \tag{5.46}$$

Because this system's output depends only on current and previous input values, we need not be concerned with initial conditions. When the input is a unit-sample, the output equals $\frac{1}{q}$ for n={0,...,q-1},then equals zero thereafter. Such systems are said to be FIR (Finite Impulse Response) because their unit sample responses have finite duration. Plotting this response (Figure 5.19) shows that the unit-sample response is a pulse of width q and height $\frac{1}{q}$. This waveform is also known as a boxcar, hence the name boxcar filter given to this system. We›ll derive its frequency response and develop its filtering interpretation in the next section. For now, note that the difference equation says that each output value equals the average of the input›s current and previous values. Thus, the output equals the running average of input's previous q values. Such a system could be used to produce the average weekly temperature (q = 7) that could be updated daily.

## 5.13 DISCRETE-TIME SYSTEMS IN THE FREQUENCY DOMAIN

As with analog linear systems, we need to find the frequency response of discrete-time systems. We used impedances to derive directly from the circuit's structure the frequency response. The only structure we have so far for a discrete-time system is the difference equation. We proceed as when we used impedances: let the input be a complex exponential signal. When we have a linear, shift-invariant system, the output should also be a complex exponential of the same frequency, changed in amplitude and phase. These amplitude and phase changes comprise the frequency response we seek. The complex exponential input signal is x(n)=Xej2πfn Note that this input occurs for all values of n. No need to worry about initial conditions here. Assume the output has a similar form: y(n)=Yej2πfn Plugging these signals into the fundamental difference equation (5.42), we have

$$Ye^{j2\pi fn} = a_1Ye^{j2\pi f(n-1)} + \cdots + a_pYe^{j2\pi f(n-p)} + b_0Xe^{j2\pi fn} + b_1Xe^{j2\pi f(n-1)} + \cdots + b_qXe^{j2\pi f(n-q)} \tag{5.47}$$

The assumed output does indeed satisfy the difference equation if the output complex amplitude is related to the input amplitude by

$$Y = \frac{b_0 + b_1e^{-(j2\pi f)} + \cdots + b_qe^{-(j2\pi qf)}}{1 - a_1e^{-(j2\pi f)} - \cdots - a_pe^{-(j2\pi pf)}}X$$

This relationship corresponds to the system's frequency response or, by another name, its transfer function. We find that any discrete-time system defined by a difference equation has a transfer function given by

$$H\left(e^{j2\pi f}\right)=\frac{b_0+b_1e^{-(j2\pi f)}+\cdots+b_qe^{-(j2\pi qf)}}{1-a_1e^{-(j2\pi f)}-\cdots-a_pe^{-(j2\pi pf)}} \tag{5.48}$$

Furthermore, because any discrete-time signal can be expressed as a superposition of complex exponential signals and because linear discrete-time systems obey the Superposition Principle, the transfer function relates the discrete-time Fourier transform of the system›s output to the input›s Fourier transform.

$$Y\left(e^{j2\pi f}\right)=X\left(e^{j2\pi f}\right)H\left(e^{j2\pi f}\right) \tag{5.49}$$

### Example 5.7

The frequency response of the simple IIR system (difference equation given in a previous example (Example 5.5)) is given by

$$H\left(e^{j2\pi f}\right)=\frac{b}{1-ae^{-(j2\pi f)}} \tag{5.50}$$

This Fourier transform occurred in a previous example; the exponential signal spectrum portrays the magnitude and phase of this transfer function. When the filter coefficient a is positive, we have a lowpass filter; negative a results in a highpass filter. The larger the coefficient in magnitude, the more pronounced the lowpass or highpass filtering.

### Example 5.8

The length-q boxcar filter (difference equation found in a previous example (Example 5.6)) has the frequency response

$$H\left(e^{j2\pi f}\right)=\frac{1}{q}\sum_{m=0}^{q-1}e^{-(j2\pi fm)} \tag{5.51}$$

This expression amounts to the Fourier transform of the boxcar signal (Figure 5.19). There we found that this frequency response has a magnitude equal to the absolute value of dsinc(πf). See the length-10 filter's frequency response. We see that boxcar filters--length-q signal averagers--have a lowpass behavior, having a cutoff frequency of $\frac{1}{q}$.

### Exercise 5.13.1

Suppose we multiply the boxcar filter's coefficients by a sinusoid: $b_m=\frac{1}{q}\cos\left(2\pi f_0 m\right)$ Use Fourier transform properties to determine the transfer function. How would you characterize this system: Does it act like a filter? If so, what kind of filter and how do you control its characteristics with the filter's coefficients?

These examples illustrate the point that systems described (and implemented) by difference equations serve as filters for discrete-time signals. The filter's order is given by the number p of denominator coefficients in the transfer function (if the system is IIR) or by the number q of numerator coefficients if the filter is FIR. When a system›s transfer function has both terms, the system is usually IIR, and its order equals p regardless of q. By selecting the coefficients and filter type, filters having virtually any frequency response desired can be designed. This design flexibility can't be found in analog systems. In the next section, we detail how analog signals can be filtered by computers, offering a much greater range of filtering possibilities than is possible with circuits.

## 5.14 FILTERING IN THE FREQUENCY DOMAIN

Because we are interested in actual computations rather than analytic calculations, we must consider the details of the discrete Fourier transform. To compute the length-N DFT, we assume that the signal has a duration less than or equal to N. Because frequency responses have an explicit frequency-domain specification in terms of filter coefficients, we don›t have a direct handle on which signal has a Fourier transform equaling a given frequency response. Finding this signal is quite easy. First of all, note that the discrete-time Fourier transform of a unit sample equals one for all frequencies. Since the input and output of linear, shift-invariant systems are related to each other by $Y\left(e^{j2\pi f}\right)=H\left(e^{j2\pi f}\right)X\left(e^{j2\pi f}\right)$, a unit-sample input, which has $X\left(e^{j2\pi f}\right)=1$ ,results in the output's Fourier transform equaling the system's transfer function.

### Exercise 5.14.1

The above statement is a very important result. Derive it yourself.

In the time-domain, the output for a unit-sample input is known as the system's unit-sample response, and is denoted by h(n). Combining the frequency-domain and time-domain interpretations of a linear, shift-invariant system's unit-sample response, we have that h(n) and the transfer function are Fourier transform pairs in terms of the discrete-time Fourier transform.

$$h(n)\leftrightarrow H\left(e^{j2\pi f}\right) \tag{5.52}$$

Returning to the issue of how to use the DFT to perform filtering, we can analytically specify the frequency response, and derive the corresponding length-N DFT by sampling the frequency response.

$$H(k)=H\left(e^{\frac{j2\pi k}{N}}\right)\ ,\ k=\{0,\ldots,N-1\} \tag{5.53}$$

Computing the inverse DFT yields a length-N signal no matter what the actual duration of the unit-sample response might be. If the unit-sample response has a duration less than or equal to N (it's a FIR filter), computing the inverse DFT of the sampled frequency response indeed yields the unit-sample response. If, however, the duration exceeds N, errors are encountered. The nature of these errors is easily explained by appealing to the Sampling Theorem. By sampling in the frequency domain, we have the potential for aliasing in the time domain (sampling in one domain, be it time or frequency, can result in aliasing in the other) unless we sample fast enough. Here, the duration of the unit-sample response determines the minimal sampling rate that prevents aliasing. For FIR systems — they by definition have finite-duration unit sample responses — the number of required DFT samples equals the unit-sample response's duration: N ≥ q

### Exercise 5.14.2

Derive the minimal DFT length for a length-q unit-sample response using the Sampling Theorem. Because sampling in the frequency domain causes repetitions of the unit-sample response in the time domain, sketch the time-domain result for various choices of the DFT length N.

### Exercise 5.14.3

Express the unit-sample response of a FIR filter in terms of difference equation coefficients. Note that the corresponding question for IIR filters is far more difficult to answer: Consider the example (Example 5.5).

For IIR systems, we cannot use the DFT to find the system's unit-sample response: aliasing of the unit-sample response will always occur. Consequently, we can only implement an IIR filter accurately in the time domain with the system›s difference equation. Frequency-domain implementations are restricted to FIR filters.

Another issue arises in frequency-domain filtering that is related to time domain aliasing, this time when we consider the output. Assume we have an input signal having duration Nx that we pass through a FIR filter having a length-q+1 unit-sample response. What is the duration of the output signal? The difference equation for this filter is

$$y(n) = b_0 x(n) + \cdots + b_q x(n-q) \tag{5.54}$$

This equation says that the output depends on current and past input values, with the input value q samples previous defining the extent of the filter's memory of past input values. For example, the output at index Nx depends on x(Nx) (which equals zero), x(Nx-1), through x(Nx-q). Thus, the output returns to zero only after the last input value passes through the filter's memory. As the input signal's last value occurs at index Nx- 1, the last nonzero output value occurs when n-q=Nx-1 or n=q+Nx-1.Thus, the output signal's duration equals q + Nx.

## Exercise 5.14.4

In words, we express this result as "The output's duration equals the input's duration plus the filter's duration minus one.". Demonstrate the accuracy of this statement. The main theme of this result is that a filter's output extends longer than either its input or its unit-sample response. Thus, to avoid aliasing when we use DFTs, the dominant factor is not the duration of input or of the unit-sample response, but of the output. Thus, the number of values at which we must evaluate the frequency response's DFT must be at least q + Nx and we must compute the same length DFT of the input. To accommodate a shorter signal than DFT length, we simply zero-pad the input: Ensure that for indices extending beyond the signal›s duration that the signal is zero. Frequency-domain filtering, as shown in Figure 5.20 below, is accomplished by storing the filter›s frequency response as the DFT H(k), computing the input's DFT X(k), multiplying them to create the output's DFT Y(k)=H(k)X(k), and computing the inverse DFT of the result to yield y(n).

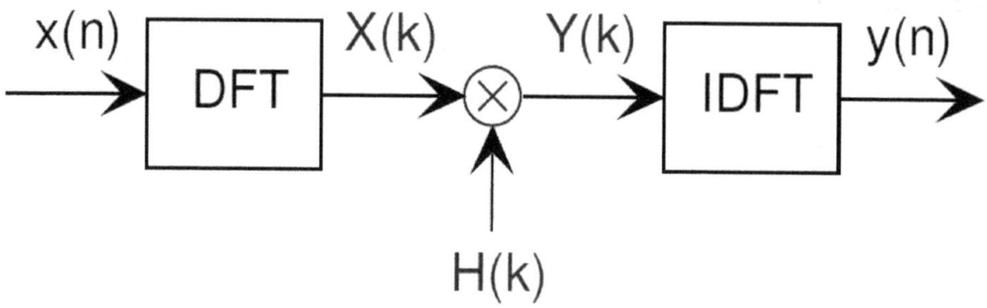


**Figure 5.20: To filter a signal in the frequency domain, first compute the DFT of the input, multiply the result by the sampled frequency response, and finally compute the inverse DFT of the product. The DFT's length must be at least the sum of the input's and unit-sample response's duration minus one. We calculate these discrete Fourier transforms using the fast Fourier transform algorithm, of course.**

Before detailing this procedure, let's clarify why so many new issues arose in trying to develop a frequency-domain implementation of linear filtering. The frequency-domain relationship between a filter's input and output is always true: $Y\left(e^{j2\pi f}\right) = H\left(e^{j2\pi f}\right) X\left(e^{j2\pi f}\right)$. The Fourier transforms in this result are discrete-time Fourier transforms; for example, $X\left(e^{j2\pi f}\right) = \sum_n x(n)\, e^{-(j2\pi f n)}$. Unfortunately, using this relationship to perform filtering is restricted to the situation when we have analytic formulas for the frequency response and the input signal. The reason why we had to "invent" the discrete Fourier transform (DFT) has the same origin: The spectrum resulting from the discrete-time Fourier transform depends on the continuous frequency variable f. That's fine for analytic calculation, but computationally we would have to make an uncountably infinite number of computations.

Note: Did you know that two kinds of infinities can be meaningfully defined? A countably infinite quantity means that it can be associated with a limiting process associated with integers. An uncountably infinite quantity cannot be so associated. The number of rational numbers is countably infinite (the numerator and denominator correspond to locating the rational by row and column; the total number so-located can be counted, voila!); the number of irrational numbers is uncountably infinite. Guess which is «bigger?»

The DFT computes the Fourier transform at a finite set of frequencies — samples the true spectrum — which can lead to aliasing in the time-domain unless we sample sufficiently fast. The sampling interval here is 1/K for a length-K DFT: faster sampling to avoid aliasing thus requires a longer transform calculation. Since the longest signal among the input, unit-sample response and output is the output, it is that signal›s duration that determines the transform length. We simply extend the other two signals with zeros (zero-pad) to compute their DFTs.

## Example 5.9

Suppose we want to average daily stock prices taken over last year to yield a running weekly average (average over five trading sessions). The filter we want is a length-5 averager (as shown in the unit-sample response (Figure 5.19)), and the input's duration is 253 (365 calendar days minus weekend days and holidays). The output duration will be 253+5-1=257253+5-1=257, and this determines the transform length we need to use. Because we want to use the FFT, we are restricted to power-of-two transform lengths. We need to choose any FFT length that exceeds the required DFT length. As it turns out, 256 is a power of two (28=256), and this length just undershoots our required length. To use frequency domain techniques, we must use length-512 fast Fourier transforms.

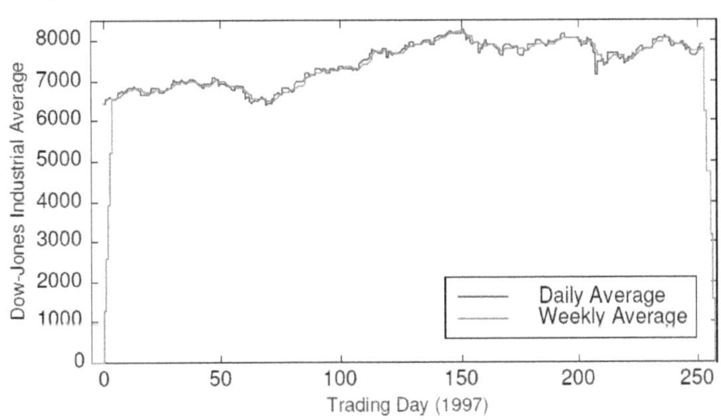


**Figure 5.21: The blue line shows the Dow Jones Industrial Average from 1997, and the red one the length-5 boxcar-filtered result that provides a running weekly of this market index. Note the "edge" effects in the filtered output.**

Figure 5.21 shows the input and the filtered output. The MATLAB programs that compute the filtered output in the time and frequency domains are:

```
Time Domain
h = [1 1 1 1 1]/5;
y = filter(h,1,[djia zeros(1,4)]);

Frequency Domain
h = [1 1 1 1 1]/5;
DJIA = fft(djia, 512);
H = fft(h, 512);
Y = H.*X;
y = ifft(Y);
```

**Note:** The filter program has the feature that the length of its output equals the length of its input. To force it to produce a signal having the proper length, the program zero-pads the input appropriately.

MATLAB's fft function automatically zero-pads its input if the specified transform length (its second argument) exceeds the signal›s length. The frequency domain result will have a small imaginary component — largest value is 2.2×10-11 — because of the inherent finite precision nature of computer arithmetic. Because of the unfortunate misfit between signal lengths and favored FFT lengths, the number of arithmetic operations in the time-domain implementation is far less than those required by the frequency domain version: 514 versus 62,271. If the input signal had been one sample shorter, the frequency-domain computations would have been more than a factor of two less (28,696), but far more than in the time-domain implementation.

An interesting signal processing aspect of this example is demonstrated at the beginning and end of the output. The ramping up and down that occurs can be traced to assuming the input is zero before it begins and after it ends. The filter "sees" these initial and final values as the difference equation passes over the input. These artifacts can be handled in two ways: we can just ignore the edge effects or the data from previous and succeeding years' last and first week, respectively, can be placed at the ends.

## 5.15 EFFICIENCY OF FREQUENCY-DOMAIN FILTERING

To determine for what signal and filter durations a time- or frequency-domain implementation would be the most efficient, we need only count the computations required by each. For the time-domain, difference-equation approach, we need (NX+q) (2(q)+1). The frequency-domain approach requires three Fourier transforms, each requiring $\frac{5K}{2}(\log_2 K)$ computations for a length-K FFT, and the multiplication of two spectra (6K computations). The output-signal-duration-determined length must be at least Nx+q. Thus, we must compare

$$(N_x+q)\,(2q+1) \leftrightarrow 6\,(N_x+q)+5\,(N_x+q)\log_2\,(N_x+q)$$

Exact analytic evaluation of this comparison is quite difficult (we have a transcendental equation to solve). Insight into this comparison is best obtained by dividing by Nx+ q.

$$2q+1 \leftrightarrow 6+5\log_2\,(N_x+q)$$

With this manipulation, we are evaluating the number of computations per sample. For any given value of the filter's order q, the right side, the number of frequency-domain computations, will exceed the left if the signal's duration is long enough. However, for filter durations greater than about 10, as long as the input is at least 10 samples, the frequency-domain approach is faster so long as the FFT's power-of-two constraint is advantageous.

The frequency-domain approach is not yet viable; what will we do when the input signal is infinitely long? The difference equation scenario fits perfectly with the envisioned digital filtering structure, but so far we have required the input to have limited duration (so that we could calculate its Fourier transform). The solution to this problem is quite simple: Section the input into frames, filter each, and add the results together. To section a signal means expressing it as a linear combination of length-Nx non-overlapping. Because the filter is linear, filtering a sum of terms is equivalent to summing the results of filtering each term.

$$\left(x\,(n)=\sum_{m=-\infty}^{\infty}x\,(n-mN_x)\right)\Rightarrow\left(y\,(n)=\sum_{m=-\infty}^{\infty}y\,(n-mN_x)\right) \tag{5.55}$$

As illustrated in Figure 5.22, note that each filtered section has a duration longer than the input. Consequently, we must literally add the filtered sections together, not just butt them together.

Computational considerations reveal a substantial advantage for a frequency-domain implementation over a time-domain one. The number of computations for a time-domain implementation essentially remains constant whether we section the input or not. Thus, the number of computations for each output is 2(q)+1. In the frequency-domain approach, computation counting changes because we need only compute the filter's frequency response H(k) once, which amounts to a fixed overhead. We need only compute two DFTs and multiply them to filter a section. Letting Nx denote a section›s length, the number of computations for a section amounts to (Nx+q) log2 (Nx+q) + 6 (Nx+q). In addition, we must add the filtered outputs together; the number of terms to add corresponds to the excess duration of the output compared with the input (q). The frequency-domain approach thus requires $\log_2(N_x+q)+6+\frac{q}{N_x+q}$ computations per output value. For even modest filter orders, the frequency-domain approach is much faster.

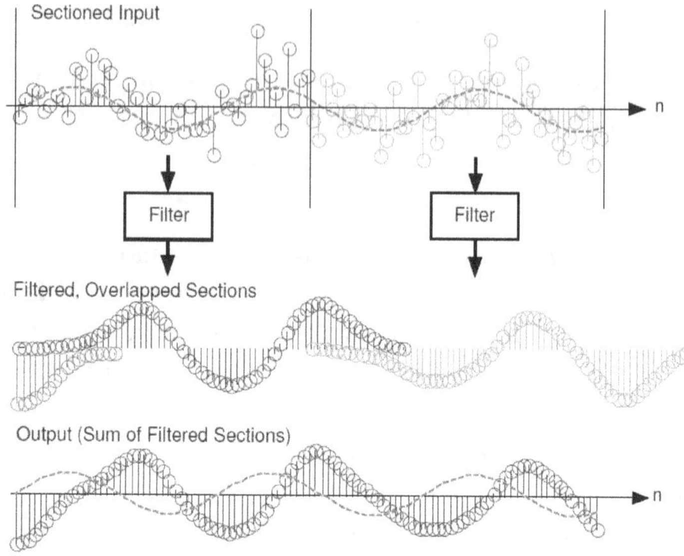


**Figure 5.22: The noisy input signal is sectioned into length-48 frames, each of which is filtered using frequency-domain techniques. Each filtered section is added to other outputs that overlap to create the signal equivalent to having filtered the entire input. The sinusoidal component of the signal is shown as the red dashed line.**

## Exercise 5.14.1

Show that as the section length increases, the frequency domain approach becomes increasingly more efficient.

Note that the choice of section duration is arbitrary. Once the filter is chosen, we should section so that the required FFT length is precisely a power of two: Choose Nx so that $N_x+q=2^L$.

Implementing the digital filter shown in the A/D block diagram (Figure 5.24) with a frequency-domain implementation requires some additional signal management not required by time-domain implementations. Conceptually, a real-time, time-domain filter could accept each sample as it becomes available, calculate the difference equation, and produce the output value, all in less than the sampling interval TS. Frequency-domain approaches don't operate on a sample-by-sample basis; instead, they operate on sections. They filter in real time by producing Nx outputs for the same number of inputs faster than NxTS. Because they generally take longer to produce an output section than the sampling interval duration, we must filter one section while accepting into memory the next section to be filtered. In programming, the operation of building up sections while computing on previous ones is known as buffering. Buffering can also be used in time-domain filters as well but isn't required.

## Example 5.10

We want to lowpass filter a signal that contains a sinusoid and a significant amount of noise. The example shown in Figure shows a portion of the noisy signal›s waveform. If it weren›t for the overlaid sinusoid, discerning the sine wave in the signal is virtually impossible. One of the primary applications of linear filters is noise removal: preserve the signal by matching filter's passband with the signal's spectrum and greatly reduce all other frequency components that may be present in the noisy signal.

A smart Rice engineer has selected a FIR filter having a unit-sample response corresponding a period-17 sinusoid: $h(n) = \frac{1}{17}\left(1 - \cos\left(\frac{2\pi n}{17}\right)\right),\ n = \{0, \ldots, 16\}$, which make q=16. Its frequency response (determined by computing the discrete Fourier transform) is shown in Figure 5.23 below. To apply, we can select the length of each section so that the frequency-domain filtering approach is maximally efficient: Choose the section length Nx so that Nx+ q is a power of two. To use a length-64 FFT, each section must be 48 samples long. Filtering with the difference equation would require 33 computations per output while the frequency domain requires a little over 16; this frequency-domain implementation is over twice as fast! .

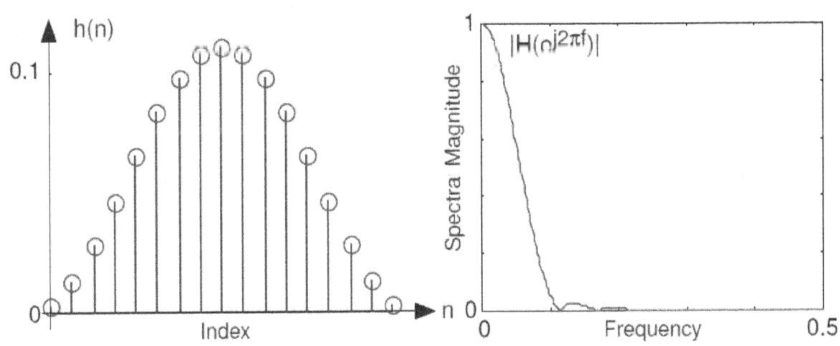


**Figure 5.23: The figure shows the unit-sample response of a length-17 Hanning filter on the left and the frequency response on the right. This filter functions as a lowpass filter having a cutoff frequency of about 0.1.**

We note that the noise has been dramatically reduced, with a sinusoid now clearly visible in the filtered output. Some residual noise remains because noise components within the filter's passband appear in the output as well as the signal.

## Exercise 5.15.2

Note that when compared to the input signal's sinusoidal component, the output's sinusoidal component seems to be delayed. What is the source of this delay? Can it be removed?

## 5.16 DISCRETE-TIME FILTERING OF ANALOG SIGNALS

Because of the Sampling Theorem, (Section 5.3.2: The Sampling Theorem), we can process, in particular filter, analog signals "with a computer" by constructing the system shown in Figure 5.24. To use this system, we are assuming that the input signal has a lowpass spectrum and can be bandlimited without affecting important signal aspects. Bandpass signals can also be filtered digitally, but require a more complicated system. Highpass signals cannot be filtered digitally. Note that the input and output filters must be analog filters; trying to operate without them can lead to potentially very inaccurate digitization.

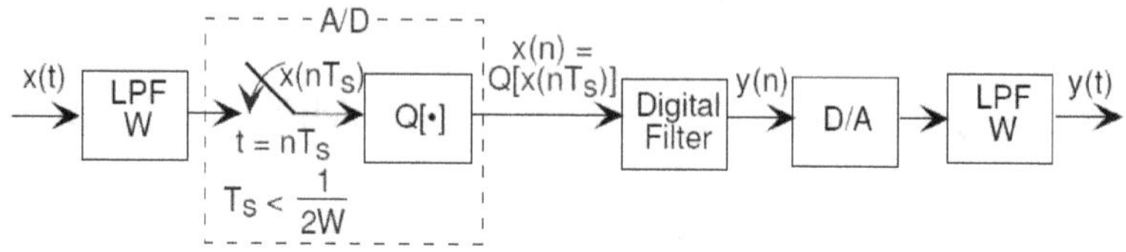


**Figure 5.24: To process an analog signal digitally, the signal x(t) must be filtered with an anti-aliasing filter (to ensure a bandlimited signal) before A/D conversion. This lowpass filter (LPF) has a cutoff frequency of W Hz, which determines allowable sampling intervals Ts. The greater the number of bits in the amplitude quantization portion Q[•] of the A/D converter, the greater the accuracy of the entire system. The resulting digital signal x(n) can now be filtered in the time-domain with a difference equation or in the frequency domain with Fourier transforms. The resulting output y(n) then drives a D/A converter and a second anti-aliasing filter (having the same bandwidth as the first one).**

Another implicit assumption is that the digital filter can operate in real time: The computer and the filtering algorithm must be sufficiently fast so that outputs are computed faster than input values arrive. The sampling interval, which is determined by the analog signal's bandwidth, thus determines how long our program has to compute each output y(n). The computational complexity for calculating each output with a difference equation (5.42) is O(p+q). Frequency domain implementation of the filter is also possible. The idea begins by computing the Fourier transform of a length-N portion of the input x(n), multiplying it by the filter›s transfer function, and computing the inverse transform of the result. This approach seems overly complex and potentially inefficient. Detailing the complexity, however, we have O(N log N) for the two transforms (computed using the FFT algorithm) and O(N) for the multiplication by the transfer function, which makes the total complexity O(N log N) for N input values. A frequency domain implementation thus requires O( log N) computational complexity for each output value. The complexities of time-domain and frequency-domain implementations depend on different aspects of the filtering: The time-domain implementation depends on the combined orders of the filter while the frequency-domain implementation depends on the logarithm of the Fourier transform's length.

It could well be that in some problems the time-domain version is more efficient (more easily satisfies the real time requirement), while in others the frequency domain approach is faster. In the latter situations, it is the FFT algorithm for computing the Fourier transforms that enables the superiority of frequency-domain implementations. Because complexity considerations only express how algorithm running-time increases with system parameter choices, we need to detail both implementations to determine which will be more suitable for any given filtering problem. Filtering with a difference equation is straightforward, and the number of computations that must be made for each output value is 2(p+q).

### Exercise 5.16.1

Derive this value for the number of computations for the general difference equation (5.42).

## 5.17 DIGITAL SIGNAL PROCESSING PROBLEMS

### Problem 5.1: Sampling and Filtering

The signal s(t) is bandlimited to 4 kHz. We want to sample it, but it has been subjected to various signal processing manipulations.

- What sampling frequency (if any works) can be used to sample the result of passing s(t) through an RC highpass filter with R = 10kΩ and C = 8nF?
- What sampling frequency (if any works) can be used to sample the derivative of s(t)?
- The signal s(t) has been modulated by an 8 kHz sinusoid having an unknown phase: the resulting signal is s(t)sin(2πf0t+φ), with f0 = 8kHz and φ=? Can the modulated signal be sampled so that the original signal can be recovered from the modulated signal regardless of the phase value φ? If so, show how and find the smallest sampling rate that can be used; if not, show why not.

### Problem 5.2: Non-Standard Sampling

Using the properties of the Fourier series can ease finding a signal's spectrum.

- Suppose a signal s(t) is periodic with period T. If ck represents the signal›s Fourier series coefficients, what are the Fourier series coefficients of $s\left(t-\frac{T}{2}\right)$?
- Find the Fourier series of the signal p(t) shown in the Figure 5.25 (Pulse Signal).
- Suppose this signal is used to sample a signal bandlimited to $\frac{1}{T}$Hz. Find an expression for and sketch the spectrum of the sampled signal.
- Does aliasing occur? If so, can a change in sampling rate prevent aliasing; if not, show how the signal can be recovered from these samples.

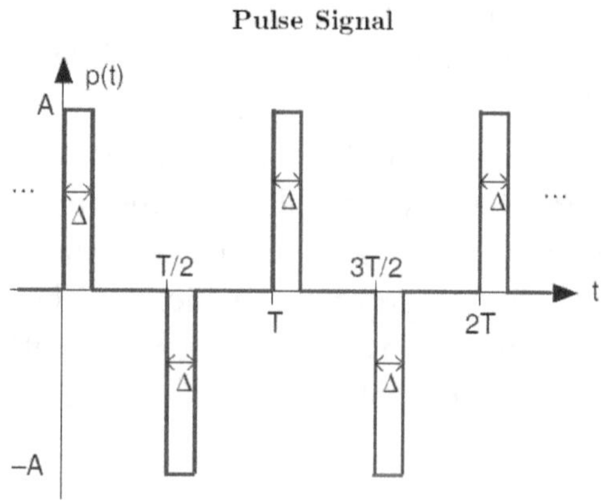


**Figure 5.25**

## Problem 5.3: A Different Sampling Scheme

A signal processing engineer from Texas A&M claims to have developed an improved sampling scheme. He multiplies the bandlimited signal by the depicted periodic pulse signal to perform sampling (Figure 5.26).

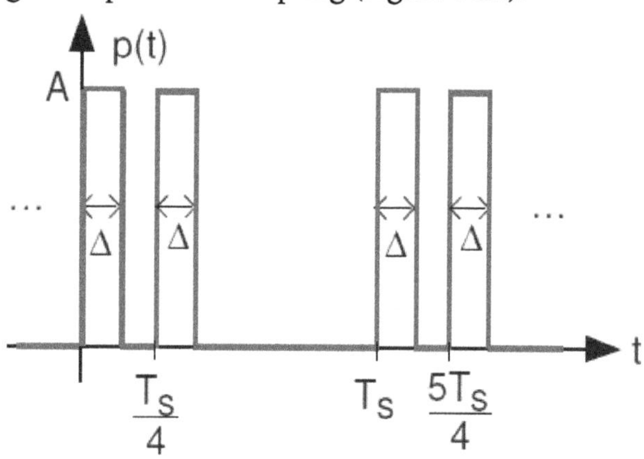


**Figure 5.26**

- Find the Fourier spectrum of this signal.
- Will this scheme work? If so, how should TS be related to the signal›s bandwidth? If not, why not?

## Problem 5.4: Bandpass Sampling

The signal s(t) has the indicated spectrum.

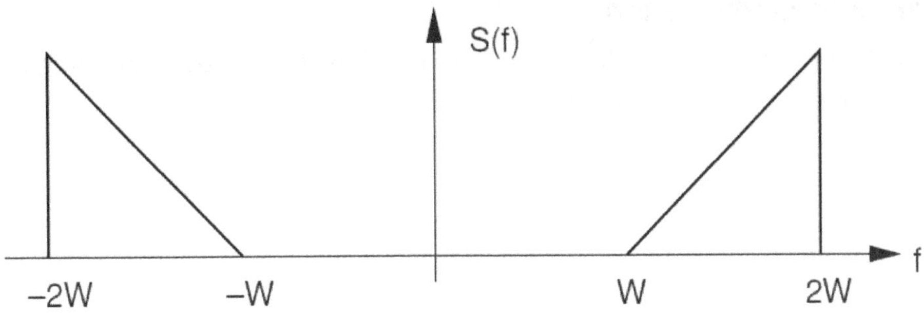


**Figure 5.27**

- What is the minimum sampling rate for this signal suggested by the Sampling Theorem?
- Because of the particular structure of this spectrum, one wonders whether a lower sampling rate could be used. Show that this is indeed the case, and find the system that reconstructs s(t) from its samples.

## Problem 5.5: Sampling Signals

If a signal is bandlimited to W Hz, we can sample it at any rate $\frac{1}{T_s} > 2W$ and recover the waveform exactly. This statement of the Sampling Theorem can be taken to mean that all information about the original signal can be extracted from the samples. While true in principle, you do have to be careful how you do so. In addition to the rms value of a signal, an important aspect of a signal is its peak value, which equals max{|s(t)|}.

- Let s(t) be a sinusoid having frequency W Hz. If we sample it at precisely the Nyquist rate, how accurately do the samples convey the sinusoid's amplitude? In other words, find the worst case example.
- How fast would you need to sample for the amplitude estimate to be within 5% of the true value?
- Another issue in sampling is the inherent amplitude quantization produced by A/D converters. Assume the maximum voltage allowed by the converter is Vmax volts and that it quantizes amplitudes to b bits. We can express the quantized sample Q(s(nTS)) as s(nTS)+ε(t), where ε(t) represents the quantization error at the nth sample. Assuming the converter rounds, how large is maximum quantization error?
- We can describe the quantization error as noise, with a power proportional to the square of the maximum error. What is the signal-to-noise ratio of the quantization error for a full-range sinusoid? Express your result in decibels.

## Problem 5.6: Hardware Error

An A/D converter has a curious hardware problem: Every other sampling pulse is half its normal amplitude (Figure 5.28).

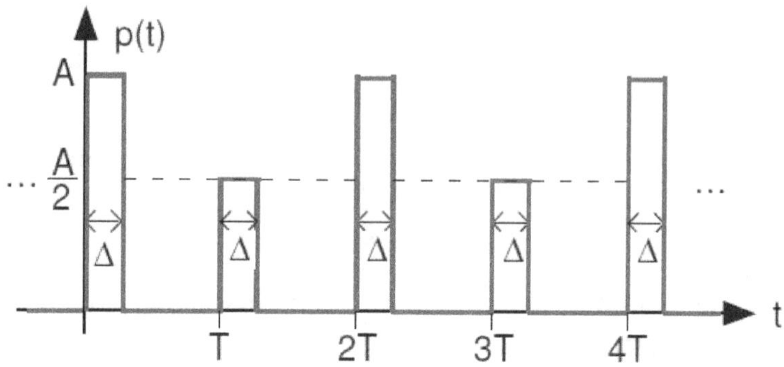


**Figure 5.28**

- Find the Fourier series for this signal.
- Can this signal be used to sample a bandlimited signal having highest frequency $W = \frac{1}{2T}$?

## Problem 5.7: Simple D/A Converter

Commercial digital-to-analog converters don't work this way, but a simple circuit illustrates how they work. Let's assume we have a B-bit converter. Thus, we want to convert numbers having a B-bit representation into a voltage proportional to that number. The first step taken by our simple converter is to represent the number by a sequence of B pulses occurring at multiples of a time interval T. The presence of a pulse indicates a "1" in the corresponding bit position, and pulse absence means a "0" occurred. For a 4-bit converter, the number 13 has the binary representation 1101 (1310=1×23+1×22+0×21+1×20) and would be represented by the depicted pulse sequence. Note that the pulse sequence is "backwards" from the binary representation. We'll see why that is.

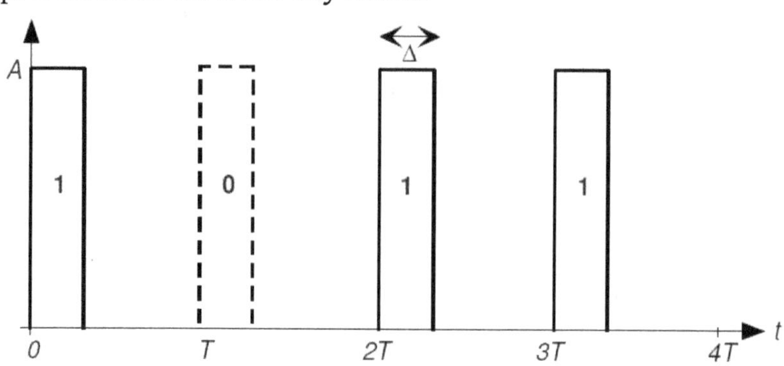


**Figure 5.29**

This signal (Figure 5.29) serves as the input to a first-order RC lowpass filter. We want to design the filter and the parameters Δ and T so that the output voltage at time 4T (for a 4-bit converter) is proportional to the number. This combination of pulse creation and filtering constitutes our simple D/A converter. The requirements are

- The voltage at time t = 4T should diminish by a factor of 2 the further the pulse occurs from this time. In other words, the voltage due to a pulse at 3T should be twice that of a pulse produced at 2T, which in turn is twice that of a pulse at T, etc.
- The 4-bit D/A converter must support a 10 kHz sampling rate.

Show the circuit that works. How do the converter's parameters change with sampling rate and number of bits in the converter?

## Problem 5.8: Discrete-Time Fourier Transforms

Find the Fourier transforms of the following sequences, where s(n) is some sequence having Fourier transform $S\left(e^{j2\pi f}\right)$.

- (-1)ns(n)
- s(n)cos(2πf0n)
- $x(n)=\begin{cases} s\left(\frac{n}{2}\right) & \text{if } n\ (\text{even}) \\ 0 & \text{if } n\ (\text{odd}) \end{cases}$
- ns(n)

## Problem 5.9: Spectra of Finite-Duration Signals

Find the indicated spectra for the following signals.

- The discrete-time Fourier transform of $s(n)=\begin{cases} \cos^2\left(\frac{\pi}{4}n\right) & \text{if } n=\{-1,0,1\} \\ 0 & \text{if otherwise} \end{cases}$
- The discrete-time Fourier transform of $s(n)=\begin{cases} n & \text{if } n=\{-2,-1,0,1,2\} \\ 0 & \text{if otherwise} \end{cases}$
- The discrete-time Fourier transform of $s(n)=\begin{cases} \sin\left(\frac{\pi}{4}n\right) & \text{if } n=\{0,\ldots,7\} \\ 0 & \text{if otherwise} \end{cases}$
- The length-8 DFT of the previous signal.

## Problem 5.10: Just Whistlin'

Sammy loves to whistle and decides to record and analyze his whistling in lab. He is a very good whistler; his whistle is a pure sinusoid that can be described by sa(t)=sin(4000t). To analyze the spectrum, he samples his recorded whistle with a sampling interval of TS=2.5×10−4 to obtain s(n)=sa(nTS). Sammy (wisely) decides to analyze a few samples at a time, so he grabs 30 consecutive, but arbitrarily chosen, samples. He calls this sequence x(n) and realizes he can write it as

$$x(n)=\sin\left(4000nT_S+\theta\right),\ \ n=\{0,\ldots,29\}$$

- Did Sammy under- or over-sample his whistle?
- What is the discrete-time Fourier transform of x(n) and how does it depend on θ?
- How does the 32-point DFT of x(n) depend on θ?

## Problem 5.11: Discrete-Time Filtering

We can find the input-output relation for a discrete-time filter much more easily than for analog filters. The key idea is that a sequence can be written as a weighted linear combination of unit samples.

- Show that $x(n) = \sum_i x(i)\,\delta(n-i)$ where δ(n) is the unit-sample.

$$\delta(n) = \begin{cases} 1 & \text{if } n = 0 \\ 0 & \text{otherwise} \end{cases}$$

- If h(n) denotes the unit-sample response—the output of a discrete-time linear, shift-invariant filter to a unit-sample input—find an expression for the output.
- In particular, assume our filter is FIR, with the unit-sample response having duration q+1. If the input has duration N, what is the duration of the filter's output to this signal?
- Let the filter be a boxcar averager: $h(n) = \frac{1}{q+1}$ for n={0,...,q} and zero otherwise. Let the input be a pulse of unit height and duration N. Find the filter's output when $N = \frac{q+1}{2}$, $q$ an odd integer.

## Problem 5.12: A Digital Filter

A digital filter has the depicted unit-sample reponse.

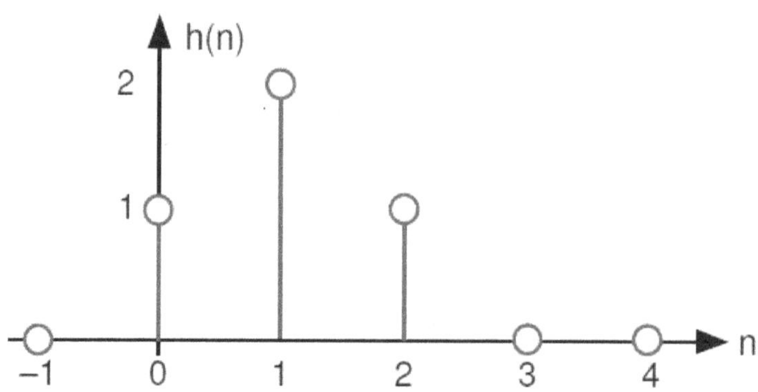


**Figure 5.30**

- What is the difference equation that defines this filter's input-output relationship?

- What is this filter's transfer function?
- What is the filter's output when the input is $\sin\left(\frac{\pi n}{4}\right)$?

## Problem 5.13: A Special Discrete-Time Filter

Consider a FIR filter governed by the difference equation

$$y(n) = \frac{1}{3}x(n+2) + \frac{2}{3}x(n+1) + x(n) + \frac{2}{3}x(n-1) + \frac{1}{3}x(n-2)$$

- Find this filter's unit-sample response.
- Find this filter's transfer function. Characterize this transfer function (i.e., what classic filter category does it fall into).
- Suppose we take a sequence and stretch it out by a factor of three.

$$x(n) = \begin{cases} s\left(\frac{n}{3}\right) & \text{if } n = 3m \quad , \quad m = \{\ldots, -1, 0, 1, \ldots\} \\ 0 & \text{otherwise} \end{cases}$$

Sketch the sequence x(n) for some example s(n). What is the filter's output to this input? In particular, what is the output at the indices where the input x(n) is intentionally zero? Now how would you characterize this system?

## Problem 5.14: Simulating the Real World

Much of physics is governed by differntial equations, and we want to use signal processing methods to simulate physical problems. The idea is to replace the derivative with a discrete-time approximation and solve the resulting differential equation. For example, suppose we have the differential equation

$$\frac{dy(t)}{dt} + ay(t) = x(t)$$

and we approximate the derivative by

$$\frac{d}{dt}y(t)\mid_{t=nT} \simeq \frac{y(nT) - y((n-1)T)}{T}$$

where T essentially amounts to a sampling interval.

- What is the difference equation that must be solved to approximate the differential equation?
- When x(t) = u(t), the unit step, what will be the simulated output?
- Assuming x(t) is a sinusoid, how should the sampling interval T be chosen so that the approximation works well?

## Problem 5.15: Derivatives

The derivative of a sequence makes little sense, but still, we can approximate it. The digital filter described by the difference equation

$$y(n) = x(n) - x(n-1)$$

resembles the derivative formula. We want to explore how well it works.

- What is this filter's transfer function?
- What is the filter's output to the depicted triangle input (Figure 5.31)?

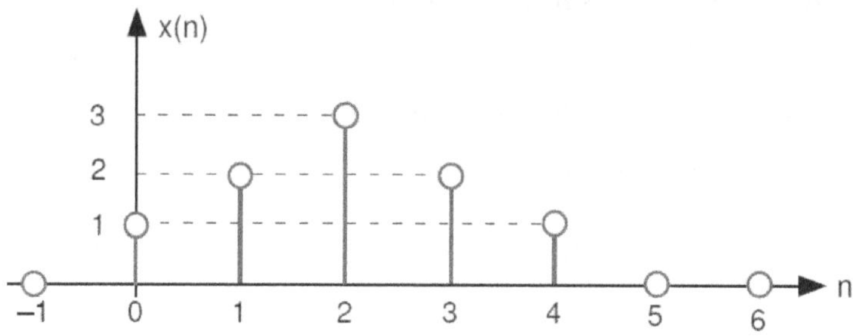


**Figure 5.31**

- **Suppose the signal x(n) is a sampled analog signal:** $x(n)=x(nT_S)$. Under what conditions will the filter act like a differentiator? In other words, when will y(n) be proportional to $\frac{d}{dt}x(t)|_{t=nT_s}$?

## Problem 5.16: The DFT

Let's explore the DFT and its properties.

- What is the length-K DFT of length-N boxcar sequence, where $N < K$?
- Consider the special case where $K = 4$. Find the inverse DFT of the product of the DFTs of two length-3 boxcars.
- If we could use DFTs to perform linear filtering, it should be true that the product of the input's DFT and the unit-sample response's DFT equals the output's DFT. So that you can use what you just calculated, let the input be a boxcar signal and the unit-sample response also be a boxcar. The result of part (b) would then be the filter's output if we could implement the filter with length-4 DFTs. Does the actual output of the boxcar-filter equal the result found in the previous part?
- What would you need to change so that the product of the DFTs of the input and unit-sample response in this case equaled the DFT of the filtered output?

## Problem 5.17: DSP Tricks

Sammy is faced with computing lots of discrete Fourier transforms. He will, of course, use the FFT algorithm, but he is behind schedule and needs to get his results as quickly as possible. He gets the idea of computing two transforms at one time by computing the transform of s(n)=s1(n)+js2(n), where s1(n) and s2(n) are two real-valued signals of which he needs to compute the spectra. The issue is whether he can retrieve the individual DFTs from the result or not.

- What will be the DFT S(k) of this complex-valued signal in terms of S1(k) and S2(k), the DFTs of the original signals?
- Sammy's friend, an Aggie who knows some signal processing, says that retrieving the wanted DFTs is easy: "Just find the real and imaginary parts of S(k) ." Show that this approach is too simplistic.
- While his friend's idea is not correct, it does give him an idea. What approach will work? Hint: Use the symmetry properties of the DFT.
- How does the number of computations change with this approach? Will Sammy's idea ultimately lead to a faster computation of the required DFTs?

## Problem 5.18: Discrete Cosine Transform (DCT)

The discrete cosine transform of a length-N sequence is defined to be

$$S_c(k) = \sum_{n=0}^{N-1} s(n) \cos\left(\frac{2\pi nk}{2N}\right)$$

Note that the number of frequency terms is 2N−1:k={0,...,2N−1}.

- Find the inverse DCT.
- Does a Parseval's Theorem hold for the DCT?
- You choose to transmit information about the signal s(n) according to the DCT coefficients. You could only send one, which one would you send?

## Problem 5.19: A Digital Filter

A digital filter is described by the following difference equation:

$$y(n) = ay(n-1) + ax(n) - x(n-1),\ a = \frac{1}{\sqrt{2}}$$

- What is this filter's unit sample response?
- What is this filter's transfer function?
- What is this filter's output when the input is $\sin\left(\frac{\pi n}{4}\right)$?

## Problem 5.20: Another Digital Filter

A digital filter is determined by the following difference equation.

$$y(n) = y(n-1) + x(n) - x(n-4)$$

- Find this filter's unit sample response.
- What is the filter's transfer function? How would you characterize this filter (lowpass, highpass, special purpose, ...)?
- Find the filter's output when the input is the sinusoid $\sin\left(\frac{\pi n}{2}\right)$.
- In another case, the input sequence is zero for n < 0, then becomes nonzero. Sammy measures the output to be y(n)=δ(n)+δ(n-1).Can his measurement be correct? In other words, is there an input that can yield this output? If so, find the input x(n) that gives rise to this output. If not, why not?

## Problem 5.21: Yet Another Digital Filter

A filter has an input-output relationship given by the difference equation

$$y(n) = \frac{1}{4}x(n) + \frac{1}{2}x(n-1) + \frac{1}{4}x(n-2)$$

- What is the filter's transfer function? How would you characterize it?
- What is the filter's output when the input equals $\cos\left(\frac{\pi n}{2}\right)$?
- What is the filter's output when the input is the depicted discrete-time square wave (Figure 5.32)?

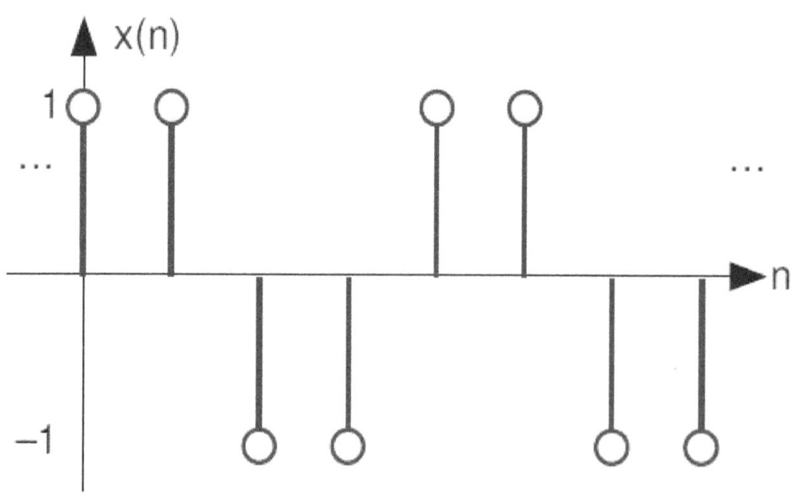


**Figure 5.32**

## Problem 5.22: A Digital Filter in the Frequency Domain

We have a filter with the transfer function operating on the input signal x(n)=δ(n)-δ(n-2) that yields the output y(n).

$$H\left(e^{j2\pi f}\right) = e^{-(j2\pi f)}\cos\left(2\pi f\right)$$

- What is the filter's unit-sample response?
- What is the discrete-Fourier transform of the output?
- What is the time-domain expression for the output?

## Problem 5.23: Digital Filters

A discrete-time system is governed by the difference equation

$$y\left(n\right) = y\left(n-1\right) + \frac{x\left(n\right) + x\left(n-1\right)}{2}$$

- Find the transfer function for this system.
- What is this system's output when the input is $\sin\left(\frac{\pi n}{2}\right)$?
- If the output is observed to be y(n)=δ(n)+δ(n-1), then what is the input?

## Problem 5.24: Digital Filtering

A digital filter has an input-output relationship expressed by the difference equation

$$y\left(n\right) = \frac{x\left(n\right) + x\left(n-1\right) + x\left(n-2\right) + x\left(n-3\right)}{4}$$

- Plot the magnitude and phase of this filter's transfer function.
- What is this filter's output when $x\left(n\right) = \cos\left(\frac{\pi n}{2}\right) + 2\sin\left(\frac{2\pi n}{3}\right)$?

## Problem 5.25: Detective Work

The signal x(n)=δ(n)-δ(n-1)

- Find the length-8 DFT (discrete Fourier transform) of this signal.
- You are told that when x(n) served as the input to a linear FIR (finite impulse response) filter, the output was y(n)=δ(n)-δ(n-1)+2δ(n-2).Is this statement true? If so, indicate why and find the system's unit sample response; if not, show why not.

## Problem 5.26

A discrete-time, shift invariant, linear system produces an output y(n)={1,-1,0,0,...} when its input x(n) equals a unit sample.

- Find the difference equation governing the system.
- Find the output when x(n)=cos(2πf0n)
- How would you describe this system's function?

## Problem 5.27: Time Reversal has Uses

A discrete-time system has transfer function H(ej2πf). A signal x(n) is passed through this system to yield the signal w(n). The time-reversed signal w(-n) is then passed through the system to yield the time-reversed output y(-n). What is the transfer function between x(n) and y(n)?

## Problem 5.28: Removing "Hum"

The slang word "hum" represents power line waveforms that creep into signals because of poor circuit construction. Usually, the 60 Hz signal (and its harmonics) are added to the desired signal. What we seek are filters that can remove hum. In this problem, the signal and the accompanying hum have been sampled; we want to design a digital filter for hum removal.

- Find filter coefficients for the length-3 FIR filter that can remove a sinusoid having digital frequency f0 from its input.
- Assuming the sampling rate is fs to what analog frequency does f0 correspond?
- A more general approach is to design a filter having a frequency response magnitude proportional to the absolute value of a cosine: |H(ej2πf)|∝|cos(πfN)|.In this way, not only can the fundamental but also its first few harmonics be removed. Select the parameter N and the sampling rate so that the frequencies at which the cosine equals zero correspond to 60 Hz and its odd harmonics through the fifth.
- Find the difference equation that defines this filter.

## Problem 5.29: Digital AM Receiver

Thinking that digital implementations are always better, our clever engineer wants to design a digital AM receiver. The receiver would bandpass the received signal, pass the result through an A/D converter, perform all the demodulation with digital signal processing systems, and end with a D/A converter to produce the analog message signal. Assume in this problem that the carrier frequency is always a large even multiple of the message signal›s bandwidth W.

- What is the smallest sampling rate that would be needed?
- Show the block diagram of the least complex digital AM receiver.
- Assuming the channel adds white noise and that a b-bit A/D converter is used, what is the output's signal-to-noise ratio?

## Problem 5.30: DFTs

A problem on Samantha's homework asks for the 8-point DFT of the discrete-time signal $\delta(n-1)+\delta(n-7)$

- What answer should Samantha obtain?
- As a check, her group partner Sammy says that he computed the inverse DFT of her answer and got $\delta(n+1)+\delta(n-1)$ Does Sammy's result mean that Samantha's answer is wrong?
- The homework problem says to lowpass-filter the sequence by multiplying its DFT by

$$H(k)=\begin{cases}1 & \text{if } k=\{0,1,7\}\\ 0 & \text{otherwise}\end{cases}$$

and then computing the inverse DFT. Will this filtering algorithm work? If so, find the filtered output; if not, why not?

## Problem 5.31: Stock Market Data Processing

Because a trading week lasts five days, stock markets frequently compute running averages each day over the previous five trading days to smooth price fluctuations. The technical stock analyst at the Buy-Lo--Sell-Hi brokerage firm has heard that FFT filtering techniques work better than any others (in terms of producing more accurate averages).

- What is the difference equation governing the five-day averager for daily stock prices?
- Design an efficient FFT-based filtering algorithm for the broker. How much data should be processed at once to produce an efficient algorithm? What length transform should be used?
- Is the analyst's information correct that FFT techniques produce more accurate averages than any others? Why or why not?

## Problem 5.32: Echoes

Echoes not only occur in canyons, but also in auditoriums and telephone circuits. In one situation where the echoed signal has been sampled, the input signal $x(n)$ emerges as $x(n)+a1x(n-n1)+a2x(n-n2)$.

- Find the difference equation of the system that models the production of echoes.
- To simulate this echo system, ELEC 241 students are asked to write the most efficient (quickest) program that has the same input-output relationship.

Suppose the duration of x(n) is 1,000 and that $a_1 = \frac{1}{2}$, $n_1 = 10$, $a_2 = \frac{1}{5}$, and n2=25. Half the class votes to just program the difference equation while the other half votes to program a frequency domain approach that exploits the speed of the FFT. Because of the undecided vote, you must break the tie. Which approach is more efficient and why?

- Find the transfer function and difference equation of the system that suppresses the echoes. In other words, with the echoed signal as the input, what system's output is the signal x(n)?

## Problem 5.33: Digital Filtering of Analog Signals

RU Electronics wants to develop a filter that would be used in analog applications, but that is implemented digitally. The filter is to operate on signals that have a 10 kHz bandwidth, and will serve as a lowpass filter.

- What is the block diagram for your filter implementation? Explicitly denote which components are analog, which are digital (a computer performs the task), and which interface between analog and digital worlds.
- What sampling rate must be used and how many bits must be used in the A/D converter for the acquired signal's signal-to-noise ratio to be at least 60 dB? For this calculation, assume the signal is a sinusoid.
- If the filter is a length-128 FIR filter (the duration of the filter's unit-sample response equals 128), should it be implemented in the time or frequency domain?
- Assuming H(ej2πf) is the transfer function of the digital filter, what is the transfer function of your system?

## Problem 5.34: Signal Compression

Because of the slowness of the Internet, lossy signal compression becomes important if you want signals to be received quickly. An enterprising 241 student has proposed a scheme based on frequency-domain processing. First of all, he would section the signal into length-N blocks, and compute its N-point DFT. He then would discard (zero the spectrum) at half of the frequencies, quantize them to b-bits, and send these over the network. The receiver would assemble the transmitted spectrum and compute the inverse DFT, thus reconstituting an N-point block.

- At what frequencies should the spectrum be zeroed to minimize the error in this lossy compression scheme?
- The nominal way to represent a signal digitally is to use simple b-bit quantization of the time-domain waveform. How long should a section be in the proposed scheme so that the required number of bits/sample is smaller than that nominally required?

- Assuming that effective compression can be achieved, would the proposed scheme yield satisfactory results?

## Solutions to Exercises in Chapter 5

### Solution to Exercise 5.2.1

For b-bit signed integers, the largest number is 2b-1−1. For b=32, we have 2,147,483,647 and for b = 64, we have 9,223,372,036,854,775,807 or about 9:2 x 1018.

### Solution to Exercise 5.2.2

In floating point, the number of bits in the exponent determines the largest and smallest representable numbers. For 32-bit floating point, the largest (smallest) numbers are 2±(127) = 1:7 x 1038 (5:9 _ 10−39). For 64-bit floating point, the largest number is about 109863.

### Solution to Exercise 5.2.3

25 = 110112 and 7 = 1112. We find that 110012 + 1112 = 1000002 = 32.

### Solution to Exercise 5.3.1

The only effect of pulse duration is to unequally weight the spectral repetitions. Because we are only concerned with the repetition centered about the origin, the pulse duration has no significant effect on recovering a signal from its samples.

### Solution to Exercise 5.3.2

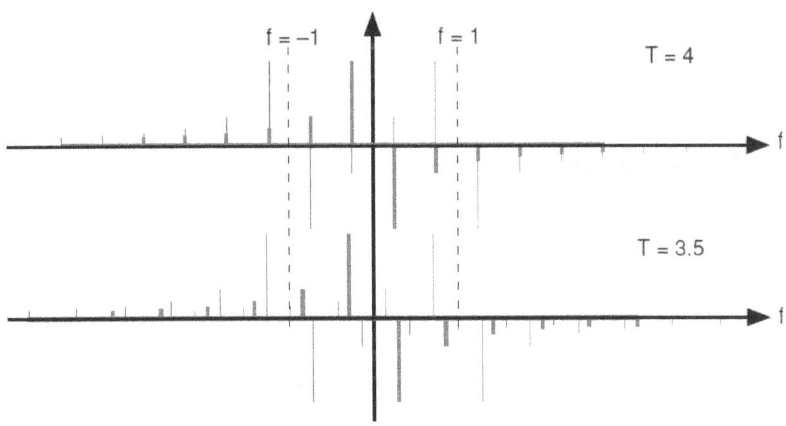


**Figure 5.33**

The square wave's spectrum is shown by the bolder set of lines centered about the origin. The dashed lines correspond to the frequencies about which the spectral repetitions (due to sampling with Ts = 1) occur. As the square wave's period decreases, the negative frequency lines move to the left and the positive frequency ones to the right.

### Solution to Exercise 5.3.3

The simplest bandlimited signal is the sine wave. At the Nyquist frequency, exactly two samples/period would occur. Reducing the sampling rate would result in fewer samples/period, and these samples would appear to have arisen from a lower frequency sinusoid.

### Solution to Exercise 5.4.1

The plotted temperatures were quantized to the nearest degree. Thus, the high temperature's amplitude was quantized as a form of A/D conversion.

### Solution to Exercise 5.4.2

The signal-to-noise ratio does not depend on the signal amplitude. With an A/D range of [–A, A]. the quantization interval $\Delta = \frac{2A}{2^B}$ and the signal's rms value (again assuming it is a sinusoid) is $\frac{A}{\sqrt{2}}$.

### Solution to Exercise 5.4.3

Solving 2–B = :001 results in B = 10 bits.

### Solution to Exercise 5.4.4

A 16-bit A/D converter yields a SNR of 6 x 16 + 10log1:5 = 97:8 dB.

### Solution to Exercise 5.6.1

$$\begin{aligned} S\left(e^{j2\pi(f+1)}\right) &= \sum_{n=-\infty}^{\infty} s(n)\, e^{-(j2\pi(f+1)n)} \\ &= \sum_{n=-\infty}^{\infty} e^{-(j2\pi n)} s(n)\, e^{-(j2\pi f n)} \\ &= \sum_{n=-\infty}^{\infty} s(n)\, e^{-(j2\pi f n)} \\ &= S\left(e^{j2\pi f}\right) \end{aligned} \quad (5.56)$$

### Solution to Exercise 5.6.2

$$\alpha \sum_{n=n_0}^{N+n_0-1} \alpha^n - \sum_{n=n_0}^{N+n_0-1} \alpha^n = \alpha^{N+n_0} - \alpha^{n_0}$$

which, after manipulation, yields the geometric sum formula.

### Solution to Exercise 5.6.3

If the sampling frequency exceeds the Nyquist frequency, the spectrum of the samples equals the analog spectrum, but over the normalized analog frequency fT. Thus, the energy in the sampled signal equals the original signal's energy multiplied by T.

### Solution to Exercise 5.7.1

This situation amounts to aliasing in the time-domain.

### Solution to Exercise 5.8.1

When the signal is real-valued, we may only need half the spectral values, but the complexity remains unchanged. If the data are complex-valued, which demands retaining all frequency values, the complexity is again the same. When only K frequencies are needed, the complexity is O (KN).

### Solution to Exercise 5.9.1

If a DFT required 1ms to compute, and signal having ten times the duration would require 100ms to compute. Using the FFT, a 1ms computing time would increase by a factor of about 10log210 = 33, a factor of 3 less than the DFT would have needed.

### Solution to Exercise 5.9.2

The upper panel has not used the FFT algorithm to compute the length-4 DFTs while the lower one has. The ordering is determined by the algorithm.

### Solution to Exercise 5.9.3

The transform can have any greater than or equal to the actual duration of the signal. We simply "pad" the signal with zero-valued samples until a computationally advantageous signal length results. Recall that the FFT is an algorithm to compute the DFT (Section 5.7). Extending the length of the signal this way merely means we are sampling the frequency axis more finely than required. To use the Cooley-Tukey algorithm, the length of the resulting zero-padded signal can be 512, 1024, etc. samples long.

### Solution to Exercise 5.10.1

Number of samples equals 1:2 x 11025 = 13230. The datarate is 11025 x 16 = 176:4 kbps. The storage required would be 26460 bytes.

### Solution to Exercise 5.10.2

The oscillations are due to the boxcar window's Fourier transform, which equals the sinc function.0

### Solution to Exercise 5.10.3

These numbers are powers-of-two, and the FFT algorithm can be exploited with these lengths. To compute a longer transform than the input signal's duration, we simply zero-pad the signal.

### Solution to Exercise 5.11.1

In discrete-time signal processing, an amplifier amounts to a multiplication, a very easy operation to perform.

### Solution to Exercise 5.12.1

The indices can be negative, and this condition is not allowed in MATLAB. To fix it, we must start the signals later in the array.

### Solution to Exercise 5.12.2

Such terms would require the system to know what future input or output values would be before the current value was computed. Thus, such terms can cause difficulties.

### Solution to Exercise 5.13.1

It now acts like a bandpass filter with a center frequency of f0 and a bandwidth equal to twice of the original lowpass filter.

### Solution to Exercise 5.14.1

The DTFT of the unit sample equals a constant (equaling 1). Thus, the Fourier transform of the output equals the transfer function.

### Solution to Exercise 5.14.2

In sampling a discrete-time signal's Fourier transform L times equally over [0; 2π) to form the DFT, the corresponding signal equals the periodic repetition of the original signal.

$$S(k) \leftrightarrow \sum_{i=-\infty}^{\infty} s(n-iL) \tag{5.57}$$

To avoid aliasing (in the time domain), the transform length must equal or exceed the signal's duration.

### Solution to Exercise 5.14.3

The difference equation for an FIR filter has the form

$$y(n) = \sum_{m=0}^{q} b_m x(n-m) \tag{5.58}$$

The unit-sample response equals

$$h(n) = \sum_{m=0}^{q} b_m \delta(n-m) \tag{5.59}$$

which corresponds to the representation described in a problem (Example 5.6) of a length- q boxcar filter.

### Solution to Exercise 5.14.4

The unit-sample response's duration is q + 1 and the signal's Nx. Thus the statement is correct.

### Solution to Exercise 5.15.1

Let N denote the input's total duration. The time-domain implementation requires a total of N(2q+1) computations, or 2q + 1 computations per input value. In the frequency domain, we split the input into $\frac{N}{N_x}$ sections, each of which requires $\log_2(N_x+q)+6+\frac{q}{N_x+q}$ per input in the section. Because we divide again by Nx to find the number of computations per input value in the entire input, this quantity decreases as Nx increases. For the time-domain implementation, it stays constant.

### Solution to Exercise 5.15.2

The delay is not computational delay here – the plot shows the first output value is aligned with the filter's first input – although in real systems this is an important consideration. Rather, the delay is due to the filter's phase shift: A phase-shifted sinusoid is equivalent to a time-delayed one: $\cos(2\pi fn-\phi) = \cos\left(2\pi f\left(n-\frac{\phi}{2\pi f}\right)\right)$. All filters have phase shifts. This delay could be removed if the filter introduced no phase shift. Such filters do not exist in analog form, but digital ones can be programmed, but not in real time. Doing so would require the output to emerge before the input arrives!

### Solution to Exercise 5.16.1

We have p+q +1 multiplications and p+q –1 additions. Thus, the total number of arithmetic operations equals 2 (p + q).

# 6 Chapter

# BASICS OF INFORMATION COMMUNICATION

## 6.1 INTRODUCTION TO INFORMATION COMMUNICATION

From the perspective of a communications engineer, signals serve as a means of conveyinginformation.Systemshavetheabilitytomodifysignals,whichconsequently impacts the information content. Data is presented in organized formats, which can be either physical or electronic. Speech is an analog signal, while computer files are composed of discrete-time signals in the form of byte sequences. However, it is important to note that the index sequences the byte location, not the time sample. Communication systems aim to transmit information without altering it, either through point-to-point communication (from one place to another), broadcast communication (from one place to many others), or multi-party communication (such as a telephone conference call or a chat room). Communication systems can be categorized as either analog, such as radio, or digital, such as computer networks. This chapter presents a unified theory that serves as the foundation for understanding the functioning of such systems. We provide a description and analysis of several systems, including both older systems such as AM radio and newer systems such as computer networks. The debate on the superiority between analog and digital communication has been resolved because to Claude Shannon's seminal research on information theory in 1948, the advancement of affordable and powerful computers, and the establishment of high-capacity communication networks. The solution lies in implementing a digital communication strategy. Typically, it is advisable to transform all signals containing information into signals that are discrete in time and quantized in amplitude. Essentially, digital signals, such as computer files (which are a specific type of symbolic signals), are in the correct format. The Sampling Theorem enables us to turn analog signals into digital signals. Shannon demonstrated that a well-designed system can successfully transmit digital information without any errors, even when the communication channel introduces noise into all transmissions. This surprising outcome has no equivalent in analog systems; AM radio will continue to be characterized by noise. The integration of these theoretical and engineering findings on communications systems has had significant implications in other domains. The audio compact disc (CD) and the digital videodisk (DVD) are today regarded as digital communication systems, incorporating communication design principles at every stage.

Go back to the fundamental model of communication (Figure 1.3: Fundamental model of the communication). Communications design begins with two fundamental considerations.

- What is the nature of the information source, and to what extent can the receiver tolerate errors in the received information?
- What are the channel's characteristics and how do they affect the transmitted signal?

In short, what are we going to send and how are we going to send it? Interestingly, digital as well as analog transmission are accomplished using analog signals, like voltages in Ethernet (an example of wireline communications) and electromagnetic radiation (wireless) in cellular telephone.

## 6.2 DIFFERENT TYPES OF COMMUNICATION CHANNELS

Electrical communications channels can be categorized as either wired or wireless channels. Wireline channels establish a physical connection between a broadcaster and receiver using a "wire," which can be a twisted pair, coaxial cable, or optic fiber. Therefore, wireline lines offer greater privacy and are significantly less susceptible to interference. Wireline channels consist of a direct link between a transmitter and a receiver, similar to a telephone's point-to-point connection. Eavesdropping on a conversation necessitates the tapping of a wire and the measurement of its voltage. Wireline channels can operate in broadcast modes, where multiple receivers are coupled to one or more transmitters. An illustrative instance of this scenario is cable television. Computer networks can be classified as either point-to-point or broadcast networks. Wireless channels have a higher level of accessibility, as the antenna of a transmitter emits a signal that can be received by any antenna that is sufficiently close. Unlike wireline channels, where the receiver only receives the signal from the transmitter, the receiver's antenna is sensitive to electromagnetic radiation from any source. This characteristic has two aspects: The smiley face symbolizes the ability of a receiver to accept transmissions from various sources, enabling receiver electronics to choose desired signals and ignore others. This allows for portable transmission and reception. On the other hand, the frowny face indicates that interference and noise are more common in wireless situations compared to wired connections. The presence of interference in a wireless communication channel reduces its flexibility.

**Note:** You will hear the term tetherless networking applied to completely wireless computer networks.

Maxwell's equations neatly summarize the physics of all electromagnetic phenomena, including circuits, radio, and optic fiber transmission.

$$\nabla \times E = -\frac{\partial (\mu H)}{\partial t} \tag{6.1}$$

$$div (\epsilon E) = \rho$$

$$\nabla \times H = \sigma E + \frac{\partial (\epsilon E)}{\partial t}$$

$$div (\mu H) = 0$$

where E is the electric field, H the magnetic field, ε dielectric permittivity, μ magnetic permeability, σ electrical conductivity, and ρ is the charge density. Kirchoff›s Laws represent special cases of these equations for circuits. We are not going to solve Maxwell›s equations here; do bear in mind that a fundamental understanding of communications channels ultimately depends on fluency with Maxwell›s equations. Perhaps the most important aspect of them is that they are linear with respect to the electrical and magnetic fields. Thus, the fields (and therefore the voltages and currents) resulting from two or more sources will add.

Note: Nonlinear electromagnetic media do exist. The equations as written here are simpler versions that apply to free-space propagation and conduction in metals. Nonlinear media are becoming increasingly important in optic fiber communications, which are also governed by Maxwell's equations.

## 6.3 WIRELINE CHANNELS

Wireline channels were initially employed for electrical communications in the mid-nineteenth century, namely for the telegraph. In this context, the channel refers to one of multiple lines that connect the transmitter to the receiver. The transmitter generates a voltage that is proportional to the message signal and then applies this voltage to the wire(s). In order for current to flow, it is necessary to have a circuit that forms a closed route. For single-wire communications, the earth serves as the pathway for the current's return. The name "ground" for the reference node in circuits actually originated in single-wire telegraphs. The earth's electrical properties exhibit significant variability. Single-wire metallic channels are incapable of transmitting high-quality signals with a bandwidth exceeding a few hundred Hertz across long distances.

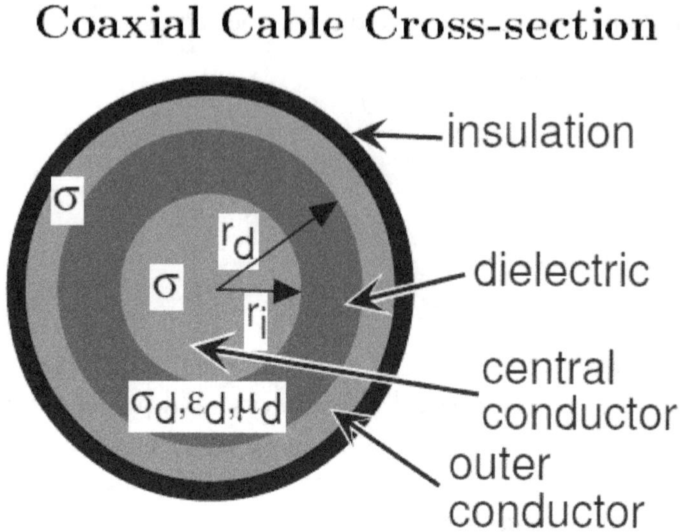


**Figure 6.1: Coaxial cable consists of one conductor wrapped around the central conductor. This type of cable supports broader bandwidth signals than twisted pair, and finds use in cable television and Ethernet.**

Consequently, most wireline channels today essentially consist of pairs of conducting wires (Figure 6.1 (Coaxial Cable Cross-Section), and the transmitter applies a message-related voltage across the pair. How these pairs of wires are physically configured greatly affects their transmission characteristics. One example is twisted pair, wherein the wires are wrapped about each other. Telephone cables are one example of a twisted pair channel. Another is coaxial cable, where a concentric conductor surrounds a central wire with a dielectric material in between. Coaxial cable, fondly called "co-ax" by engineers, is what Ethernet uses as its channel. In either case, wireline channels form a dedicated circuit between transmitter and receiver. As we shall find subsequently, several transmissions can share the circuit by amplitude modulation techniques; commercial cable TV is an example. These information-carrying circuits are designed so that interference from nearby electromagnetic sources is minimized. Thus, by the time signals arrive at the receiver, they are relatively interference- and noise-free.

Both twisted pair and co-ax are examples of transmission lines, which all have the circuit model shown in Figure 6.2 (Circuit Model for a Transmission Line) for an infinitesimally small length. This circuit model arises from solving Maxwell›s equations for the particular transmission line geometry.

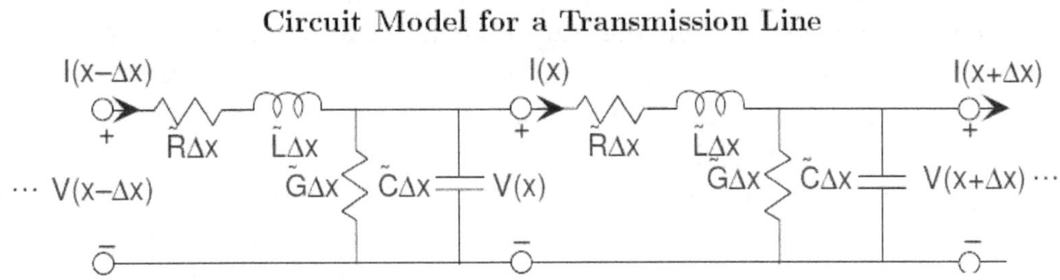


**Figure 6.2: The so-called distributed parameter model for two-wire cables has the depicted circuit model structure. Element values depend on geometry and the properties of materials used to construct the transmission line.**

The series resistance comes from the conductor used in the wires and from the conductor's geometry. The inductance and the capacitance derive from transmission line geometry, and the parallel conductance from the medium between the wire pair. Note that all the circuit elements have values expressed by the product of a constant times a length; this notation represents that element values here have per-unit-length units. For example, the series resistance $\tilde{R}$ has units of ohms/meter. For coaxial cable, the element values depend on the inner conductor's radius ri, the outer radius of the dielectric rd, the conductivity of the conductors σ, and the conductivity σd, dielectric constant εd, and magnetic permittivity μd of the dielectric as

$$\tilde{R}=\frac{1}{2\pi\delta\sigma}\left(\frac{1}{r_d}+\frac{1}{r_i}\right) \tag{6.2}$$

$$\tilde{C}=\frac{2\pi\epsilon_d}{\ln\left(\frac{r_d}{r_i}\right)}$$

$$\tilde{G}=\frac{2\left(\pi,\sigma_d\right)}{\ln\left(\frac{r_d}{r_i}\right)}$$

$$\tilde{L}=\frac{\mu_d}{2\pi}\ln\left(\frac{r_d}{r_i}\right)$$

For twisted pair, having a separation d between the conductors that have conductivity σ and common radius r and that are immersed in a medium having dielectric and magnetic properties, the element values are then

$$\tilde{R}=\frac{1}{\pi r\delta\sigma} \tag{6.3}$$

$$\tilde{C}=\frac{\pi\epsilon}{\operatorname{arccosh}\left(\frac{d}{2r}\right)}$$

$$\tilde{G}=\frac{\pi\sigma}{\operatorname{arccosh}\left(\frac{d}{2r}\right)}$$

$$\tilde{L}=\frac{\mu}{\pi}\left(\frac{\delta}{2r}+\operatorname{arccosh}\left(\frac{d}{2r}\right)\right)$$

The voltage between the two conductors and the current flowing through them will depend on distance x along the transmission line as well as time. We express this dependence as v(x,t) and i(x,t). When we place a sinusoidal source at one end of the transmission line, these voltages and currents will also be sinusoidal because the transmission line model consists of linear circuit elements. As is customary in analyzing linear circuits, we express voltages and currents as the real part of complex exponential signals, and write circuit variables as a complex amplitude—here dependent on distance—times a complex exponential: v(x,t)=R(V(x) and i(x,t)=Re (I(x)ej2πft). Using the transmission line circuit model, we find from KCL, KVL, and v-i relations the equations governing the complex amplitudes.

## KCL at Center Node

$$I\left(x\right)=I\left(x-\Delta\left(x\right)\right)-V\left(x\right)\left(\tilde{G}+j2\pi f\,\tilde{C}\right)\Delta\left(x\right) \tag{6.4}$$

V-I relation for RL series

$$V\left(x\right)-V\left(x+\Delta\left(x\right)\right)=I\left(x\right)\left(\tilde{R}+j2\pi f\,\tilde{L}\right)\Delta\left(x\right) \tag{6.5}$$

Rearranging and taking the limit Δ(x)→0 yields the so-called transmission line equations.

$$\frac{d}{dx}I\left(x\right)=-\left(\left(\tilde{G}+j2\pi f\,\tilde{C}\right)V\left(x\right)\right) \tag{6.6}$$

$$\frac{d}{dx}V\left(x\right)=-\left(\left(\tilde{R}+j2\pi f\,\tilde{L}\right)I\left(x\right)\right)$$

By consolidating these equations, we may derive a singular equation that dictates the variation of the complex amplitude of voltage or current with respect to position along the transmission line. By differentiating the second equation and substituting the first equation into the obtained result, we obtain the equation that governs the voltage.

$$\frac{d^2}{dx^2}V\left(x\right)=\left(\tilde{G}+j2\pi f\,\tilde{C}\right)\left(\tilde{R}+j2\pi f\,\tilde{L}\right)V\left(x\right) \tag{6.7}$$

This equation's solution is

$$V\left(x\right)=V_{+}e^{-(\gamma x)}+V_{-}e^{\gamma x} \tag{6.8}$$

Calculating its second derivative and comparing the result with our equation for the voltage can check this solution.

$$\begin{aligned}\frac{d^2}{dx^2}V\left(x\right) &= \gamma^2\left(V_{+}e^{-(\gamma x)}+V_{-}e^{\gamma x}\right)\\ &= \gamma^2 V\left(x\right)\end{aligned} \tag{6.9}$$

Our solution works so long as the quantity γ satisfies

$$\begin{aligned}\gamma &= \pm\left(\sqrt{\left(\widetilde{G}+j2\pi f\,\widetilde{C}\right)\left(\widetilde{R}+j2\pi f\,\widetilde{L}\right)}\right)\\ &= \pm\left(a\left(f\right)+jb\left(f\right)\right)\end{aligned} \tag{6.10}$$

Thus, γ depends on frequency, and we express it in terms of real and imaginary parts as indicated. The quantities V+ and V- are constants determined by the source and physical considerations. For example, let the spatial origin be the middle of the transmission line model in Figure 6.2. Because the circuit model contains simple circuit elements, physically possible solutions for voltage amplitude cannot increase with distance along the transmission line. Expressing γ in terms of its real and imaginary parts in our solution shows that such increases are a (mathematical) possibility. V(x)=V+e(-(a+jb))x +V-e(a+jb)x The voltage cannot increase without limit; because a(f) is always positive, we must segregate the solution for negative and positive x. The first term will increase exponentially for x<0 unless V+=0 in this region; a similar result applies to V- for x>0. These physical constraints give us a cleaner solution.

$$V\left(x\right)=\begin{cases}V_{+}e^{(-(a+jb))x} & \text{if } x>0\\ V_{-}e^{(a+jb)x} & \text{if } x<0\end{cases} \tag{6.11}$$

This solution suggests that voltages (and currents too) will decrease exponentially along a transmission line. The space constant, also known as the attenuation constant, is the distance over which the voltage decreases by a factor of $\frac{1}{e}$. It equals the reciprocal of a(f), which depends on frequency, and is expressed by manufacturers in units of dB/m.

The presence of the imaginary part of γ, b(f), also provides insight into how transmission lines work. Because the solution for x>0 is proportional to e-(ibx) , we know that the voltage's complex amplitude will vary sinusoidally in space. The complete solution for the voltage has the form

$$v\left(x,t\right)=Re\left(V_{+}e^{-(ax)}e^{j(2\pi ft-bx)}\right) \tag{6.12}$$

The complex exponential portion has the form of a propagating wave. If we could take a snapshot of the voltage (take its picture at t=t1), we would see a sinusoidally varying waveform along the transmission line. One period of this variation, known as the wavelength, equals $\lambda=\frac{2\pi}{b}$. If we were to take a second picture at some later time t=t2, we would also see a sinusoidal voltage. Because

$$2\pi f t_2 - bx = 2\pi f\left(t_1 + t_2 - t_1\right) - bx = 2\pi f t_1 - b\left(x - \frac{2\pi f}{b}\left(t_2 - t_1\right)\right)$$

the second waveform appears to be the first one, but delayed—shifted to the right—in space. Thus, the voltage appeared to move to the right with a speed equal to $\frac{2\pi f}{b}$ (assuming b > 0). We denote this propagation speed by c, and it equals

$$c = \left|\frac{2\pi f}{\mathrm{Im}\left(\sqrt{\left(\tilde{G} + j2\pi f\,\tilde{C}\right)\left(\tilde{R} + j2\pi f\,\tilde{L}\right)}\right)}\right| \tag{6.13}$$

In the high-frequency region where $j2\pi f\,\tilde{L} \gg \tilde{R}$ and $j2\pi f\,\tilde{C} \gg \tilde{G}$, the quantity under the radical simplifies to $-4\left(\pi^2, f^2, \tilde{L}, \tilde{C}\right)$, and we find the propagation speed to be

$$\underset{f\to\infty}{\text{limit}}\; c = \frac{1}{\sqrt{\tilde{L}\tilde{C}}}$$

For typical coaxial cable, this propagation speed is a fraction (one-third to two-thirds) of the speed of light.

## Exercise 6.3.1

Find the propagation speed in terms of physical parameters for both the coaxial cable and twisted pair examples.

By using the second of the transmission line equation (6.6), we can solve for the current's complex amplitude. Considering the spatial region x>0, for example, we find that

$$\frac{d}{dx}V(x) = -\left(\gamma V(x)\right) = \left(-\left(\tilde{R} + j2\pi f\,\tilde{L}\right)\right) I(x)$$

which means that the ratio of voltage and current complex amplitudes does not depend on distance.

$$\begin{aligned}\frac{V(x)}{I(x)} &= \sqrt{\frac{\tilde{R} + j2\pi f\tilde{L}}{\tilde{G} + j2\pi f\tilde{C}}} \\ &= Z_0\end{aligned} \tag{6.15}$$

The quantity Z0 is known as the transmission line's characteristic impedance. Note that when the signal frequency is sufficiently high, the characteristic impedance is real, which means the transmission line appears resistive in this high-frequency regime.

$$\underset{f\to\infty}{\text{limit}}\; Z_0 = \sqrt{\frac{\tilde{L}}{\tilde{C}}} \tag{6.16}$$

Typical values for characteristic impedance are 50 and 75 Ω.

A related transmission line is the optic fiber. Here, the electromagnetic field is light, and it propagates down a cylinder of glass. In this situation, we don't have two conductors—in fact we have none—and the energy is propagating in what corresponds to the dielectric material of the coaxial cable. Optic fiber communication has exactly the same properties as other transmission lines: Signal strength decays exponentially according to the fiber's space constant and propagates at some speed less than light would in free space. From the encompassing view of Maxwell's equations, the only difference is the electromagnetic signal's frequency. Because no electric conductors are present and the fiber is protected by an opaque "insulator," optic fiber transmission is interference-free.

### Exercise 6.3.2

From tables of physical constants, find the frequency of a sinusoid in the middle of the visible light range. Compare this frequency with that of a mid-frequency cable television signal.

To summarize, we use transmission lines for high-frequency wireline signal communication. In wireline communication, we have a direct, physical connection—a circuit—between transmitter and receiver. When we select the transmission line characteristics and the transmission frequency so that we operate in the high-frequency regime, signals are not filtered as they propagate along the transmission line: The characteristic impedance is real-valued—the transmission line's equivalent impedance is a resistor—and all the signal's components at various frequencies propagate at the same speed. Transmitted signal amplitude does decay exponentially along the transmission line. Note that in the high-frequency regime the space constant is approximately zero, which means the attenuation is quite small.

### Exercise 6.3.3

What is the limiting value of the space constant in the high frequency regime?

## 6.4 WIRELESS CHANNELS

Wireless channels utilize the prediction derived from Maxwell's equation that electromagnetic fields propagate in free space in a manner similar to light. Applying a voltage to an antenna generates an electromagnetic field that spreads in all directions. The amount of power flowing in each direction is influenced by the antenna's geometry. This field then induces electric currents in the antenna of the receiver. The shape of an antenna controls the strength of the electric field generated by a voltage at a specific frequency. Generally, the primary determinant is the proportion between the size of the antenna and the wavelength of the field. The fundamental equation that establishes the relationship between frequency and wavelength for a propagating wave is

$$\lambda f = c$$

Thus, wavelength and frequency are inversely related: High frequency corresponds to small wavelengths. For example, a 1 MHz electromagnetic field has a wavelength of 300 m. Antennas that are of a same size or distance from the ground as the wavelength are able to emit fields with the highest efficiency. Therefore, when the frequency decreases, the size of the antenna must increase. Since the majority of information transmissions are baseband signals, which means they have energy concentrated at low frequencies, they need to undergo modulation in order to be transferred via wireless channels at higher frequencies.

The decrease in signal strength as the receiver moves away from the transmitter in most antenna-based wireless systems may be understood by examining the way the radiated power changes with distance from the transmitting antenna. An antenna emits a specific quantity of power into the surrounding environment, and ideally this power spreads out equally in all directions without any loss. The total power, obtained by integrating the radiated power throughout the surface of a sphere centered at the transmitter, is constant regardless of the sphere's radius. This requirement arises due to the principle of energy conservation. Thus, if p(d) represents the power integrated with respect to direction at a distance d from the antenna, the total power will be p(d)4πd2. For this quantity to be a constant, we must have

$$p(d) \propto \frac{1}{d^2}$$

which means that the received signal amplitude AR must be proportional to the transmitter›s amplitude AT and inversely related to distance from the transmitter.

$$A_R = \frac{kA_T}{d}$$

for some value of the constant k. Hence, as the distance between the transmitter and the receiver increases, the strength of the received signal diminishes. While the attenuation in wireline channels can be regulated by physical characteristics and the selection of transmission frequency, the inverse-distance attenuation in wireless channels remains constant regardless of the frequency.

### Exercise 6.4.1

Why don't signals attenuate according to the inverse-square law in a conductor? What is the difference between the wireline and wireless cases?

The speed of propagation is governed by the dielectric constant μ0 and magnetic permeability ε0 of free space.

$$\begin{aligned} c &= \frac{1}{\sqrt{\mu_0 \epsilon_0}} \\ &= 3 \times 10^8 \mathrm{m/s} \end{aligned} \quad (6.18)$$

Known familiarly as the speed of light, it sets an upper limit on how fast signals can propagate from one place to another. Because signals travel at a finite speed, a receiver senses a transmitted signal only after a time delay inversely related to the propagation speed:

$$\Delta(t) = \frac{d}{c}$$

At the speed of light, a signal travels across the United States in 16 ms, a reasonably small time delay. If a lossless (zero space constant) coaxial cable connected the East and West coasts, this delay would be two to three times longer because of the slower propagation speed.

## 6.5 LINE-OF-SIGHT TRANSMISSION

Transmission across long distances, regardless of the type of channel, experiences issues with signal weakening. The Circuit Models module (Section 6.3) investigates the losses that occur in wireline channels. It examines how repeaters can be used to increase the distance between the transmitter and receiver, compensating for the passive losses imposed by the wireline channel. In wireless channels, there is both radiation loss (p. 231) and the possibility that one antenna may not have line of sight to another due to the curvature of the earth.

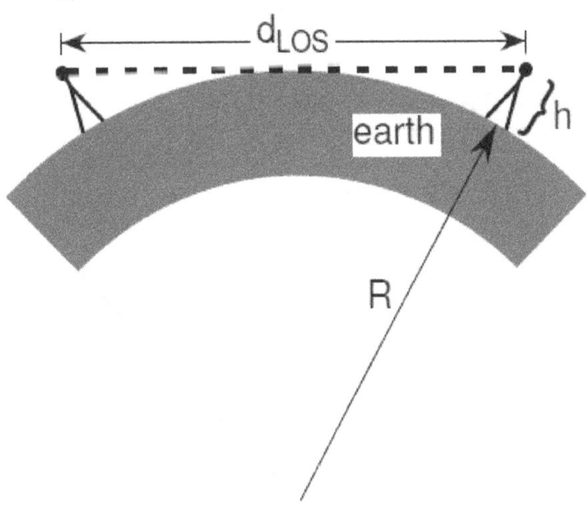


**Figure 6.3: Two antennae are shown each having the same height. Line-of-sight transmission means the transmitting and receiving antennae can "see" each other as shown. The maximum distance at which they can see each other, dLOS, occurs when the sighting line just grazes the earth's surface.**

At the usual radio frequencies, propagating electromagnetic energy does not follow the earth's surface. Line-of-sight communication has the transmitter and receiver antennas in visual contact with each other. Assuming both antennas have height

$$d_{\text{LOS}} = 2\sqrt{2hR + h^2} \simeq 2\sqrt{2Rh} \tag{6.19}$$

where R is the earth's radius (6.38×106 m).

### Exercise 6.5.1

Derive the expression of line-of-sight distance using only the Pythagorean Theorem. Generalize it to the case where the antennas have different heights (as is the case with commercial radio and cellular telephone). What is the range of cellular telephone where the handset antenna has essentially zero height?

### Exercise 6.5.2

Can you imagine a situation wherein global wireless communication is possible with only one transmitting antenna? In particular, what happens to wavelength when carrier frequency decreases?

Using a 100 m antenna would provide line-of-sight transmission over a distance of 71.4 km. Using such very tall antennas would provide wireless communication within a town or between closely spaced population centers. Consequently, networks of antennas sprinkle the countryside (each located on the highest hill possible) to provide long-distance wireless communications: Each antenna receives energy from one antenna and retransmits to another. This kind of network is known as a relay network.

## 6.6 THE IONOSPHERE AND COMMUNICATIONS

If we were limited to line-of-sight communications, long distance wireless communication, like ship-to-shore communication, would be impossible. At the turn of the century, Marconi, the inventor of wireless telegraphy, boldly tried such long distance communication without any evidence — either empirical or theoretical — that it was possible. When the experiment worked, but only at night, physicists scrambled to determine why (using Maxwell's equations, of course). It was Oliver Heaviside, a mathematical physicist with strong engineering interests, who hypothesized that an invisible electromagnetic "mirror" surrounded the earth. What he meant was that at optical frequencies (and others as it turned out), the mirror was transparent, but at the frequencies Marconi used, it reflected electromagnetic radiation back to earth. He had predicted the existence of the ionosphere, a plasma that encompasses the earth at altitudes hi between 80 and 180 km that reacts to solar radiation: It becomes transparent at Marconi's frequencies during the day, but becomes a mirror at night when solar radiation diminishes. The maximum distance along the earth's surface that can be reached by a single ionospheric reflection is $2R\arccos\left(\frac{R}{R+h_i}\right)$, which ranges between 2,010 and 3,000 km when we substitute minimum and maximum ionospheric altitudes. This distance does not span the United States or cross the Atlantic; for transatlantic communication, at least two reflections would be required. The communication delay encountered with a single reflection in this channel is $\frac{2\sqrt{2Rh_i+h_i^2}}{c}$, which ranges between 6.8 and 10 ms, again a small time interval.

## 6.7 COMMUNICATION WITH SATELLITES

Satellites are essential for worldwide wireless connectivity. In this process, terrestrial stations broadcast signals to satellites in space, which then enhance the signal and transmit it back to the Earth. Satellites will traverse the sky unless they are in geosynchronous orbits, when their orbital period around the equator precisely equals the earth's rotation period of one day. If the TV satellite were not in geosynchronous orbit, the homeowner would need to constantly readjust their antenna. According to Newton's equations, the time T required for one complete orbit of a body is determined by the distance R from the center of the Earth.

$$R = \sqrt[3]{\frac{GMT^2}{4\pi^2}} \tag{6.20}$$

where G is the gravitational constant and M the earth's mass. Calculations yield R = 42200 km, which corresponds to an altitude of 35700 km. This altitude greatly exceeds that of the ionosphere, requiring satellite transmitters to use frequencies that pass through it. Of great importance in satellite communications is the transmission delay. The time for electromagnetic fields to propagate to a geosynchronous satellite and return is 0.24 s, a significant delay.

### Exercise 6.7.1

In addition to delay, the propagation attenuation encountered in satellite communication far exceeds what occurs in ionospheric-mirror based communication. Calculate the attenuation incurred by radiation going to the satellite (one-way loss) with that encountered by Marconi (total going up and down). Note that the attenuation calculation in the ionospheric case, assuming the ionosphere acts like a perfect mirror, is not a straightforward application of the propagation loss formula (p. 231).

## 6.8 NOISE AND INTERFERENCE

As previously stated, communications can be affected to different extents by interference and noise. It is necessary to provide a more accurate definition of these quantities and explain their distinctions. Interference refers to signals that are created by humans. Telephone lines can experience power-line interference, which in the United States is typically a distorted sinusoidal signal with a frequency of 60 Hz. Adjacent-cell phone talks can interfere with cellular telephone channels that use the same signal frequency. The issue with such interference is that it occupies the identical frequency range as the targeted transmission signal and possesses a comparable structure.

## Exercise 6.8.1

Suppose interference occupied a different frequency band; how would the receiver remove it?

We use the notation i(t) to represent interference. Because interference has man-made structure, we can write an explicit expression for it that may contain some unknown aspects (how large it is, for example).

Noise signals have little structure and arise from both human and natural sources. Satellite channels are subject to deep space noise arising from electromagnetic radiation pervasive in the galaxy. Thermal noise plagues all electronic circuits that contain resistors. Thus, in receiving small amplitude signals, receiver amplifiers will most certainly add noise as they boost the signal›s amplitude. All channels are subject to noise, and we need a way of describing such signals despite the fact we can›t write a formula for the noise signal like we can for interference. The most widely used noise model is white noise. It is defined entirely by its frequency-domain characteristics.

- White noise has constant power at all frequencies.
- At each frequency, the phase of the noise spectrum is totally uncertain: It can be any value in between 0 and 2π, and its value at any frequency is unrelated to the phase at any other frequency.
- When noise signals arising from two different sources add, the resultant noise signal has a power equal to the sum of the component powers.

Because of the emphasis here on frequency-domain power, we are led to define the power spectrum. Because of Parseval's Theorem, we define the power spectrum Ps(f) of a non-noise signal s(t) to be the magnitude-squared of its Fourier transform.

$$P_s(f) \equiv (|S(f)|)^2 \quad (6.21)$$

Integrating the power spectrum over any range of frequencies equals the power the signal contains in that band. Because signals must have negative frequency components that mirror positive frequency ones, we routinely calculate the power in a spectral band as the integral over positive frequencies multiplied by two.

$$\text{Power in } [f_1, f_2] = 2\int_{f_1}^{f_2} P_s(f)\,df \quad (6.22)$$

Using the notation n(t) to represent a noise signal›s waveform, we define noise in terms of its power spectrum. For white noise, the power spectrum equals the constant $\frac{N_0}{2}$. With this definition, the power in a frequency band equals N0(f2-f1)

When we pass a signal through a linear, time-invariant system, the output's spectrum equals the product of the system›s frequency response and the input›s spectrum. Thus, the power spectrum of the system›s output is given by

$$P_y(f) = (|H(f)|)^2 P_x(f) \tag{6.23}$$

This result applies to noise signals as well. When we pass white noise through a filter, the output is also a noise signal but with power spectrum $(|H(f)|)^2 \frac{N_0}{2}$.

## 6.9 CHANNEL MODELS

Both wireline and wireless channels exhibit similar features, enabling us to employ a unified model to understand how the channel impacts transmitted signals.

- The transmitted signal is usually not filtered by the channel.
- The signal can be attenuated.
- The signal propagates through the channel at a speed equal to or less than the speed of light, which means that the channel delays the transmission.
- The channel may introduce additive interference and/or noise.

Letting $\alpha$ represent the attenuation introduced by the channel, the receiver›s input signal is related to the transmitted one by

$$r(t) = \alpha x(t - \tau) + i(t) + n(t) \tag{6.24}$$

This expression corresponds to the system model for the channel shown in Figure 6.4 below. In this book, we shall assume that the noise is white.

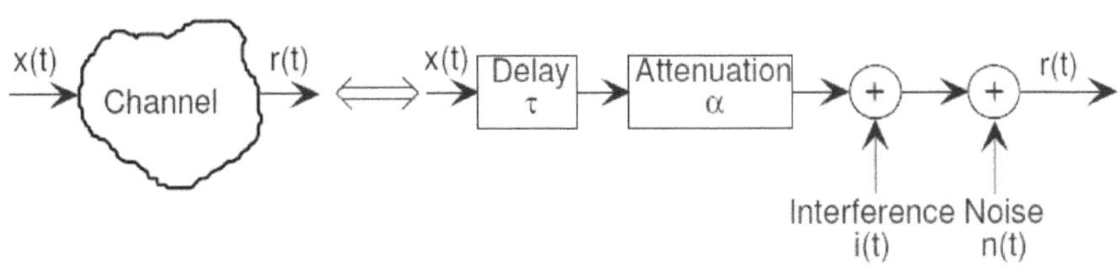


**Figure 6.4: The channel component of the fundamental model of communication (Figure 1.3: Fundamental model of communication) has the depicted form. The attenuation is due to propagation loss. Adding the interference and noise is justified by the linearity property of Maxwell's equations.**

### Exercise 6.9.1

Is this model for the channel linear?

As expected, the signal that emerges from the channel is corrupted, but does contain the transmitted signal. Communication system design begins with detailing the channel model, then developing the transmitter and receiver that best compensate for the channel's corrupting behavior. We characterize the channel's quality by the signal-to-interference ratio (SIR) and the signal-to-noise ratio (SNR). The ratios are computed according to the relative power of each within the transmitted signal's bandwidth. Assuming the signal x(t)'s spectrum spans the frequency interval [fl,fu], these ratios can be expressed in terms of power spectra.

$$\mathrm{SIR} = \frac{2\alpha^2 \int_0^\infty P_x(f)\,df}{2\int_{f_l}^{f_u} P_i(f)\,df} \tag{6.25}$$

$$\mathrm{SNR} = \frac{2\alpha^2 \int_0^\infty P_x(f)\,df}{N_0(f_u - f_l)} \tag{6.26}$$

In most cases, the interference and noise powers do not vary for a given receiver. Variations in signal-to-interference and signal-to-noise ratios arise from the attenuation because of transmitter-to-receiver distance variations.

## 6.10 BASEBAND COMMUNICATION

Analog communication systems are employed to transmit analog message signals, such as music, speech, and television. The transmission and reception of analog signals via analog methods inevitably leads to a noisy received signal, given that the channel introduces noise, which is quite likely. Baseband communication is the most basic type of analog communication.

**Point of Interest:** We use analog communication techniques for analog message signals, like music, speech, and television. Transmission and reception of analog signals using analog results in an inherently noisy received signal (assuming the channel adds noise, which it almost certainly does).

Here, the transmitted signal equals the message times a transmitter gain.

$$x(t) = Gm(t) \tag{6.27}$$

An example, which is somewhat out of date, is the wireline telephone system. You don't use baseband communication in wireless systems simply because low-frequency signals do not radiate well. The receiver in a baseband system can't do much more than filter the received signal to remove out-of-band noise (interference is small in wireline channels). Assuming the signal occupies a bandwidth of W Hz (the signal's spectrum extends from zero to W), the receiver applies a lowpass filter having the same bandwidth, as shown in Figure 6.5.

**Figure 6.5: The receiver for baseband communication systems is quite simple: a lowpass filter having the same bandwidth as the signal**

We use the signal-to-noise ratio of the receiver's output $\hat{m}(t)$ to evaluate any analog-message communication system. Assume that the channel introduces an attenuation α and white noise of spectral height $\frac{N_0}{2}$. The filter does not affect the signal component—we assume its gain is unity—but does filter the noise, removing frequency components above W Hz. In the filter's output, the received signal power equals α2G2 power (m) and the noise power N0W, which gives a signal-to-noise ratio of

$$\mathrm{SNR}_{baseband} = \frac{\alpha^2 G^2 \mathrm{power}\,(m)}{N_0 W} \tag{6.28}$$

The signal power power (m) will be proportional to the bandwidth W; thus, in baseband communication the signal-to-noise ratio varies only with transmitter gain and channel attenuation and noise level.

## 6.11 MODULATED COMMUNICATION

In order to transmit analog message signals, modulation is necessary, particularly for wireless channels such as commercial radio and television, as well as wireline systems like cable television. The spectrum of the transmitted signal is at significantly higher frequencies than the frequencies occupied by the signal.

Point of Interest: We use analog communication techniques for analog message signals, like music, speech, and television. Transmission and reception of analog signals using analog results in an inherently noisy received signal (assuming the channel adds noise, which it almost certainly does).

The key idea of modulation is to affect the amplitude, frequency or phase of what is known as the carrier sinusoid. Frequency modulation (FM) and less frequently used phase modulation (PM) are not discussed here; we focus on amplitude modulation (AM). The amplitude modulated message signal has the form

$$x(t) = A_c\,(1 + m(t))\cos(2\pi f_c t) \tag{6.29}$$

where fc is the carrier frequency and Ac the carrier amplitude. Also, the signal's amplitude is assumed to be less than one:|m(t)|<1|. From our previous exposure to amplitude modulation (see the Fourier Transform example (Example 4.5)), we know that the transmitted signal's spectrum occupies the frequency range [fc−W,fc+W], assuming the signal's bandwidth is W Hz (see the figure (Figure 6.6)). The carrier frequency is usually much larger than the signal's highest frequency: fc >> W, which means that the transmitter antenna and carrier frequency are chosen jointly during the design process.

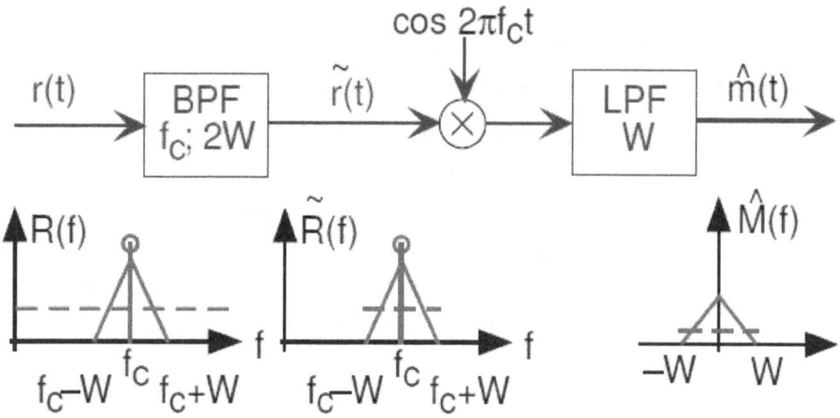


**Figure 6.6: The AM coherent receiver along with the spectra of key signals is shown for the case of a triangular-shaped signal spectrum. The dashed line indicates the white noise level. Note that the filters' characteristics — cutoff frequency and center frequency for the bandpass filter — must be match to the modulation and message parameters.**

Ignoring the attenuation and noise introduced by the channel for the moment, reception of an amplitude modulated signal is quite easy (see Problem 4.20). The so-called coherent receiver multiplies the input signal by a sinusoid and lowpass-filters the result (Figure 6.6).

$$\begin{aligned}\hat{m}(t) &= \mathrm{LPF}\left(x(t)\cos(2\pi f_c t)\right)\\ &= \mathrm{LPF}\left(A_c(1+m(t))\cos^2(2\pi f_c t)\right)\end{aligned} \tag{6.30}$$

Because of our trigonometric identities, we know that

$$\cos^2(2\pi f_c t) = \frac{1}{2}\left(1+\cos(2\pi 2 f_c t)\right) \tag{6.31}$$

At this point, the message signal is multiplied by a constant and a sinusoid at twice the carrier frequency. Multiplication by the constant term returns the message signal to baseband (where we want it to be!) while multiplication by the double-frequency term yields a very high frequency signal. The lowpass filter removes this high-frequency signal, leaving only the baseband signal. Thus, the received signal is

$$\hat{m}(t) = \frac{A_c}{2}(1 + m(t)) \tag{6.32}$$

### Exercise 6.11.1

This derivation relies solely on the time domain; derive the same result in the frequency domain. You won't need the trigonometric identity with this approach.

Because it is so easy to remove the constant term by electrical means—we insert a capacitor in series with the receiver's output—we typically ignore it and concentrate on the signal portion of the receiver's output when calculating signal-to-noise ratio.

## 6.12 SIGNAL-TO-NOISE RATIO OF AN AMPLITUDE-MODULATED SIGNAL

When we consider the much more realistic situation when we have a channel that introduces attenuation and noise, we can make use of the just-described receiver's linear nature to directly derive the receiver's output. The attenuation affects the output in the same way as the transmitted signal: It scales the output signal by the same amount. The white noise, on the other hand, should be filtered from the received signal before demodulation. We must thus insert a bandpass filter having bandwidth 2W and center frequency fc : This filter has no effect on the received signal-related component, but does remove out-of-band noise power. As shown in the triangular-shaped signal spectrum, we apply coherent receiver to this filtered signal, with the result that the demodulated output contains noise that cannot be removed: It lies in the same spectral band as the signal.

As we derive the signal-to-noise ratio in the demodulated signal, let's also calculate the signal-to-noise ratio of the bandpass filter's output $\tilde{r}(t)$.The signal component of $\tilde{r}(t)$.equals $\alpha A_c m(t)\cos(2\pi f_c t)$. This signal's Fourier transform equals

$$\frac{\alpha A_c}{2}(M(f + f_c) + M(f - f_c)) \tag{6.33}$$

making the power spectrum,

$$\frac{\alpha^2 A_c{}^2}{4}\left((|M(f + f_c)|)^2 + (|M(f - f_c)|)^2\right) \tag{6.34}$$

### Exercise 6.12.1

If you calculate the magnitude-squared of the first equation, you don't obtain the second unless you make an assumption. What is it?

Thus, the total signal-related power in $\tilde{r}(t)$ is $\frac{\alpha^2 A_c{}^2}{2}\text{power}(m)$. The noise power equals the integral of the noise power spectrum; because the power spectrum is constant over the transmission band, this integral equals the noise amplitude N0 times the filter›s bandwidth 2W. The so-called received signal-to-noise ratio — the signal-to-noise ratio after the de rigeur front-end bandpass filter and before demodulation — equals

$$\mathrm{SNR}_r = \frac{\alpha^2 {A_c}^2 \mathrm{power}\,(m)}{4N_0W} \tag{6.35}$$

The demodulated signal $\hat{m}\,(t) = \frac{\alpha A_c m(t)}{2} + n_{\mathrm{out}}\,(t)$. Clearly, the signal power equals $\frac{\alpha^2 {A_c}^2 \mathrm{power}(m)}{4}$. To determine the noise power, we must understand how the coherent demodulator affects the bandpass noise found in $\tilde{r}\,(t)$. Because we are concerned with noise, we must deal with the power spectrum since we don't have the Fourier transform available to us. Letting Pñ(f) denote the power spectrum of $\tilde{r}\,(t)$'s noise component, the power spectrum after multiplication by the carrier has the form

$$\frac{P(f+f_c)+P(f-f_c)}{4} \tag{6.36}$$

The delay and advance in frequency indicated here results in two spectral noise bands falling in the low-frequency region of lowpass filter's passband. Thus, the total noise power in this filter's output equals $2 \cdot \frac{N_0}{2} \cdot W \cdot 2 \cdot \frac{1}{4} = \frac{N_0 W}{2}$. The signal-to-noise ratio of the receiver's output thus equals

$$\begin{aligned} \mathrm{SNR}_{\hat{m}} &= \frac{\alpha^2 {A_c}^2 \mathrm{power}(m)}{2N_0W} \\ &= 2\mathrm{SNR}_r \end{aligned} \tag{6.37}$$

Let's break down the components of this signal-to-noise ratio to better appreciate how the channel and the transmitter parameters affect communications performance. Better performance, as measured by the SNR, occurs as it increases.

- More transmitter power — increasing Ac — increases the signal-to-noise ratio proportionally.
- The carrier frequency fc has no effect on SNR, but we have assumed that fc>>W.
- The signal bandwidth W enters the signal-to-noise expression in two places: implicitly through the signal power and explicitly in the expression›s denominator. If the signal spectrum had a constant amplitude as we increased the bandwidth, signal power would increase proportionally. On the other hand, our transmitter enforced the criterion that signal amplitude was constant. Signal amplitude essentially equals the integral of the magnitude of the signal's spectrum.

Note: This result isn't exact, but we do know that $m\,(0) = \int_{-\infty}^{\infty} M\,(f)\,df$ .

Enforcing the signal amplitude specification means that as the signal's bandwidth increases we must decrease the spectral amplitude, with the result that the signal power remains constant. Thus, increasing signal bandwidth does indeed decrease the signal-to-noise ratio of the receiver's output.

- Increasing channel attenuation – moving the receiver farther from the transmitter – decreases the signal-to-noise ratio as the square. Thus, signal-to-noise ratio decreases as distance-squared between transmitter and receiver.
- Noise added by the channel adversely affects the signal-to-noise ratio.

In summary, amplitude modulation provides an effective means for sending a bandlimited signal from one place to another. For wireline channels, using baseband or amplitude modulation makes little difference in terms of signal-to-noise ratio. For wireless channels, amplitude modulation is the only alternative. The one AM parameter that does not affect signal-to-noise ratio is the carrier frequency fc: We can choose any value we want so long as the transmitter and receiver use the same value. However, suppose someone else wants to use AM and chooses the same carrier frequency. The two resulting transmissions will add, and both receivers will produce the sum of the two signals. What we clearly need to do is talk to the other party, and agree to use separate carrier frequencies. As more and more users wish to use radio, we need a forum for agreeing on carrier frequencies and on signal bandwidth. On earth, this forum is the government. In the United States, the Federal Communications Commission (FCC) strictly controls the use of the electromagnetic spectrum for communications. Separate frequency bands are allocated for commercial AM, FM, cellular telephone (the analog version of which is AM), short wave (also AM), and satellite communications.

### Exercise 6.12.2

Suppose all users agree to use the same signal bandwidth. How closely can the carrier frequencies be while avoiding communications crosstalk? What is the signal bandwidth for commercial AM? How does this bandwidth compare to the speech bandwidth?

## 6.13 DIGITAL COMMUNICATION

Effective, error-free transmission of a sequence of bits—a bit stream {b(0),b(1),...} —is the goal here. Our findings indicate that analog techniques, such as amplitude modulation, consistently produce a received signal that contains both noise and the message signal when noise is introduced by the channel. There is significant variation among digital communication schemes. After determining the method of encoding bits into analog signals for transmission over wireline or wireless channels, we will proceed to devise a technique for appending communication bits to the message bits, which will significantly minimize errors caused by the channel. Conceptually, it is possible for digital communication errors to be completely eliminated, notwithstanding the presence of noise introduced by the channel.

We represent a bit by associating one of two specific analog signals with the bit's value. Thus, if b(n)=0, we transmit the signal s0(t); if b(n)=1, send s1(t). These two signals form the signal set specifically tailored for digital communication, taking into consideration the channel and bit stream. Almost always, these signals share a finite time T, which is referred to as the bit interval. The kind of signals we utilize ultimately determines the quality of bit reception. Remarkably, both baseband and modulated signal sets can achieve equivalent performance. Additional factors influence the impact of signal set selection on the performance of digital communication.

### Exercise 6.13.1

What is the expression for the signal arising from a digital transmitter sending the bit stream b(n),n={...,-1,0,1,...} using the signal set s0(t), s1(t), each signal of which has duration T?

## 6.14 BINARY PHASE SHIFT KEYING

A commonly used example of a signal set consists of pulses that are negatives of each other (Figure 6.7).

$$s_0(t) = Ap_T(t) \tag{6.38}$$

$$s_1(t) = -(Ap_T(t))$$

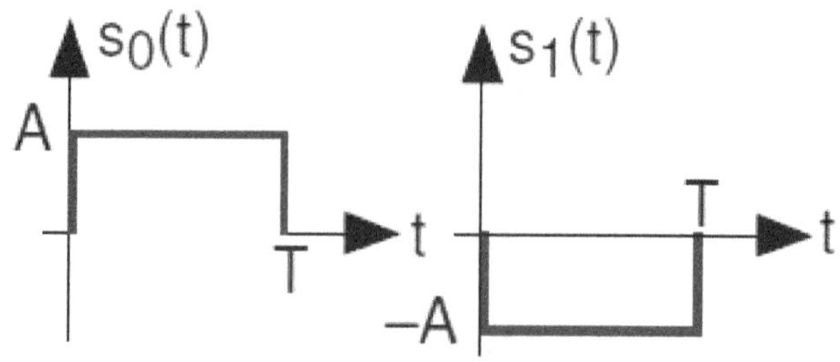


**Figure 6.7**

Here, we have a baseband signal set suitable for wireline transmission. The entire bit stream b(n) is represented by a sequence of these signals. Mathematically, the transmitted signal has the form

$$x(t) = \sum_{nn} (-1)^{b(n)} A p_T (t - nT) \quad (6.39)$$

and graphically Figure 6.8, shows what a typical transmitted signal might be.

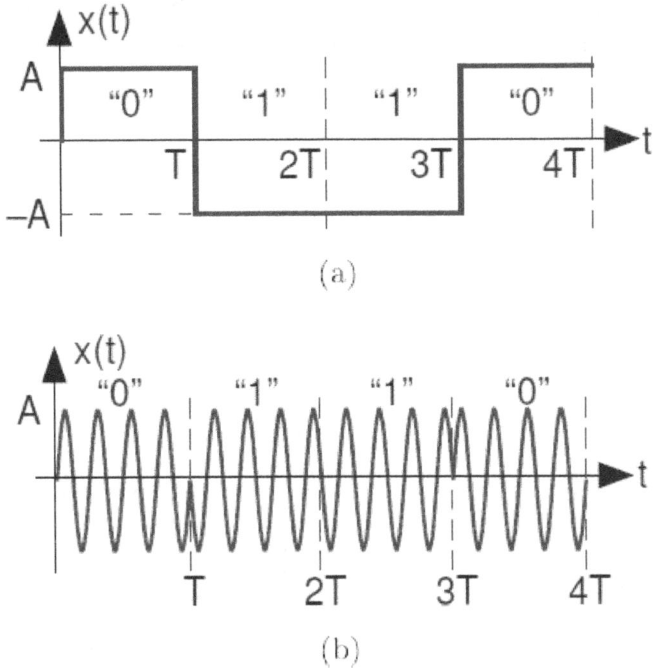


**Figure 6.8: The upper plot shows how a baseband signal set for transmitting the bit sequence.**

This way of representing a bit stream – changing the bit changes the sign of the transmitted signal---is known as binary phase shift keying and abbreviated BPSK. The name comes from concisely expressing this popular way of communicating digital information. The word «binary» is clear enough (one binary-valued quantity is transmitted during a bit interval). Changing the sign of sinusoid amounts to changing – shifting – the phase by π (although we don't have a sinusoid yet). The word "keying" reflects back to the first electrical communication system, which happened to be digital as well: the telegraph.

The data rate R of a digital communication system is how frequently an information bit is transmitted. In this example it equals the reciprocal of the bit interval: $R = \frac{1}{T}$. Thus, for a 1 Mbps (megabit per second) transmission, we must have T=1μs.

The choice of signals to represent bit values is arbitrary to some degree. Clearly, we do not want to choose signal set members to be the same; we couldn't distinguish bits if we did so. We could also have made the negative-amplitude pulse represent a 0 and the positive one a 1. This choice is indeed arbitrary and will have no effect on performance assuming the receiver knows which signal represents which bit. As in all communication systems, we design transmitter and receiver together.

A simple signal set for both wireless and wireline channels amounts to amplitude modulating a baseband signal set (more appropriate for a wireline channel) by a carrier having a frequency harmonic with the bit interval.

$$s_0(t) = Ap_T(t)\sin\left(\frac{2\pi kt}{T}\right) \tag{6.40}$$

$$s_1(t) = -\left(Ap_T(t)\sin\left(\frac{2\pi kt}{T}\right)\right)$$

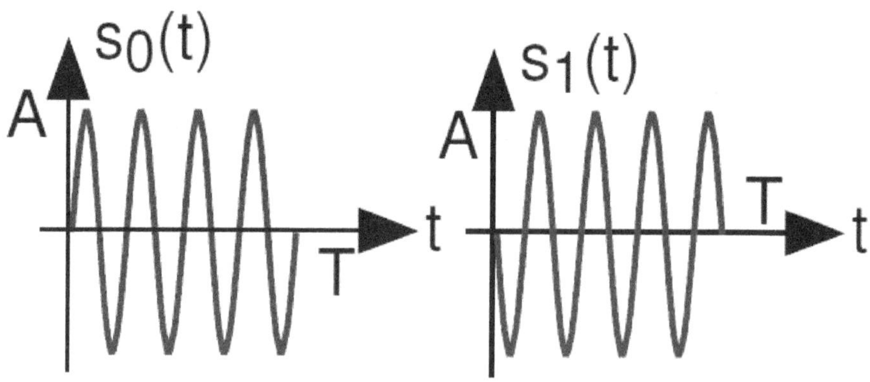


Figure 6.9

### Exercise 6.14.1

What is the value of k in this example?

This signal set is also known as a BPSK signal set. We'll show later that indeed both signal sets provide identical performance levels when the signal-to-noise ratios are equal.

### Exercise 6.14.2

Write a formula, in the style of the baseband signal set as shown in the beginning of this section, for the transmitted signal as shown in the plot of the baseband signal set that emerges when we use this modulated signal.

What is the transmission bandwidth of these signal sets? We need only consider the baseband version as the second is an amplitude-modulated version of the first. The bandwidth is determined by the bit sequence. If the bit sequence is constant—always 0 or always 1—the transmitted signal is a constant, which has zero bandwidth. The worst-case—bandwidth consuming—bit sequence is the alternating one shown in Figure 6.10. In this case, the transmitted signal is a square wave having a period of 2T.

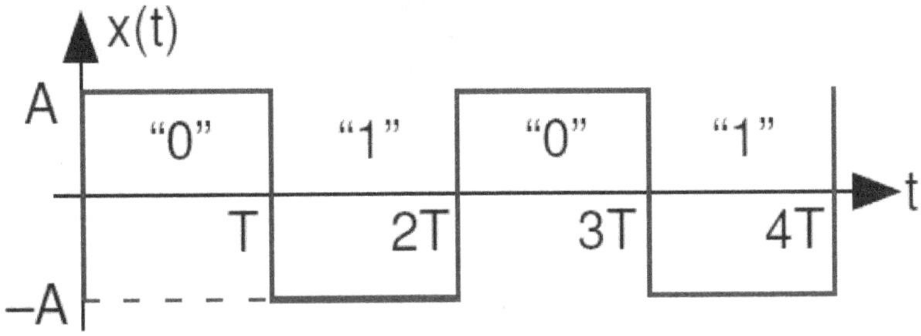


**Figure 6.10: Here we show the transmitted waveform corresponding to an alternating bit sequence**

From our work in Fourier series, we know that this signal's spectrum contains odd-harmonics of the fundamental, which here equals $\frac{1}{2T}$. Thus, strictly speaking, the signal's bandwidth is infinite. In practical terms, we use the 90%-power bandwidth to assess the effective range of frequencies consumed by the signal. The first and third harmonics contain that fraction of the total power, meaning that the effective bandwidth of our baseband signal is $\frac{3}{2T}$or, expressing this quantity in terms of the datarate, $\frac{3R}{2}$. Thus, a digital communications signal requires more bandwidth than the datarate: a 1 Mbps baseband system requires a bandwidth of at least 1.5 MHz. Listen carefully when someone describes the transmission bandwidth of digital communication systems: Did they say "megabits" or "megahertz"?

## Exercise 6.14.3

Show that indeed the first and third harmonics contain 90% of the transmitted power. If the receiver uses a front-end filter of bandwidth $\frac{3}{2T}$, what is the total harmonic distortion of the received signal?

## Exercise 6.14.4

What is the 90% transmission bandwidth of the modulated signal set?

## 6.15 FREQUENCY SHIFT KEYING

In frequency-shift keying (FSK), the bit affects the frequency of a carrier sinusoid.

$$s_0(t) = Ap_T(t)\sin(2\pi f_0 t) \tag{6.41}$$

$$s_1(t) = Ap_T(t)\sin(2\pi f_1 t)$$

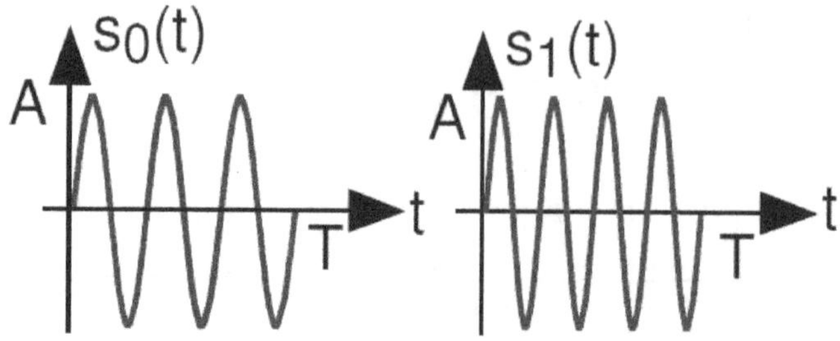


**Figure 6.11**

The frequencies f0, f1 are usually harmonically related to the bit interval. In the depicted example, $f_0 = \frac{3}{T}$ and $f_1 = \frac{4}{T}$.As can be seen from the transmitted signal for our example bit stream (Figure 6.12) the transitions at bit interval boundaries are smoother than those of BPSK.

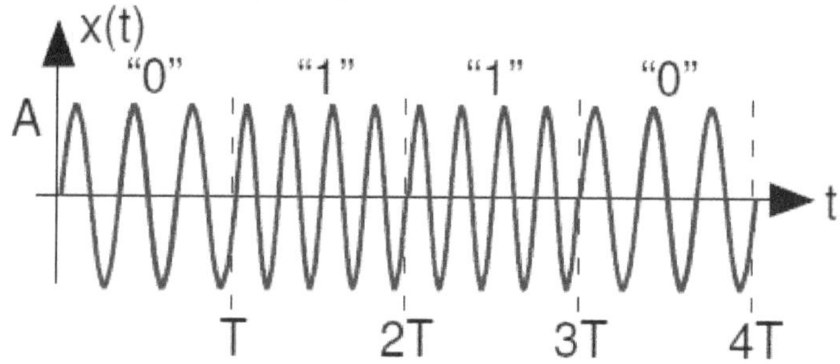


**Figure 6.12: This plot shows the FSK waveform for same bitstream used in the BPSK example (Figure 6.8).**

To determine the bandwidth required by this signal set, we again consider the alternating bit stream. Think of it as two signals added together: The first comprised of the signal s0(t), the zero signal, s0(t), zero, etc., and the second having the same structure but interleaved with the first and containing s1(t) (Figure 6.13).

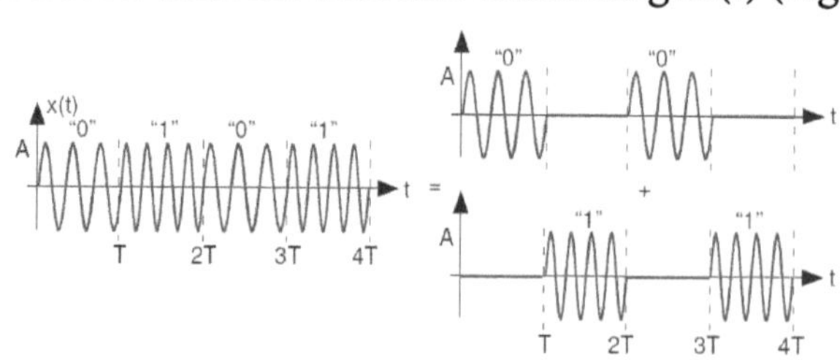


**Figure 6.13: The depicted decomposition of the FSK - modulated alternating bit stream into its frequency components simplifies the calculation of its bandwidth.**

Each component can be thought of as a fixed-frequency sinusoid multiplied by a square wave of period 2T that alternates between one and zero. This baseband square wave has the same Fourier spectrum as our BPSK example, but with the addition of the constant term c0. This quantity's presence changes the number of Fourier series terms required for the 90% bandwidth: Now we need only include the zero and first harmonics to achieve it. The bandwidth thus equals, with $f_0 < f_1$, $f_1 + \frac{1}{2T} - (f_0 - \frac{1}{2T}) = f_1 - f_0 + \frac{1}{T}$.

If the two frequencies are harmonics of the bit-interval duration, $f_0 = \frac{k_0}{T}$ and $f_1 = \frac{k_1}{T}$ with $k_1 > k_0$,the bandwidth equals $\frac{k_1 + -k_0 + 1}{T}$. If the difference between harmonic numbers is 1, then the FSK bandwidth is smaller than the BPSK bandwidth. If the difference is 2, the bandwidths are equal and larger differences produce a transmission bandwidth larger than that resulting from using a BPSK signal set.

## 6.16 DIGITAL COMMUNICATION RECEIVERS

The receiver interested in the transmitted bit stream must perform two tasks when received waveform r(t) begins.

- It must determine when bit boundaries occur: The receiver needs to synchronize with the transmitted signal. Because transmitter and receiver are designed in concert, both use the same value for the bit interval T. Synchronization can occur because the transmitter begins sending with a reference bit sequence, known as the preamble. This reference bit sequence is usually the alternating sequence as shown in the square wave example and in the FSK example (Figure 6.13). The receiver knows what the preamble bit sequence is and uses it to determine when bit boundaries occur. This procedure amounts to what in digital hardware as self-clocking signaling: The receiver of a bit stream must derive the clock — when bit boundaries occur — from its input signal. Because the receiver usually does not determine which bit was sent until synchronization occurs, it does not know when during the preamble it obtained synchronization. The transmitter signals the end of the preamble by switching to a second bit sequence. The second preamble phase informs the receiver that data bits are about to come and that the preamble is almost over.
- Once synchronized and data bits are transmitted, the receiver must then determine every T seconds what bit was transmitted during the previous bit interval. We focus on this aspect of the digital receiver because this strategy is also used in synchronization.

The receiver for digital communication is known as a matched filter.

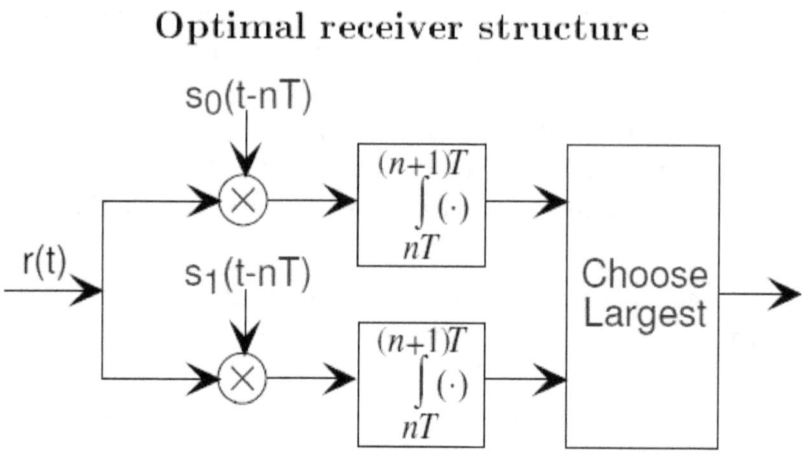


**Figure 6.14: The optimal receiver structure for digital communication faced with additive white noise channels is the depicted matched filter**

This receiver, shown in Figure 6.14, multiplies the received signal by each of the possible members of the transmitter signal set, integrates the product over the bit interval, and compares the results. Whichever path through the receiver yields the largest value corresponds to the receiver's decision as to what bit was sent during the previous bit interval. For the next bit interval, the multiplication and integration begins again, with the next bit decision made at the end of the bit interval. Mathematically, the received value of b(n), which we label $\hat{b}(n)$, is given by:

$$\hat{b}(n) = \underset{i}{argmax} \int_{nT}^{(n+1)T} r(t)\, s_i(t)\, dt \tag{6.42}$$

You may not have seen the $\underset{i}{argmax}$ notation before. $max_i\{i, \cdot\}$ yields the maximum value of its argument with respect to the index i. $\underset{i}{argmax}$ equals the value of the index that yields the maximum. Note that the precise numerical value of the integrator's output does not matter; what does matter is its value relative to the other integrator's output.

Let's assume a perfect channel for the moment: The received signal equals the transmitted one. If bit 0 were sent using the baseband BPSK signal set, the integrator outputs would be

$$\int_{nT}^{(n+1)T} r(t)\, s_0(t)\, dt = A^2 T \tag{6.43}$$

$$\int_{nT}^{(n+1)T} r(t)\, s_1(t)\, dt = -\left(A^2 T\right)$$

If bit 1 were sent,

$$\int_{nT}^{(n+1)T} r(t)\, s_0(t)\, dt = -\left(A^2T\right) \tag{6.44}$$

$$\int_{nT}^{(n+1)T} r(t)\, s_1(t)\, dt = A^2T$$

### Exercise 6.16.1

Can you develop a receiver for BPSK signal sets that requires only one multiplier-integrator combination?

### Exercise 6.16.2

What is the corresponding result when the amplitude-modulated BPSK signal set is used?

Clearly, this receiver would always choose the bit correctly. Channel attenuation would not affect this correctness; it would only make the values smaller, but all that matters is which is largest.

## 6.17 DIGITAL COMMUNICATION IN THE PRESENCE OF NOISE

When we incorporate additive noise into our channel model, so that r(t)=αsi(t)+n(t), errors can creep in. If the transmitter sent bit 0 using a BPSK signal set (Section 6.14), the integrators' outputs in the matched filter receiver would be:

$$\int_{nT}^{(n+1)T} r(t)\, s_0(t)\, dt = \alpha A^2T + \int_{nT}^{(n+1)T} n(t)\, s_0(t)\, dt \tag{6.45}$$

$$\int_{nT}^{(n+1)T} r(t)\, s_1(t)\, dt = \alpha A^2T + \int_{nT}^{(n+1)T} n(t)\, s_1(t)\, dt$$

It is the quantities containing the noise terms that cause errors in the receiver's decision-making process. Because they involve noise, the values of these integrals are random quantities drawn from some probability distribution that vary erratically from bit interval to bit interval. Because the noise has zero average value and has an equal amount of power in all frequency bands, the values of the integrals will hover about zero. What is important is how much they vary. If the noise is such that its integral term is more negative than αA2T, then the receiver will make an error, deciding that the transmitted zero-valued bit was indeed a one. The probability that this situation occurs depends on three factors:

- **Signal Set Choice** — The difference between the signal-dependent terms in the integrators› outputs (equations above) defines how large the noise term must be for an incorrect receiver decision to result. What affects the probability of such errors occurring is the energy in the difference of the received signals in comparison to the noise term›s variability. The signal-difference energy equals

$$\int_0^T (s_1(t) - s_0(t))^2 dt$$

For our BPSK baseband signal set, the difference-signal-energy term is 4α2A4T2.

- **Variability of the Noise Term** — We quantify variability by the spectral height of the white noise $\frac{N_0}{2}$ added by the channel.
- **Probability Distribution of the Noise Term** — The value of the noise terms relative to the signal terms and the probability of their occurrence directly affect the likelihood that a receiver error will occur. For the white noise we have been considering, the underlying distributions are Gaussian. Deriving the following expression for the probability the receiver makes an error on any bit transmission is complicated but can be found at here and here.

$$\begin{aligned} p_e &= Q\left(\sqrt{\frac{\int_0^T (s_1(t)-s_0(t))^2 dt}{2N_0}}\right) \\ &= Q\left(\sqrt{\frac{2\alpha^2 A^2 T}{N_0}}\right) \text{ for the BPSK case} \end{aligned} \tag{6.46}$$

Here Q(•) is the integral $Q(x) = \frac{1}{\sqrt{2\pi}} \int_x^\infty e^{-\frac{\alpha^2}{2}} d\alpha$. This integral has no closed form expression, but it can be accurately computed. As Figure 6.15 illustrates, Q(•) is a decreasing, very nonlinear function.

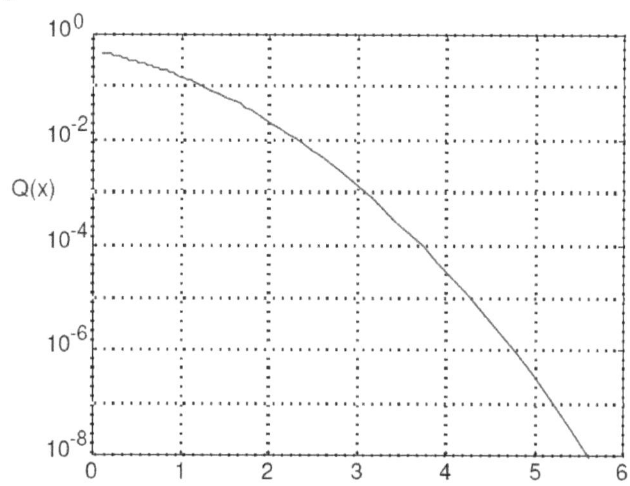


**Figure 6.15: The function Q(x) is plotted in semilogarithmic coordinates. Note that it decreases very rapidly for small increases in its arguments. For example, when x increases from 4 to 5, Q(x) decreases by a factor of 100.**

The term A2T equals the energy expended by the transmitter in sending the bit; we label this term Eb. We arrive at a concise expression for the probability the matched filter receiver makes a bit-reception error.

$$p_e = Q\left(\sqrt{\frac{2\alpha^2 E_b}{N_0}}\right) \tag{6.47}$$

Figure 6.16 shows how the receiver's error rate varies with the signal-to-noise ratio $\frac{\alpha^2 E_b}{N_0}$.

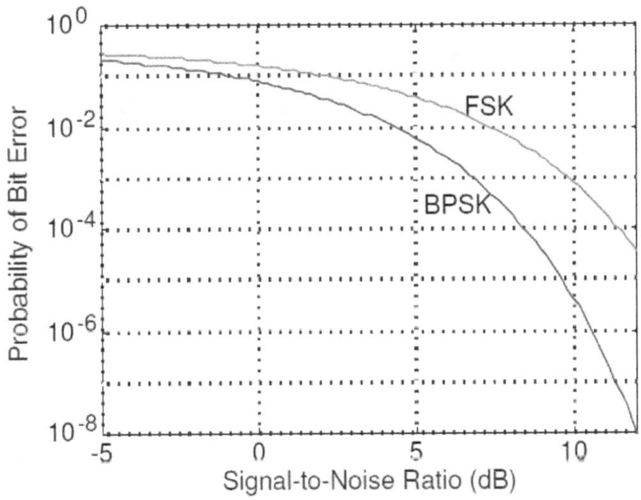


**Figure 6.16: The probability that the matched-filter receiver makes an error on any bit transmission is plotted against the signal-to-noise ratio of the received signal. The upper curve shows the performance of the FSK signal set, the lower (and therefore better) one the BPSK signal set.**

### Exercise 6.17.1

Derive the probability of error expression for the modulated BPSK signal set, and show that its performance identically equals that of the baseband BPSK signal set.

## 6.18 DIGITAL COMMUNICATION SYSTEM PROPERTIES

Results from the Receiver Error module (Section 6.17) reveals several properties digital communication systems.

- As the received signal becomes increasingly noisy, whether due to increased distance from the transmitter (smaller α) or to increased noise in the channel (larger N0), the probability the receiver makes an error approaches 1/2. In such situations, the receiver performs only slightly better than the "receiver" that ignores what was transmitted and merely guesses what bit was transmitted. Consequently, it becomes almost impossible to communicate information when digital channels become noisy.

- As the signal-to-noise ratio increases, performance gains--smaller probability of error pe- can be easily obtained. At a signal-to-noise ratio of 12 dB, the probability the receiver makes an error equals 10-8. In words, one out of one hundred million bits will, on the average, be in error.
- Once the signal-to-noise ratio exceeds about 5 dB, the error probability decreases dramatically. Adding 1 dB improvement in signal-to-noise ratio can result in a factor of 10 smaller pe.
- Signal set choice can make a significant difference in performance. All BPSK signal sets, baseband or modulated, yield the same performance for the same bit energy. The BPSK signal set does perform much better than the FSK signal set once the signal-to-noise ratio exceeds about 5 dB.

### Exercise 6.18.1

Derive the expression for the probability of error that would result if the FSK signal set were used.

The matched-filter receiver provides impressive performance once adequate signal-to-noise ratios occur. You might wonder whether another receiver might be better. The answer is that the matched-filter receiver is optimal: No other receiver can provide a smaller probability of error than the matched filter regardless of the SNR. Furthermore, no signal set can provide better performance than the BPSK signal set, where the signal representing a bit is the negative of the signal representing the other bit. The reason for this result rests in the dependence of probability of error pe on the difference between the noise-free integrator outputs: For a given Eb, no other signal set provides a greater difference.

How small should the error probability be? Out of N transmitted bits, on the average Npe bits will be received in error. Do note the phrase "on the average" here: Errors occur randomly because of the noise introduced by the channel, and we can only predict the probability of occurrence. Since bits are transmitted at a rate R, errors occur at an average frequency of Rpe. Suppose the error probability is an impressively small number like 10-6. Data on a computer network like Ethernet is transmitted at a rate R = 100Mbps, which means that errors would occur roughly 100 per second. This error rate is very high, requiring a much smaller pe to achieve a more acceptable average occurrence rate for errors occurring. Because Ethernet is a wireline channel, which means the channel noise is small and the attenuation low, obtaining very small error probabilities is not difficult. We do have some tricks up our sleeves, however, that can essentially reduce the error rate to zero without resorting to expending a large amount of energy at the transmitter. We need to understand digital channels (Section 6.19) and Shannon›s Noisy Channel Coding Theorem (Section 6.30).

## 6.19 DIGITAL CHANNELS

Let's review how digital communication systems work within the Fundamental Model of Communication (Figure 1.3: Fundamental Model of communication). As shown in Figure 6.17 (DigMC), the message is a single bit. The entire analog transmission/reception system, which is discussed in Digital Communication (Section 6.13), Signal Sets, BPSK Signal Set, Transmission Bandwidth, Frequency Shift Keying (Section 6.15), Digital Communication Receivers (Section 6.16). Factors in Receiver Errors (Section 6.17), Digital Communication System Properties, and Error Probability, can be lumped into a single known as the digital channel.

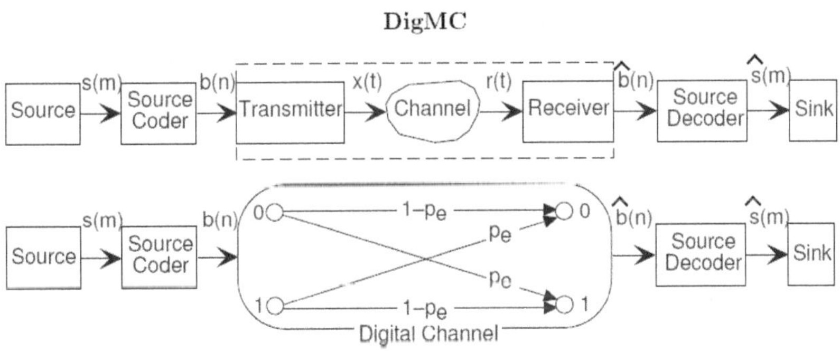


**Figure 6.17: The steps in transmitting digital information are shown in the upper system, the Fundamental Model of Communication. The symbolic-valued signal s(m) forms the message, and it is encoded into a bit sequence b(n). The indices differ because more than one bit/symbol is usually required to represent the message by a bitstream. Each bit is represented by an analog signal, transmitted through the (unfriendly) channel, and received by a matched-filter receiver. From the received bitstream $\hat{b}(n)$ the received symbolic-valued signal $\hat{s}(m)$ is derived.The lower block diagram shows an equivalent system wherein the analog portions are combined and modeled by a transition diagram, which shows how each transmitted bit could be received. For example, transmitting a 0 results in the reception of a 1 with probability pe (an error) or a 0 with probability 1- pe (no error).**

Transition diagrams are used to depict digital channels. These diagrams show the output alphabet symbols that are produced for each conceivable transmission signal, as well as the probability associated with different reception possibilities. The probabilities of transitions originating from the same symbol must collectively add up to a total of one. The transition diagram illustrated in the binary symmetric channel depicts the reception of sent bits for the matched-filter receiver and the observed signal sets. The probability of error, pe, is the only parameter of the digital channel. It represents the combined effect of signal set choice, channel parameters, and the matched-filter receiver. Using this straightforward yet totally precise model, we can focus on the process of receiving bits.

## 6.20 ENTROPY

The formulation of communication theory is most effective for signals with symbolic values. In 1948, Claude Shannon released The Mathematical Theory of Communication, which served as the foundation for digital communication. He demonstrated the efficacy of probabilistic models in analyzing signals with symbolic values, enabling him to accurately measure the amount of information contained in a signal.

In the simplest signal model, each symbol can occur at index n with a probability Pr[ak], k={1,...,K}. What this model says is that for each signal value a K-sided coin is flipped (note that the coin need not be fair). For this model to make sense, the probabilities must be numbers between zero and one and must sum to one.

$$0 \leq Pr\left[a_k\right] \leq 1 \tag{6.48}$$

$$\sum_{k=1}^{K} Pr\left[a_k\right] = 1 \tag{6.49}$$

This coin-flipping model presupposes that symbols occur independently of preceding or following symbols, which is an inaccurate assumption for typed text. Although this probabilistic model is overly simplistic, the concepts we discuss here are applicable even when more precise, yet still probabilistic, models are employed. The fundamental measure that defines a symbolic-valued signal is the entropy of its set of symbols.

$$H\left(A\right) = -\sum_{kk} Pr\left[a_k\right] \log_2 Pr\left[a_k\right] \tag{6.50}$$

Entropy is measured in bits since we utilize the base-2 logarithm. In order to comprehend this definition, it is imperative that we pay particular attention to symbols that have a probability of zero of happening. We define 0log2K as the logarithm of K to the base 2, where K is a symbol that never occurs with a probability of zero. The minimal value is obtained when there is only one symbol present, with a probability of one, while the remaining symbols have a probability of zero.

### Exercise 6.20.1

Derive the maximum-entropy results, both the numeric aspect (entropy equals log2K) and the theoretical one (equally likely symbols maximize entropy). Derive the value of the minimum entropy alphabet.

**Example 6.1**

A four-symbol alphabet has the following probabilities.

$$Pr\left[a_0\right] = \frac{1}{2}$$

$$Pr\left[a_1\right] = \frac{1}{4}$$

$$Pr\left[a_2\right] = \frac{1}{8}$$

$$Pr\left[a_3\right] = \frac{1}{8}$$

Note that these probabilities sum to one as they should. As $\frac{1}{2} = 2^{-1}$, $\log_2 \frac{1}{2} = -1$. The entropy of this alphabet equals

$$\begin{aligned} H\left(A\right) &= -\left(\tfrac{1}{2}\log_2\tfrac{1}{2} + \tfrac{1}{4}\log_2\tfrac{1}{4} + \tfrac{1}{8}\log_2\tfrac{1}{8} + \tfrac{1}{8}\log_2\tfrac{1}{8}\right) \\ &= -\left(\tfrac{1}{2} - 1 + \tfrac{1}{4} - 2 + \tfrac{1}{8} - 3 + \tfrac{1}{8} - 3\right) \\ &= 1.75\text{bits} \end{aligned} \tag{6.51}$$

## 6.21 SOURCE CODING THEOREM

The importance of the entropy of an alphabet lies in its ability to be represented using a sequence of bits. Bit sequences are the fundamental currency in digital communications, serving as the universal method for expressing signals with symbolic values. A codebook is used to translate symbols into bit sequences and vice versa. It is essentially a table that correlates symbols with certain bit sequences. When constructing this table, it is essential to assign a distinct bit sequence to each symbol to ensure accurate conversion between symbol and bit sequences.

Point of Interest: You may be conjuring the notion of hiding information from others when we use the name codebook for the symbol-to-bit-sequence table. There is no relation to cryptology, which comprises mathematically provable methods of securing information. The codebook terminology was developed during the beginnings of information theory just after World War II.

Digital communication refers to the transmission of symbolic-valued signals from one location to another, as discussed in Section 6.13. When confronted with the issue, such as uploading a file over the Internet, our initial step is to encode each character as a sequence of bits. In order to expedite the file transmission, we aim to minimize the number of bits used. Nevertheless, it is imperative to avoid using an insufficient number of bits that would render the receiver incapable of discerning the identity of each character in the bit sequence. As an illustration, we may allocate one bit for each character. The file transmission would be rapid, yet futile, due to the codebook's propensity to generate errors. Shannon proved in his monumental work what we call today the Source Coding Theorem. Let B(ak) denote the number of bits used to represent the symbol ak. The average number of bits $B\bar{(A)}$ required to represent the entire alphabet equals $\sum_{k=1}^{K} B(a_k)\, Pr[a_k]$. The Source Coding Theorem states that the average number of bits needed to accurately represent the alphabet need only to satisfy

$$H(A) \leq B\bar{(A)} < H(A) + 1 \tag{6.52}$$

Therefore, the entropy of the alphabet indicates precisely how many bits, on average, are required to transmit the alphabet. The lower the entropy of an alphabet, the less number of bits needed for digital transmission of files represented in that alphabet.

## Example 6.2

A four-symbol alphabet has the following probabilities.

$$Pr[a_0] = \frac{1}{2}$$

$$Pr[a_1] = \frac{1}{4}$$

$$Pr[a_2] = \frac{1}{8}$$

$$Pr[a_3] = \frac{1}{8}$$

and an entropy of 1.75 bits (Example 6.1). Let's see if we can find a codebook for this four-letter alphabet that satisfies the Source Coding Theorem. The simplest code to try is known as the simple binary code: convert the symbol's index into a binary number and use the same number of bits for each symbol by including leading zeros where necessary.

$$a_0 \leftrightarrow 00 a_1 \leftrightarrow 01 a_2 \leftrightarrow 10 a_3 \leftrightarrow 11 \tag{6.53}$$

Whenever the number of symbols in the alphabet is a power of two (as in this case), the average number of bits $B\bar{(A)}$ which equals 2 in this case. Because the entropy equals 1.75 bits, the simple binary code indeed satisfies the Source Coding Theorem—we are within one bit of the entropy limit—but you might wonder if you can do better. If we choose a codebook with differing number of bits for the symbols, a smaller average number of bits can indeed be obtained. The idea is to use shorter bit sequences for the symbols that occur more often. One codebook like this is

$$a_0 \leftrightarrow 0 a_1 \leftrightarrow 10 a_2 \leftrightarrow 110 a_3 \leftrightarrow 111 \tag{6.54}$$

Now $B\bar{(A)} = 1 \cdot \times \frac{1}{2} + 2 \cdot \times \frac{1}{4} + 3 \cdot \times \frac{1}{8} + 3 \cdot \times \frac{1}{8} = 1.75$. We can reach the entropy limit! The simple binary code is, in this case, less efficient than the unequal-length code. Using the efficient code, we can transmit the symbolic-valued signal having this alphabet 12.5% faster. Furthermore, we know that no more efficient codebook can be found because of Shannon's Theorem.

## 6.22 COMPRESSION AND THE HUFFMAN CODE

Shannon's Source Coding Theorem (6.52) has additional applications in the realm of data compression. Our objective is to efficiently encode a signal source, which can be a computer file or an image, by utilizing the minimum number of bits. Compression algorithms that assign symbols to sequences of bits are classified as lossless if they comply with the Source Coding Theorem. On the other hand, they are considered lossy if they use less bits than the entropy of the alphabet. Using a lossy compression technique means that it is not possible to accurately recover a signal with symbolic values from its compressed version. One may inquire about the rationale behind intentionally creating errors, yet lossy compression techniques are frequently utilized when the advantages of effectively representing the signal outweigh the significance of the faults.

According to Shannon's Source Coding Theorem, symbolic-valued signals need, on average, a minimum of H(A) bits to encode each of their values. These values are symbols selected from the alphabet A. In Section 6.21, the module on the Source Coding Theorem explains that utilizing a fixed rate source coder, which generates a constant amount of bits per symbol, may not be the optimal method for encoding symbols into bits. Not included in that discussion is a method for devising an effective source encoder: one that is certain to generate the minimum number of bits per symbol on average. The source coder is not distinct, however the Huffman source coding algorithm is one method that does achieve that limitation.

**Point of Interest:** In the early years of information theory, the race was on to be the first to find a provably maximally efficient source coding algorithm. The race was won by then MIT graduate student David Huffman in 1954, who worked on the problem as a project in his information theory course. We›re pretty sure he received an "A."

- Create a vertical table for the symbols, the best ordering being in decreasing order of probability.

- Form a binary tree to the right of the table. A binary tree always has two branches at each node. Build the tree by merging the two lowest probability symbols at each level, making the probability of the node equal to the sum of the merged nodes' probabilities. If more than two nodes/symbols share the lowest probability at a given level, pick any two; your choice won't affect $\bar{B}(A)$.
- At each node, label each of the emanating branches with a binary number. The bit sequence obtained from passing from the tree's root to the symbol is its Huffman code.

## Example 6.3

The simple four-symbol alphabet used in the Entropy (Example 6.1) and Source Coding modules has a four-symbol alphabet with the following probabilities, and an entropy of 1.75 bits (Example 6.1). This alphabet has the Huffman coding tree shown in Figure 6.18 (Huffman Coding Tree).

$$Pr[a_0] = \frac{1}{2}$$

$$Pr[a_1] = \frac{1}{4}$$

$$Pr[a_2] = \frac{1}{8}$$

$$Pr[a_3] = \frac{1}{8}$$

Huffman Coding Tree

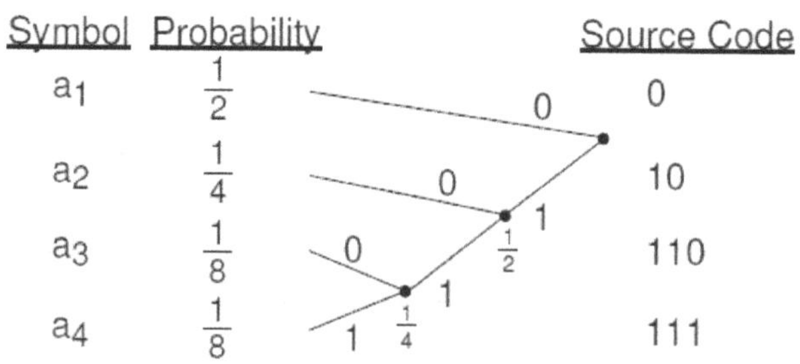


**Figure 6.18: We form a Huffman code for a four-letter alphabet having the indicated probabilities of occurrence. The binary tree created by the algorithm extends to the right, with the root node (the one at which the tree begins) defining the codewords. The bit sequence obtained by traversing the tree from the root to the symbol defines that symbol's binary code.**

The code thus obtained is not unique as we could have labeled the branches coming out of each node differently. The average number of bits required to represent this alphabet equals 1.75 bits, which is the Shannon entropy limit for this source alphabet. If we had the symbolic-valued signal s(m)={a2,a3,a1,a4,a1,a2,...}, Our Huffman code would produce the bitstream b(n)=101100111010...

If the alphabet probabilities were different, clearly a different tree, and therefore different code, could well result. Furthermore, we may not be able to achieve the entropy limit. If our symbols had the probabilities $Pr[a_1] = \frac{1}{2}$, $Pr[a_2] = \frac{1}{4}$, $Pr[a_3] = \frac{1}{5}$, and $Pr[a_4] = \frac{1}{20}$, the average number of bits/symbol resulting from the Huffman coding algorithm would equal 1.75 bits. However, the entropy limit is 1.68 bits. The Huffman code does satisfy the Source Coding Theorem—its average length is within one bit of the alphabet's entropy—but you might wonder if a better code existed. David Huffman showed mathematically that no other code could achieve a shorter average code than his. We can't do better.

### Exercise 6.22.1

Derive the Huffman code for this second set of probabilities, and verify the claimed average code length and alphabet entropy.

## 6.23 SUBTLIES OF CODING

In the Huffman code, the bit sequences that represent individual symbols can have differing lengths so the bitstream index m does not increase in lock step with the symbol-valued signal's index n. To capture how often bits must be transmitted to keep up with the source's production of symbols, we can only compute averages. If our source code averages $B\bar{(A)}$ bits/symbol and symbols are produced at a rate R, the average bit rate equals $B\bar{(A)}R$, and this quantity determines the bit interval duration T.

### Exercise 6.23.1

Calculate the relation between T and the average bit rate $B\bar{(A)}R$ is:

A subtlety of source coding is whether we need "commas" in the bitstream. When we use an unequal number of bits to represent symbols, how does the receiver determine when symbols begin and end? If you created a source code that required a separation marker in the bitstream between symbols, it would be very inefficient since you are essentially requiring an extra symbol in the transmission stream.

**Note:** A good example of this need is the Morse Code: Between each letter, the telegrapher needs to insert a pause to inform the receiver when letter boundaries occur.

As shown in this example (Example 6.3), no commas are placed in the bitstream, but you can unambiguously decode the sequence of symbols from the bitstream. Huffman showed that his (maximally efficient) code had the prefix property: No code for a symbol began another symbol›s code. Once you have the prefix property, the bitstream is partially self-synchronizing: Once the receiver knows where the bitstream starts, we can assign a unique and correct symbol sequence to the bitstream.

### Exercise 6.23.2

Sketch an argument that prefix coding, whether derived from a Huffman code or not, will provide unique decoding when an unequal number of bits/symbol are used in the code.

However, having a prefix code does not guarantee total synchronization: After hopping into the middle of a bitstream, can we always find the correct symbol boundaries? The self-synchronization issue does mitigate the use of efficient source coding algorithms.

### Exercise 6.23.3

Show by example that a bitstream produced by a Huffman code is not necessarily self-synchronizing. Are fixed-length codes self synchronizing?

Another issue is bit errors induced by the digital channel; if they occur (and they will), synchronization can easily be lost even if the receiver started "in synch" with the source. Despite the small probabilities of error offered by good signal set design and the matched filter, an infrequent error can devastate the ability to translate a bitstream into a symbolic signal. We need ways of reducing reception errors without demanding that pe be smaller.

### Example 6.4

The first electrical communications system—the telegraph—was digital. When first deployed in 1844, it communicated text over wireline connections using a binary code—the Morse code—to represent individual letters. To send a message from one place to another, telegraph operators would tap the message using a telegraph key to another operator, who would relay the message on to the next operator, presumably getting the message closer to its destination. In short, the telegraph relied on a network not unlike the basics of modern computer networks. To say it presaged modern communications would be an understatement. It was also far ahead of some needed technologies, namely the Source Coding Theorem. The Morse code, shown in the table below, was not a prefix code. To separate codes for each letter, Morse code required that a space—a pause—be inserted between each letter. In information theory, that space counts as another code letter, which means that the Morse code encoded text with a three-letter source code: dots, dashes and space. The resulting source code is not within a bit of entropy, and is grossly inefficient (about 25%). Figure 6.19 shows a Huffman code for English text, which as we know is efficient.

**Morse and Huffman Code Table**

| | % | Morse Code | Huffman Code |
|---|---|---|---|
| A | 6.22 | .- | 1011 |
| B | 1.32 | -... | 010100 |
| C | 3.11 | -.-. | 10101 |
| D | 2.97 | -.. | 01011 |
| E | 10.53 | . | 001 |
| F | 1.68 | ..-. | 110001 |
| G | 1.65 | . | 110000 |
| H | 3.63 | .... | 11001 |
| I | 6.14 | .. | 1001 |
| J | 0.06 | . | 01010111011 |
| K | 0.31 | -.- | 01010110 |
| L | 3.07 | .-.. | 10100 |
| M | 2.48 | | 00011 |
| N | 5.73 | -. | 0100 |
| O | 6.06 | | 1000 |
| P | 1.87 | . . | 00000 |
| Q | 0.10 | .- | 0101011100 |
| R | 5.87 | .-. | 0111 |
| S | 5.81 | ... | 0110 |
| T | 7.68 | - | 1101 |
| U | 2.27 | ..- | 00010 |
| V | 0.70 | ...- | 0101010 |
| W | 1.13 | . | 000011 |
| X | 0.25 | -..- | 010101111 |
| Y | 1.07 | -. | 000010 |
| Z | 0.06 | .. | 0101011101011 |

**Figure 6.19: Morse and Huffman Codes for American-Roman Alphabet. The % column indicates the average probability (expressed in percent) of the letter occurring in English. The entropy H (A) of the this source is 4.14 bits. The average Morse codeword length is 2.5 symbols. Adding one more symbol for the letter separator and converting to bits yields an average codeword length of 5.56 bits. The average Hu_man codeword length is 4.35 bits.**

## 6.24 CHANNEL CODING

We have the ability to partially rectify mistakes produced by the recipient using merely the error-laden data stream that comes out of the digital channel.

The concept is for the transmitter to broadcast not just the bits obtained from the source coder's symbols, but also extra bits obtained from the coder's bit stream. The error correcting bits aid the receiver in identifying whether an error has transpired in the data bits (the crucial bits) or in the error-correction bits. Instead of utilizing the communication model (Figure 6.17: DigMC) demonstrated before, the transmitter incorporates a channel coder prior to analog modulation, while the receiver employs the matching channel decoder (Figure 6.20). The diagram displayed represents the Fundamental Model of Digital Communication.

**Figure 6.20: To correct errors that occur in the digital channel, a channel coder and decoder are added to the communication system. Properly designed channel coding can greatly reduce the probability (from the uncoded value of pe) that a data bit b(n) is received incorrectly even when the probability of c(l) be received in error remains pe or becomes larger. This system forms the Fundamental Model of Digital Communication.**

According to Shannon's Noisy Channel Coding Theorem (Section 6.30), as long as the data transmission rate is not too high, there are error correction codes available that can fix all the bit mistakes caused by the channel. Regrettably, Shannon did not exhibit an error correcting code that would accomplish this extraordinary accomplishment; indeed, no one has discovered such a code. Shannon's result demonstrates its existence; it appears that there is always additional work to be done. Regardless, this should not hinder our examination of frequently employed error correcting codes, which are not only present in all digital communication systems, but also in CDs and bar codes utilized in commerce.

## 6.25 REPETITION CODES

Perhaps the simplest error correcting code is the repetition code.

Here, the transmitter sends the data bit several times, an odd number of times in fact. Because the error probability pe is always less than $\frac{1}{2}$, we know that more of the bits should be correct rather than in error. Simple majority voting of the received bits (hence the reason for the odd number) determines the transmitted bit more accurately than sending it alone. For example, let's consider the three-fold repetition code: for every bit b(n) emerging from the source coder, the channel coder produces three. Thus, the bit stream emerging from the channel coder c(l) has a data rate three times higher than that of the original bit stream b(n). The coding table illustrates when errors can be corrected and when they can't by the majority-vote decoder.

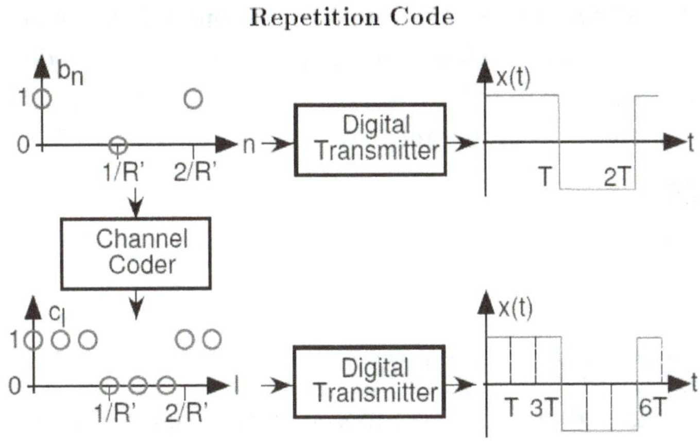


**Figure 6.21: The upper portion depicts the result of directly modulating the bit stream b(n) into a transmitted signal x(t) using a baseband BPSK signal set. R' is the datarate produced by the source coder. If that bit stream passes through a (3,1) channel coder to yield the bit stream c(l), the resulting transmitted signal requires a bit interval T three times smaller than the uncoded version. This reduction in the bit interval means that the transmitted energy/bit decreases by a factor of three, which results in an increased error probability in the receiver.**

## CODING TABLE

Table 6.1: In this example, the transmitter encodes 0 as 000. The channel creates an error (changing a 0 into a 1) that with probability pe. The first column lists all possible received datawords and the second the probability of each dataword being received. The last column shows the results of the majority-vote decoder. When the decoder produces 0, it successfully corrected the errors introduced by the channel (if there were any; the top row corresponds to the case in which no errors occurred). The error probability of the decoders is the sum of the probabilities when the decoder produces 1.

| Code | Probability | Bit |
|---|---|---|
| 000 | (1-pe)3 | 0 |
| 001 | pe (1-pe)2 | 0 |
| 010 | pe (1-pe)2 | 0 |
| 011 | pe2 (1-pe) | 1 |
| 100 | pe (1-pe)2 | 0 |
| 101 | pe2 (1-pe) | 1 |
| 110 | pe2 (1-pe) | 1 |
| 111 | pe3 | 1 |

Thus, if one bit of the three bits is received in error, the receiver can correct the error; if more than one error occurs, the channel decoder announces the bit is 1 instead of transmitted value of 0. Using this repetition code, the probability $\hat{b}(n) \neq 0$ equals 3pe2×(1-pe)+pe3. This probability of a decoding error is always less than pe , the uncoded value, so long as $p_e < \frac{1}{2}$.

### Exercise 6.25.1

Demonstrate mathematically that this claim is indeed true. Is

$$3{p_e}^2 \times (1 - p_e) + {p_e}^3 \leq p_e?$$

## 6.26 BLOCK CHANNEL CODING

Because of the higher datarate imposed by the channel coder, the probability of bit error occurring in the digital channel increases relative to the value obtained when no channel coding is used. The bit interval duration must be reduced by $\frac{K}{N}$ in comparison to the no-channel-coding situation, which means the energy per bit Eb goes down by the same amount. The bit interval must decrease by a factor of three if the transmitter is to keep up with the data stream, as illustrated here (Figure 6.21: Repetition Code).

Point of Interest: It is unlikely that the transmitter's power could be increased to compensate. Such is the sometimes-unfriendly nature of the real world.

Because of this reduction, the error probability pe of the digital channel goes up. The question thus becomes does channel coding really help: Is the effective error probability lower with channel coding even though the error probability for each transmitted bit is larger? The answer is no: Using a repetition code for channel coding cannot ultimately reduce the probability that a data bit is received in error. The ultimate reason is the repetition code's inefficiency: transmitting one data bit for every three transmitted is too inefficient for the amount of error correction provided.

### Exercise 6.26.1

Using MATLAB, calculate the probability a bit is received incorrectly with a three-fold repetition code. Show that when the energy per bit Eb is reduced by 1/3 that this probability is larger than the no-coding probability of error.

The repetition code (p. 258) represents a special case of what is known as block channel coding. For every K bits that enter the block channel coder, it inserts an additional N-K error-correction bits to produce a block of N bits for transmission. We use the notation (N,K) to represent a given block code's parameters. In the three-fold repetition code (p. 258), K=1 and N=3. A block code's coding efficiency E equals the ratio $\frac{K}{N}$, and quantifies the overhead introduced by channel coding. The rate at which bits must be transmitted again changes: So-called data bits b(n) emerge from the source coder at an average rate $\bar{B(A)}$ and exit the channel at a rate $\frac{1}{E}$ higher. We represent the fact that the bits sent through the digital channel operate at a different rate by using the index l for the channel-coded bit stream c(l). Note that the blocking (framing) imposed by the channel coder does not correspond to symbol boundaries in the bit stream b(n), especially when we employ variable-length source codes.

Does any error-correcting code reduce communication errors when real-world constraints are taken into account? The answer now is yes. To understand channel coding, we need to develop first a general framework for channel coding, and discover what it takes for a code to be maximally efficient: Correct as many errors as possible using the fewest error correction bits as possible (making the efficiency $\frac{K}{N}$ as large as possible).

## 6.27 ERROR-CORRECTING CODES: HAMMING DISTANCE

So-called linear codes create error-correction bits by combining the data bits linearly. The phrase «linear combination» means here single-bit binary arithmetic.

**Table 6.2**

| $0 \oplus 0 = 0$ | $1 \oplus 1 = 0$ | $0 \oplus 1 = 1$ | $1 \oplus 0 = 1$ |
|---|---|---|---|
| $0 \cdot 0 = 0$ | $1 \cdot 1 = 1$ | $0 \cdot 1 = 0$ | $1 \cdot 0 = 0$ |

For example, let's consider the specific (3, 1) error correction code described by the following coding table and, more concisely, by the succeeding matrix expression.

$$c(1) = b(1)$$
$$c(2) = b(1)$$
$$c(3) = b(1)$$

Or

$$c = Gb$$

Where

$$G = \begin{pmatrix} 1 \\ 1 \\ 1 \end{pmatrix}$$
$$c = \begin{pmatrix} c(1) \\ c(2) \\ c(3) \end{pmatrix}$$
$$b = \begin{pmatrix} b(1) \end{pmatrix}$$

The length-K (in this simple example K=1) block of data bits is represented by the vector b, and the length-N output block of the channel coder, known as a codeword, by c. The generator matrix G defines all block-oriented linear channel coders.

As we consider other block codes, the simple idea of the decoder taking a majority vote of the received bits won't generalize easily. We need a broader view that takes into account the distance between codewords. A length-N codeword means that the receiver must decide among the 2N possible datawords to select which of the 2K codewords was actually transmitted. As shown in Figure 6.22, we can think of the datawords geometrically. We define the Hamming distance between binary datawords c1 and c2, denoted by d (c1; c2) to be the minimum number of bits that must be "flipped" to go from one word to the other. For example, the distance between codewords is 3 bits. In our table of binary arithmetic, we see that adding a 1 corresponds to flipping a bit. Furthermore, subtraction and addition are equivalent. We can express the Hamming distance as

$$d\left(c_1, c_2\right) = \operatorname{sum}\left(c_1 \oplus c_2\right) \tag{6.55}$$

## Exercise 6.27.1

Show that adding the error vector col [1,0,...,0] to a codeword flips the codeword's leading bit and leaves the rest unaffected.

The probability of one bit being flipped anywhere in a codeword is $Np_e(1-p_e)^{N-1}$. The number of errors the channel introduces equals the number of ones in e; the probability of any particular error vector decreases with the number of errors.

To perform decoding when errors occur, we want to find the codeword (one of the filled circles in Figure 6.22) that has the highest probability of occurring: the one closest to the one received. Note that if a dataword lies a distance of 1 from two codewords, it is impossible to determine which codeword was actually sent. This criterion means that if any two codewords are two bits apart, then the code cannot correct the channel-induced error. Thus, to have a code that can correct all single-bit errors, codewords must have a minimum separation of three. Our repetition code has this property.

Introducing code bits increases the probability that any bit arrives in error (because bit interval durations decrease). However, using a well-designed error-correcting code corrects bit reception errors. Do we win or lose by using an error-correcting code? The answer is that we can win if the code is well-designed. The (3,1) repetition code demonstrates that we can lose (Exercise 6.26.1). To develop good channel coding, we need to develop first a general framework for channel codes and discover what it takes for a code to be maximally efficient: Correct as many errors as possible using the fewest error correction bits as possible (making the efficiency $\frac{K}{N}$ as large as possible.) We also need a systematic way of finding the codeword closest to any received dataword. A much better code than our (3,1) repetition code is the following (7,4) code.

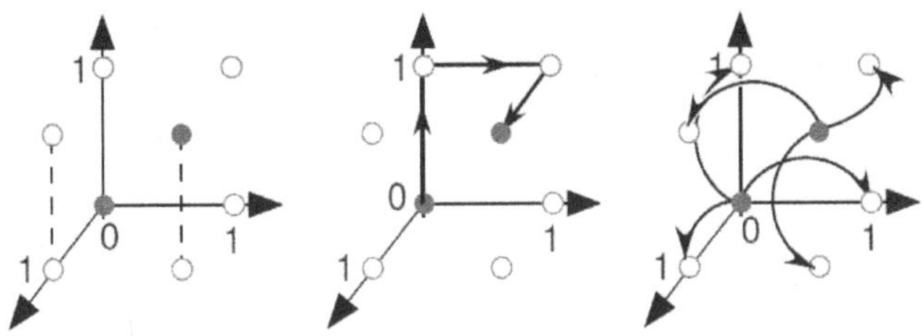

**Figure 6.22: In a (3,1) repetition code, only 2 of the possible 8 three-bit data blocks are codewords. We can represent these bit patterns geometrically with the axes being bit positions in the data block. In the left plot, the filled circles represent the codewords [0 0 0] and [1 1 1], the only possible codewords. The unfilled ones correspond to the transmission. The center plot shows that the distance between codewords is 3. Because distance corresponds to flipping a bit, calculating the Hamming distance geometrically means following the axes rather than going "as the crow flies". The right plot shows the datawords that result when one error occurs as the codeword goes through the channel. The three datawords are unit distance from the original codeword. Note that the received dataword groups do not overlap, which means the code can correct all single-bit errors.**

$$c(1) = b(1)$$
$$c(2) = b(2)$$
$$c(3) = b(3)$$
$$c(4) = b(4)$$
$$c(5) = b(1) \oplus b(2) \oplus b(3)$$
$$c(6) = b(2) \oplus b(3) \oplus b(4)$$
$$c(7) = b(1) \oplus b(2) \oplus b(4)$$

where the generator matrix is

$$G = \begin{pmatrix} 1 & 0 & 0 & 0 \\ 0 & 1 & 0 & 0 \\ 0 & 0 & 1 & 0 \\ 0 & 0 & 0 & 1 \\ 1 & 1 & 1 & 0 \\ 0 & 1 & 1 & 1 \\ 1 & 1 & 0 & 1 \end{pmatrix}$$

In this (7,4) code, 24 = 16 of the 27 = 128 possible blocks at the channel decoder correspond to error-free transmission and reception.

Error correction amounts to searching for the codeword c closest to the received block $\hat{c}$ in terms of the Hamming distance between the two. The error correction capability of a channel code is limited by how close together any two error-free blocks are. Bad codes would produce blocks close together, which would result in ambiguity when assigning a block of data bits to a received block. The quantity to examine, therefore, in designing code error correction codes is the minimum distance between codewords.

$$d_{\min} = \min\left(d\left(c_i, c_j\right)\right) \quad , \quad c_i \neq c_j \tag{6.56}$$

To have a channel code that can correct all single-bit errors, dmin≥3.

### Exercise 6.27.2

Suppose we want a channel code to have an error-correction capability of n bits. What must the minimum Hamming distance between codewords dmin be?

How do we calculate the minimum distance between codewords? Because we have 2K codewords, the number of possible unique pairs equals 2K−1(2K−1), which can be a large number. Recall that our channel coding procedure is linear, with c=Gb. Therefore ci⊕cj=G(bi⊕bj). Because bi⊕bj always yields another block of data bits, we find that the difference between any two codewords is another codeword! Thus, to find dmin we need only compute the number of ones that comprise all non-zero codewords. Finding these codewords is easy once we examine the coder›s generator matrix. Note that the columns of G are codewords (why is this?), and that all codewords can be found by all possible pairwise sums of the columns. To find dmin, we need only count the number of bits in each column and sums of columns. For our example (7, 4), G's first column has three ones, the next one four, and the last two three. Considering sums of column pairs next, note that because the upper portion of G is an identity matrix, the corresponding upper portion of all column sums must have exactly two bits. Because the bottom portion of each column differs from the other columns in at least one place, the bottom portion of a sum of columns must have at least one bit. Triple sums will have at least three bits because the upper portion of G is an identity matrix. Thus, no sum of columns has fewer than three bits, which means that dmin = 3, and we have a channel coder that can correct all occurrences of one error within a received 7-bit block.

## 6.28 ERROR-CORRECTING CODES: CHANNEL DECODING

Because the idea of channel coding has merit (so long as the code is efficient), let's develop a systematic procedure for performing channel decoding. One way of checking for errors is to try recreating the error correction bits from the data portion of the received block $\hat{c}$. Using matrix notation, we make this calculation by multiplying the received block $\hat{c}$ by the matrix H known as the parity check matrix. It is formed from the generator matrix G by taking the bottom, error-correction portion of G and attaching to it an identity matrix. For our (7,4) code,

$$H = \begin{bmatrix} \underbrace{\begin{matrix} 1 & 1 & 1 & 0 \\ 0 & 1 & 1 & 1 \\ 1 & 1 & 0 & 1 \end{matrix}}_{\text{Lower portion of } G} & \underbrace{\begin{matrix} 1 & 0 & 0 \\ 0 & 1 & 0 \\ 0 & 0 & 1 \end{matrix}}_{\text{Identity}} \end{bmatrix} \tag{6.57}$$

The parity check matrix thus has size (N-K) × N, and the result of multiplying this matrix with a received word is a length- (N-K) binary vector. If no digital channel errors occur—we receive a codeword so that $\hat{c} = c$ then $H\hat{c} = 0$. For example, the first column of G,(1,0,0,0,1,0,1)T is a codeword. Simple calculations show that multiplying this vector by H results in a length- (N-K) zero-valued vector.

## Exercise 6.28.1

Show that HC = 0 for all the columns of G. In other words, show that HG = 0 an (N-K )× N matrix of zeroes. Does this property guarantee that all codewords also satisfy HC = 0?

When the received bits $\hat{c}$ do not form a codeword, $H\hat{c}$ does not equal zero, indicating the presence of one or more errors induced by the digital channel. Because the presence of an error can be mathematically written as $\hat{c} = c \oplus e$, with e a vector of binary values having a 1 in those positions where a bit error occurred.

## Exercise 6.28.2

Show that adding the error vector (1,0,...,0)T to a codeword flips the codeword's leading bit and leaves the rest unaffected.

Consequently, $H\hat{c} = Hc \oplus e = He$. Because the result of the product is a length- (N-K) vector of binary values, we can have 2N-K - 1 non-zero values that correspond to non-zero error patterns e. To perform our channel decoding,

- compute (conceptually at least) $H\hat{c}$;
- if this result is zero, no detectable or correctable error occurred
- if non-zero, consult a table of length- (N-K) binary vectors to associate them with the minimal error pattern that could have resulted in the non-zero result; then
- add the error vector thus obtained to the received vector $\hat{c}$ to correct the error (because $c \oplus e \oplus e = c$)

- Select the data bits from the corrected word to produce the received bit sequence $\hat{b}(n)$.

The phrase minimal in the third item raises the point that a double (or triple or quadruple ...) error occurring during the transmission/reception of one codeword can create the same received word as a single-bit error or no error in another codeword. For example, (1,0,0,0,1,0,1)T and (0,1,0,0,1,1,1)T are both codewords in the example (7,4) code. The second results when the first one experiences three bit errors (first, second, and sixth bits). Such an error pattern cannot be detected by our coding strategy, but such multiple error patterns are very unlikely to occur. Our receiver uses the principle of maximum probability: An error-free transmission is much more likely than one with three errors if the bit-error probability pe is small enough.

### Exercise 6.28.3

How small must pe be so that a single-bit error is more likely to occur than a triple-bit error?

## 6.29 ERROR-CORRECTING CODES: HAMMING CODES

For the (7,4) example, we have 2N-K -1 =7 error patterns that can be corrected. We start with single-bit error patterns, and multiply them by the parity check matrix. If we obtain unique answers, we are done; if two or more error patterns yield the same result, we can try double-bit error patterns. In our case, single-bit error patterns give a unique result.

### Parity Check Matrix

**Table 6.3**

| e | He |
|---|---|
| 1000000 | 101 |
| 0100000 | 111 |
| 0010000 | 110 |
| 0001000 | 011 |
| 0000100 | 100 |
| 0000010 | 010 |
| 0000001 | 001 |

This corresponds to our decoding table: We associate the parity check matrix multiplication result with the error pattern and add this to the received word. If more than one error occurs (unlikely though it may be), this "error correction" strategy usually makes the error worse in the sense that more bits are changed from what was transmitted.

As with the repetition code, we must question whether our (7,4) code's error correction capability compensates for the increased error probability due to the necessitated reduction in bit energy. Figure 6.23 shows that if the signal-to-noise ratio is large enough channel coding yields a smaller error probability. Because the bit stream emerging from the source decoder is segmented into four-bit blocks, the fair way of comparing coded and uncoded transmission is to compute the probability of block error: the probability that any bit in a block remains in error despite error correction and regardless of whether the error occurs in the data or in coding buts. Clearly, our (7,4) channel code does yield smaller error rates, and is worth the additional systems required to make it work.

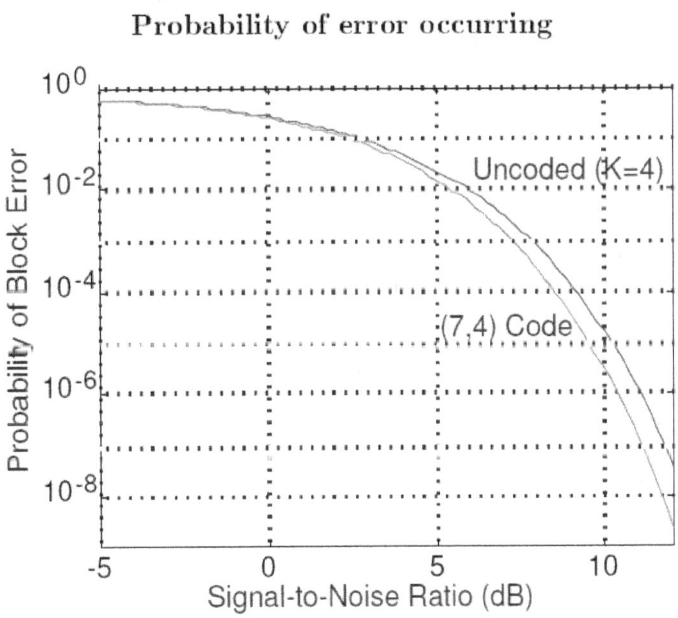


**Figure 6.23: The probability of an error occurring in transmitted K=4 data bits equals 1-(1-pe)4 as (1-pe)4 equals the probability that the four bits are received without error. The upper curve displays how this probability of an error anywhere in the four-bit block varies with the signal-to-noise ratio. When a (7,4) single-bit error correcting code is used, the transmitter reduced the energy it expends during a single-bit transmission by 4/7, appending three extra bits for error correction. Now the probability of any bit in the seven-bit block being in error after error correction equals 1-(1-pe›)7-(7pe›) (1-pe›)6 where pe› is the probability of a bit error occurring in the channel when channel coding occurs. Here (7pe›) (1-pe›)6 equals the probability of exactly on in seven bits emerging from the channel in error; The channel decoder corrects this type of error, and all data bits in the block are received correctly.**

Note that our (7,4) code has the length and number of data bits that perfectly fits correcting single bit errors. This pleasant property arises because the number of error patterns that can be corrected, 2N-K -1, equals the codeword length N. Codes that have 2N-K -1 = N are known as Hamming codes, and the following table provides the parameters of these codes. Hamming codes are the simplest single-bit error correction codes, and the generator/parity check matrix formalism for channel coding and decoding works for them.

### Hamming Codes

**Table 6.4**

| N | K | E (efficiency) |
|---|---|---|
| 3 | 1 | 0.33 |
| 7 | 4 | 0.57 |
| 15 | 11 | 0.73 |
| 31 | 26 | 0.84 |
| 63 | 57 | 0.90 |
| 127 | 120 | 0.94 |

Unfortunately, for such large blocks, the probability of multiple-bit errors can exceed the number of single-bit errors unless the channel single-bit error probability pe is very small. Consequently, we need to enhance the code's error correcting capability by adding double as well as single-bit error correction.

### Exercise 6.29.1

What must the relation between N and K be for a code to correct all single- and double-bit errors with a "perfect fit"?

## 6.30 NOISY CHANNEL CODING THEOREM

As the size of the block increases, a greater amount of mistake correction will be required. Are there any scripts that have the ability to rectify all errors? Claude Shannon's discovery of information theory is considered a remarkable accomplishment, as it provides a solution to this difficulty. The outcome is presented in two complimentary forms: the Noisy Channel Coding Theorem and its opposite.

### 6.30.1 Noisy Channel Coding Theorem

Let E represent the efficiency of an error-correcting code, which is defined as the ratio of the number of data bits to the total number of bits utilized to represent them. If the efficiency of the digital channel is lower than its capacity, it is possible to find an error-correcting code that, as its length rises, reduces the likelihood of errors occurring in the decoded block to zero.

$$\lim_{N\to\infty} Pr\,[\text{block error}] = 0 \quad , \quad E < C \tag{6.58}$$

## 6.30.2 Converse to the Noisy Channel Coding Theorem

If E > C, the probability of an error in a decoded block must approach one regardless of the code that might be chosen.

$$\lim_{N\to\infty} Pr\,[\text{block error}] = 1 \tag{6.59}$$

These results mean that it is possible to transmit digital information over a noisy channel (one that introduces errors) and receive the information without error if the code is sufficiently inefficient compared to the channel›s characteristics. Generally, a channel›s capacity changes with the signal-to-noise ratio: As one increases or decreases, so does the other. The capacity measures the overall error characteristics of a channel—the smaller the capacity the more frequently errors occur—and an overly efficient error-correcting code will not build in enough error correction capability to counteract channel errors.

This result astounded communication engineers when Shannon published it in 1948. Analog communication always yields a noisy version of the transmitted signal; in digital communication, error correction can be powerful enough to correct all errors as the block length increases. The key for this capability to exist is that the code's efficiency be less than the channel's capacity. For a binary symmetric channel, the capacity is given by

$$C = 1 + p_e \log_2 p_e - 1\log_2(1 - p_e)\ \text{bits/transmission} \tag{6.60}$$

Figure 6.24 (capacity of a channel) shows how capacity varies with error probability. For example, our (7,4) Hamming code has an efficiency of 0.57, and codes having the same efficiency but longer block sizes can be used on additive noise channels where the signal-to-noise ratio exceeds 0dB.

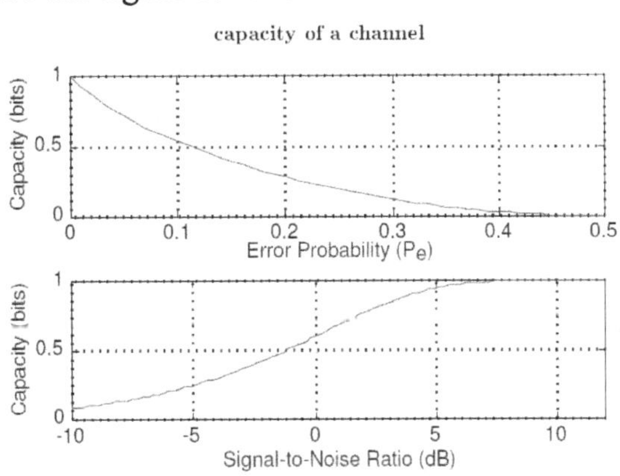


**Figure 6.24 The capacity per transmission through a binary symmetric channel is plotted as a function of the digital channel's error probability (upper) and as a function of the signal-to-noise ratio for a BPSK signal set (lower)**

## 6.31 CAPACITY OF A CHANNEL

In addition to the Noisy Channel Coding Theorem and its converse, Shannon also derived the capacity for a bandlimited (to W Hz) additive white noise channel. For this case, the signal set is unrestricted, even to the point that more than one bit can be transmitted each "bit interval." Instead of constraining channel code efficiency, the revised Noisy Channel Coding Theorem states that some error-correcting code exists such that as the block length increases, error-free transmission is possible if the source coder's datarate, $B\overline{(A)}\,R$, is less than capacity.

$$C = W\log_2(1 + \text{SNR}) \ \text{bits/s}$$

This result sets the maximum datarate of the source coder's output that can be transmitted through the bandlimited channel with no error. 1 Shannon's proof of his theorem was very clever, and did not indicate what this code might be; it has never been found. Codes such as the Hamming code work quite well in practice to keep error rates low, but they remain greater than zero. Until the "magic" code is found, more important in communication system design is the converse. It is indicated that if the pace at which your data is transmitted is higher than the capacity of the system, errors will become too many to handle, regardless of the type of channel coding employed. Capacity calculations are performed to determine the underlying constraints on transmission rates.

### Exercise 6.31.1

The first definition of capacity applies only for binary symmetric channels, and represents the number of bits/transmission. The second result states capacity more generally, having units of bits/second. How would you convert the first definition's result into units of bits/second?

### Exercise 6.5

The telephone channel has a bandwidth of 3 kHz and a signal-to-noise ratio exceeding 30 dB (at least they promise this much). The maximum data rate a modem can produce for this wireline channel and hope that errors will not become rampant is the capacity.

$$\begin{aligned} C &= 3\times 10^3\log_2\left(1+10^3\right) \\ &= 29.901 \ \text{kbps} \end{aligned} \tag{6.62}$$

Thus, the so-called 33 kbps modems operate right at the capacity limit.

Note that the data rate allowed by the capacity can exceed the bandwidth when the signal-to-noise ratio exceeds 0 dB. Our results for BPSK and FSK indicated the bandwidth they require exceeds $\frac{1}{T}$. What kind of signal sets might be used to achieve capacity? Modem signal sets send more than one bit/transmission using a number, one of the most popular of which is multi-level signaling. Here, we can transmit several bits during one transmission interval by representing bit by some signal's amplitude. For example, two bits can be sent with a signal set comprised of a sinusoid with amplitudes of ±(A) and $\pm\left(\frac{A}{2}\right)$.

## 6.32 COMPARISON OF ANALOG AND DIGITAL COMMUNICATION

Analog communication systems, such as amplitude modulation (AM) radio, are capable of transmitting a bandlimited analog signal in a cost-effective manner. This can be done either for point-to-point communication between two locations or for broadcasting from one point to multiple locations. While not depicted in this illustration, the coherent receiver (Figure 6.6) offers the maximum achievable signal-to-noise ratio for the demodulated message. An examination (Section 6.12) of this receiver therefore reveals that there will always be some remaining inaccuracy in the output of an analog system. While analog systems may be cheaper than digital systems for certain applications, digital systems provide significantly higher efficiency, superior performance, and greater flexibility.

- **Efficiency:** The Source Coding Theorem allows quantification of just how complex a given message source is and allows us to exploit that complexity by source coding (compression). In analog communication, the only parameters of interest are message bandwidth and amplitude. We cannot exploit signal structure to achieve a more efficient communication system.
- **Performance:** Because of the Noisy Channel Coding Theorem, we have a specific criterion by which to formulate error-correcting codes that can bring us as close to error-free transmission as we might want. Even though we may send information by way of a noisy channel, digital schemes are capable of error-free transmission while analog ones cannot overcome channel disturbances; see this problem (Problem 6.15) for a comparison.
- **Flexibility:** Digital communication systems can transmit real-valued discrete-time signals, which could be analog ones obtained by analog-to-digital conversion, and symbolic-valued ones (computer data, for example). Any signal that can be transmitted by analog means can be sent by digital means, with the only issue being the number of bits used in A/D conversion (how accurately do we need to represent signal amplitude). Images can be sent by analog means (commercial television), but better communication performance occurs when we use digital systems (HDTV). In addition to digital communication's ability to transmit a wider variety of signals than analog systems, point-to-point digital systems can be organized into global (and beyond as well) systems that provide efficient and flexible information transmission. Computer networks, explored in the next section, are what

we call such systems today. Even analog-based networks, such as the telephone system, employ modern computer networking ideas rather than the purely analog systems of the past.

Hence, due to the enhanced velocity of digital computers, the advancement of progressively effective algorithms, and the capability to connect computers to establish a communications infrastructure, digital communication has become the optimal selection for numerous scenarios.

## 6.33 COMMUNICATION NETWORKS

The Fundamental Model of Communications is developed by communication networks. The model depicted in Figure 6.25 accurately represents point-to-point communications, where the connection between the transmitter and receiver is direct and exclusive, without any interference from other sources. A contemporary illustration of this means of communication is the modem, which establishes a connection between a personal computer and an information server through a telephone line. The key aspect, some would say flaw, of this model is that the channel is dedicated: Only one communications link through the channel is allowed for all time. Irrespective of whether we are using a wired or wireless channel, the bandwidth for communication is valuable. If it can be shared without causing a significant decline in communication performance, as measured by the signal-to-noise ratio for analog signal transmission and the bit-error probability for digital transmission, it would be highly beneficial.

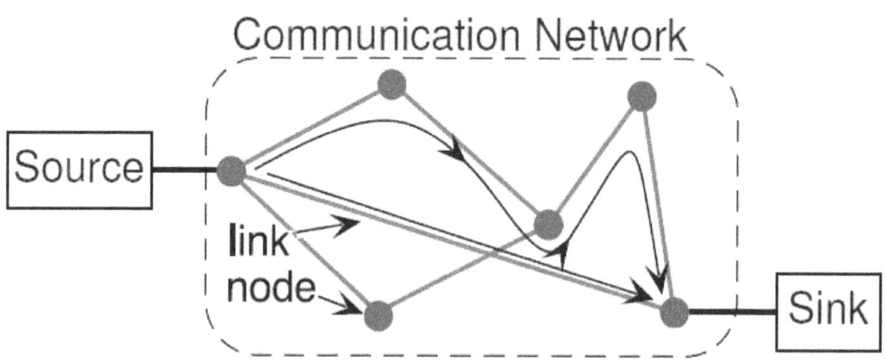


**Figure 6.25: The prototypical communications network—whether it be the postal service, cellular telephone, or the Internet—consists of nodes interconnected by links. Messages formed by the source are transmitted within the network by dynamic routing. Two routes are shown. The longer one would be used if the direct link were disabled or congested.**

The idea of a network first emerged with perhaps the oldest form of organized communication: the postal service. Most communication networks, even modern ones, share many of its aspects.

- A user writes a letter, serving in the communications context as the message source.
- This message is sent to the network by delivery to one of the network's public entry points. Entry points in the postal case are mailboxes, post offices, or your friendly mailman or mailwoman picking up the letter.
- The communications network delivers the message in the most efficient (timely) way possible, trying not to corrupt the message while doing so.
- The message arrives at one of the network's exit points, and is delivered to the recipient (what we have termed the message sink).

## Exercise 6.33.1

Develop the network model for the telephone system, making it as analogous as possible with the postal service-communications network metaphor.

What is most interesting about the network system is the ambivalence of the message source and sink about how the communications link is made. What they do care about is message integrity and communications efficiency. Furthermore, today's networks use heterogeneous links. Communication paths that form the Internet use wireline, optical fiber, and satellite communication links.

The first electrical communications network was the telegraph. Here the network consisted of telegraph operators who transmitted the message efficiently using Morse code and routed the message so that it took the shortest possible path to its destination while taking into account internal network failures (downed lines, drunken operators). From today's perspective, the fact that this nineteenth century system handled digital communications is astounding. Morse code, which assigned a sequence of dots and dashes to each letter of the alphabet, served as the source coding algorithm. The signal set consisted of a short and a long pulse. Rather than a matched filter, the receiver was the operator's ear, and he wrote the message (translating from received bits to symbols).

**Note:** Because of the need for a comma between dot-dash sequences to define letter (symbol) boundaries, the average number of bits/symbol, as described in Subtleties of Coding, exceeded the Source Coding Theorem's upper bound.

Internally, communication networks are equipped with point-to-point communication linkages between network nodes, which are accurately characterized by the Fundamental Model of Communications. However, numerous messages utilize the communications channel between nodes by employing time-domain multiplexing. Instead of the continuous manner of communication suggested in the provided Model, message sequences are transmitted, utilizing the channel's capacity at the same time. From a broader perspective, the network is responsible for directing messages by determining which nodes and links to utilize, based on destination information (i.e., the address), which is typically distinct from the message content. Network routing inherently involves dynamic processes. The message's whole route is determined by the network as it handles the message. Nodes that relay the message have an understanding of the most optimal path at the time of transmission. It is important to understand that there is no all-powerful router that sees the entire network and decides in advance the route for every message. Undoubtedly, the postal system employs dynamic routing, which takes into account factors such as non-functioning and excessively congested lines. Routing occurs in the telephone system during the initiation of a call, and the route becomes established once the phone begins to ring. Modern communication networks aim to attain optimal efficiency and reliability in delivering information.

## 6.34 MESSAGE ROUTING

When it comes to electrical networks, the majority of analog networks do not effectively utilize communication lines due to the challenge, if not impossibility, of achieving genuinely dynamic routing. In radio networks, similar as commercial television, each station is allocated a certain section of the electromagnetic spectrum, which cannot be shared with other stations or utilized in any way other than what is regulated. The telephone network is highly adaptable, but once a call is established, the route through the network remains unchanged. The users of the pathway exert authority over its utilization and may not optimize its efficiency, as evidenced by prolonged pauses during individual contemplation, for instance. Telephone network clients would be highly displeased if the telephone company temporarily interrupted the connection so that another person may utilize it. A connection that remains constant during the communication session, such as through a network, is referred to as a circuit-switched connection. In the 1960s, it became evident that digital communication was not only technically superior, but also required a new method than point-to-point for the many kinds of communication such as computer login, file transfer, and electronic mail. The concept of computer networks originated during this time, giving rise to what was once known as the ARPANET and is now referred to as the Internet. Computer networks enhance the fundamental network architecture by splitting communications into smaller units known as packets (see to Figure 6.26). The reason behind the network's decision to enforce smaller transmissions was that transferring huge files would excessively utilize network resources across the entire route. Additionally, due to the lengthy transmission time, a communication breakdown could necessitate the retransmission of the entire file. The network can effectively manage congestion by establishing individual packets with unique addresses that are routed independently of one another. The analogy might be described as follows: instead of sending a lengthy letter in the envelope you supply, the postal service opens the envelope, places each page in an individual envelope, and addresses each envelope separately based on the address provided on your envelope. These individual envelopes are then sent separately. It is essential for the network to ensure that the order of packets (page numbering) is preserved, and the network's exit point must reconstruct the original message properly.

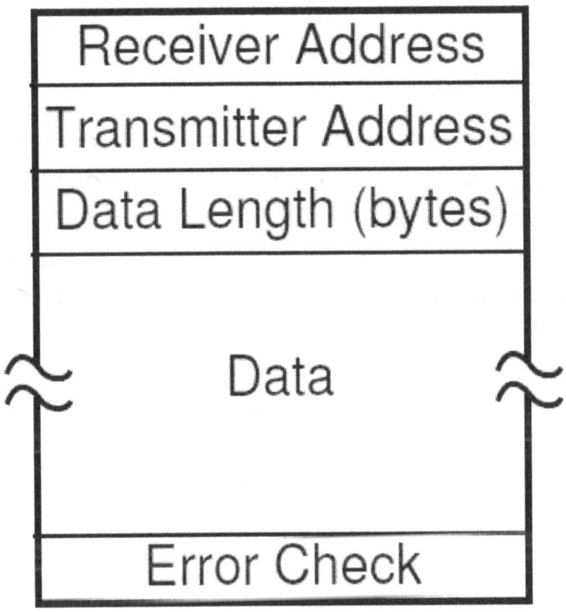


**Figure 6.26: Long messages, such as files, are broken into separate packets, then transmitted over computer networks. A packet, like a letter, contains the destination address, the return address (transmitter address), and the data. The data includes the message part and a sequence number identifying its order in the transmitted message.**

Communications networks are now classified based on whether they employ packet-based transmission or not. A system such as the telephone network is referred to as circuit switched, meaning that the network establishes a set path that remains unchanged for the whole duration of the message. Circuit switching offers the benefit of allowing customers to utilize the allocated capacity in any desired manner once the route has been established. A major drawback of this system is that users may not effectively utilize their capacity, resulting in congestion of network lines and nodes. Packet-switched networks constantly monitor network usage and route messages accordingly. Therefore, messages can be sent with efficiency, but the network cannot provide a certain quantity of capacity to the consumers.

## 6.35 NETWORK ARCHITECTURES AND INTERCONNECTION

The network's design, as depicted in Figure 6.25, exemplifies the characteristics of wide area networks (WANs). The nodes, as well as the users, are dispersed across vast geographical areas. The term "long" lacks a specific definition and is used to indicate a wide range of variations in communication lines. The Internet is undeniably the most extensive Wide Area Network (WAN), covering the entire globe and even extending beyond it. Local area networks, also known as LANs, utilize a solitary communication link and apply specialized routing. Ethernet is widely recognized as the most well-known Local Area Network (LAN).

LANs connect to other LANs and to wide area networks through special nodes known as gateways (Figure 6.27). In the Internet, a computer's address consists of a four byte sequence, which is known as its IP address (Internet Protocol address). An example address is 128.42.4.32: each byte is separated by a period. The first two bytes specify the computer's domain (here Rice University). Computers are also addressed by a more human-readable form: a sequence of alphabetic abbreviations representing institution, type of institution, and computer name. A given computer has both names (128.42.4.32 is the same as soma.rice.edu). Data transmission on the Internet requires the numerical form. So-called name servers translate between alphabetic and numerical forms, and the transmitting computer requests this translation before the message is sent to the network.

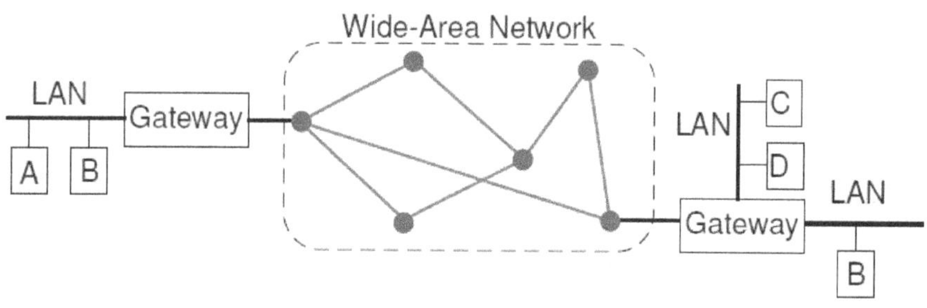


**Figure 6.27: The gateway serves as an interface between local area networks and the Internet. The two shown here translate between LAN and WAN protocols; one of these also interfaces between two LANs, presumably because together the two LANs would be geographically too dispersed.**

## 6.36 ETHERNET

Ethernet utilizes a solitary length of coaxial wire as its communication medium (Figure 6.28). This connection functions as the "ethereal" conduit through which all digital data traverse. Computers establish an electrical connection to the coaxial cable via a transceiver. This device has the ability to monitor the voltage between the core conductor and the shield, as well as apply a voltage to it. Conceptually, it comprises of two operational amplifiers, with one generating a voltage that represents a stream of bits (for transmitting data) and the other functioning as an amplifier for Ethernet voltage signals (for receiving data). The signal set for Ethernet is similar to the one depicted in BPSK Signal Sets, where one signal is the exact opposite of the other. Computers are connected in parallel, creating the circuit architecture for Ethernet as depicted in Figure 6.29.

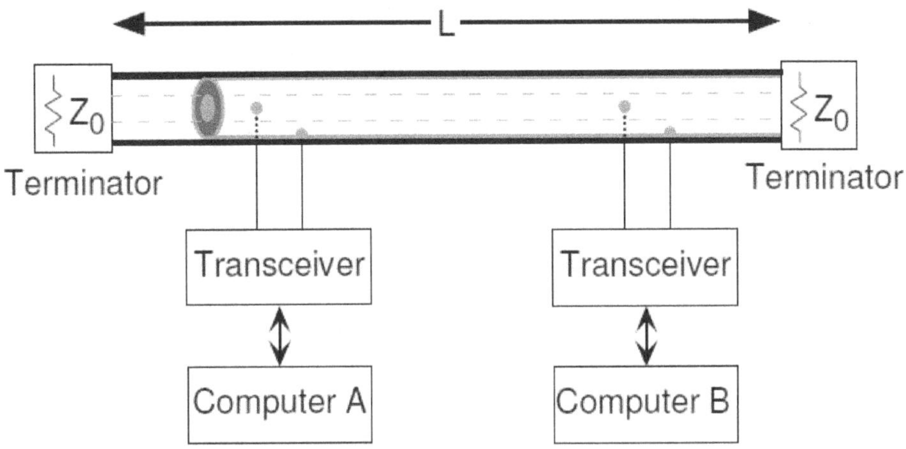


**Figure 6.28: The Ethernet architecture consists of a single coaxial cable terminated at either end by a resistor having a value equal to the cable's characteristic impedance. Computers attach to the Ethernet through an interface known as a transceiver because it sends as well as receives bit streams represented as analog voltages.**

### Exercise 6.36.1

From the viewpoint of a transceiver's sending op-amp, what is the load it sees and what is the transfer function between this output voltage and some other transceiver's receiving circuit? Why should the output resistor Rout be large?

There is no computer that has greater power or control than any other computer to determine the timing and method of sending signals. In the abs ence of scheduling authority, it is natural to question how one computer may communicate data to another without experiencing significant interference from other computers transmitting simultaneously. The key advancement of Ethernet is the utilization of a random-access technique for computers to arrange themselves. This approach is based on the principle that all packets sent across the coaxial cable can be received by all transceivers, regardless of the intended receiver computer. Ethernet is a communication technology that immediately enables broadcast. Every computer follows a series of processes to transmit a packet.

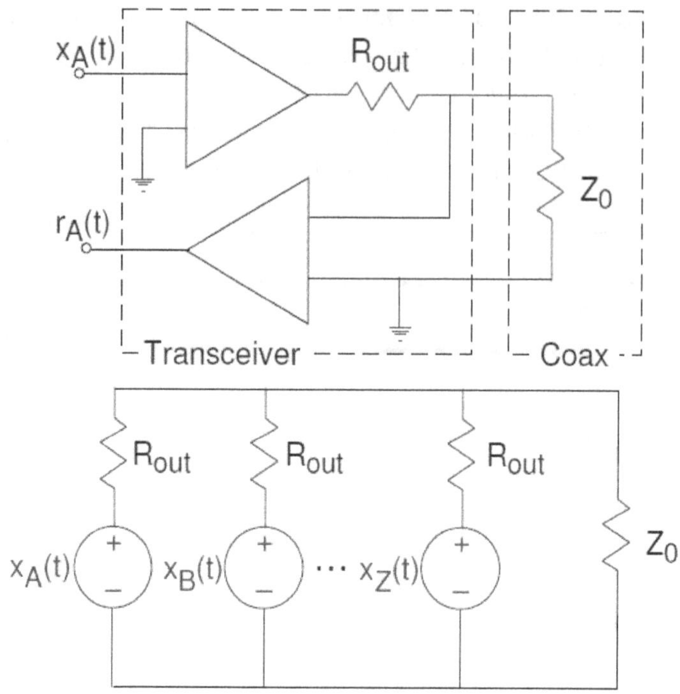


**Figure 6.29: The top circuit expresses a simplified circuit model for a transceiver. The output resistance Rout must be much larger than Z0 so that the sum of the various transmitter voltages add to create the Ethernet conductor-to-shield voltage that serves as the received signal r(t) for all transceivers. In this case, the equivalent circuit shown in the bottom circuit applies.**

- The computer senses the voltage across the cable to determine if some other computer is transmitting.
- If another computer is transmitting, wait until the transmissions finish and go back to the first step. If the cable has no transmissions, begin transmitting the packet.
- If the receiver portion of the transceiver determines that no other computer is also sending a packet, continue transmitting the packet until completion.
- On the other hand, if the receiver senses interference from another computer's transmissions, immediately cease transmission, waiting a random amount of time to attempt the transmission again (go to step 1) until only one computer transmits and the others defer. The condition wherein two (or more) computers' transmissions interfere with others is known as a collision.

The reason two computers waiting to transmit may not sense the other's transmission immediately arises because of the finite propagation speed of voltage signals through the coaxial cable. The longest time any computer must wait to determine if its transmissions do not encounter interference is $\frac{2L}{c}$, where L is the coaxial cable's length. The maximum-length-specification for Ethernet is 1 km. Assuming a propagation speed of 2/3 the speed of light, this time interval is more than 10 µs. As analyzed in Problem 22 (Problem 6.31), the number of these time intervals required to resolve the collision is, on the average, less than two!

### Exercise 6.36.2

Why does the factor of two enter into this equation? (Consider the worst-case situation of two transmitting computers located at the Ethernet's ends.)

Thus, despite not having separate communication paths among the computers to coordinate their transmissions, the Ethernet random access protocol allows computers to communicate without only a slight degradation in efficiency, as measured by the time taken to resolve collisions relative to the time the Ethernet is used to transmit information.

A subtle consideration in Ethernet is the minimum packet size Pmin. The time required to transmit such packets equals $\frac{P_{\min}}{C}$, where C is the Ethernet's capacity in bps. Ethernet now comes in two different types, each with individual specifications, the most distinguishing of which is capacity: 10 Mbps and 100 Mbps. If the minimum transmission time is such that the beginning of the packet has not propagated the full length of the Ethernet before the end-of-transmission, it is possible that two computers will begin transmission at the same time and, by the time their transmissions cease, the other's packet will not have propagated to the other. In this case, computers in-between the two will sense a collision, which renders both computer's transmissions senseless to them, without the two transmitting computers knowing a collision has occurred at all! For Ethernet to succeed, we must have the minimum packet transmission time exceed twice the voltage propagation time: $\frac{P_{\min}}{C} > \frac{2L}{c}$ or

$$P_{\min} > \frac{2LC}{c}$$

Thus, for the 10 Mbps Ethernet having a 1 km maximum length specification, the minimum packet size is 200 bits.

### Exercise 6.36.3

The 100 Mbps Ethernet was designed more recently than the 10 Mbps alternative. To maintain the same minimum packet size as the earlier, slower version, what should its length specification be? Why should the minimum packet size remain the same?

## 6.37 COMMUNICATION PROTOCOLS

The intricate process of transmitting information in a computer network, which includes ensuring the dependable transfer of individual bits across a channel, directing the information to the appropriate destination within the operating system of the receiving computers, necessitates a comprehensive framework for organizing the delivery of information. There is no one set of rules that can meet all the requirements imposed by communication channels and network organization for transmitting information. Random access concerns, which can occur in Ethernet networks, are not found in wide-area networks like the Internet. A protocol is a defined set of regulations that dictate the manner in which information is transmitted. When using the telephone network, the procedure entails lifting the receiver, listening for a dial tone, dialing a number with a predetermined amount of digits, waiting for the phone to ring, and greeting the caller with "hello".

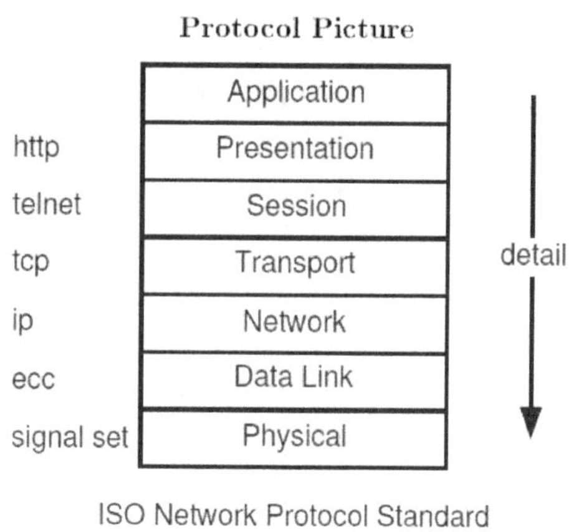


**Figure 6.30: Protocols are organized according to the level of detail required for information transmission. Protocols at the lower levels (shown toward the bottom) concern reliable bit transmission. Higher level protocols concern how bits are organized to represent information, what kind of information is defined by bit sequences, what software needs the information, and how the information is to be interpreted. Bodies such as the IEEE (Institute for Electronics and Electrical Engineers) and the ISO (International Standards Organization) define standards such as this. Despite being a standard, it does not constrain protocol implementation so much that innovation and competitive individuality are ruled out.**

In radio broadcasting, the station employs either amplitude or frequency modulation techniques, utilizing a designated carrier frequency and transmission bandwidth. To listen to the station, you only need to switch on the radio and adjust the tuning to the desired frequency. From a technical standpoint, it is not possible to employ a single protocol or a specific collection of protocols for every communication scenario. However, communication engineers have discovered that there is a shared element that connects the organization of different protocols. This intricate architecture of information dissemination permeates all contemporary networks at now. A layered, hierarchical protocol organization has been defined as the networking standard. Figure 6.30 (Protocol Picture) illustrates that protocols are categorized based on their function and level of information.

The categorization of information transmission, modification, and interpretation has a direct impact on the organization of communication systems and the specific functions fulfilled by software systems. Previously, this organizational structure was not considered in the same manner. However, it now dictates the thought process of communication engineers when it comes to all communication systems, ranging from radio to the Internet.

### Exercise 6.37.1

How do the various aspects of establishing and maintaining a telephone conversation fit into this layered protocol organization?

We now explicitly state whether we are working in the physical layer (signal set design, for example), the data link layer (source and channel coding), or any other layer. IP abbreviates Internet protocol, and governs gateways (how information is transmitted between networks having different internal organizations). TCP (transmission control protocol) governs how packets are transmitted through a wide-area network such as the Internet. Telnet is a protocol that concerns how a person at one computer logs on to another computer across a network. A moderately high level protocol such as telnet, is not concerned with what data links (wireline or wireless) might have been used by the network or how packets are routed. Rather, it establishes connections between computers and directs each byte (presumed to represent a typed character) to the appropriate operation system component at each end. It is not concerned with what the characters mean or what programs the person is typing to. That aspect of information transmission is left to protocols at higher layers.

The World Wide Web was recently established through the development of a significant set of protocols. These protocols are autonomous from the Internet. The Internet ensures the efficient and complete transmission of messages, without regard to their content. An HTTP frame contains messages and specifies how the data should be handled. The rapid evolution of the Web, built upon a mostly unchanged Internet, exemplifies the ability to organize information transfer without excessively limiting the specifics.

## 6.38 INFORMATION COMMUNICATION PROBLEMS

### Problem 6.1: Signals on Transmission Lines

A modulated signal needs to be sent over a transmission line having a characteristic impedance of $Z0 = 50\Omega$. So that the signal does not interfere with signals others may be transmitting, it must be bandpass filtered so that its bandwidth is 1 MHz and centered at 3.5 MHz. The filter's gain should be one in magnitude. An op-amp filter (Figure 6.31) is proposed.

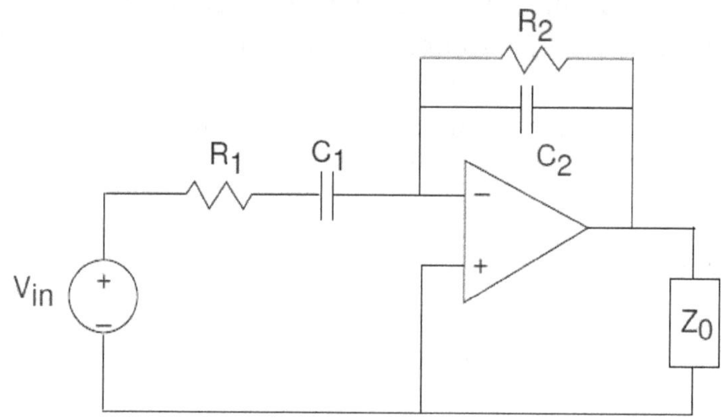


**Figure 6.31**

- What is the transfer function between the input voltage and the voltage across the transmission line?
- Find values for the resistors and capacitors so that design goals are met.

### Problem 6.2: Noise in AM Systems

The signal $\hat{s}(t)$ emerging from an AM communication system consists of two parts: the message signal, s(t), and additive noise. The plot (Figure 6.32) shows the message spectrum S(f) and noise power spectrum PN(f). The noise power spectrum lies completely within the signal's band, and has a constant value there of $\frac{N_0}{2}$.

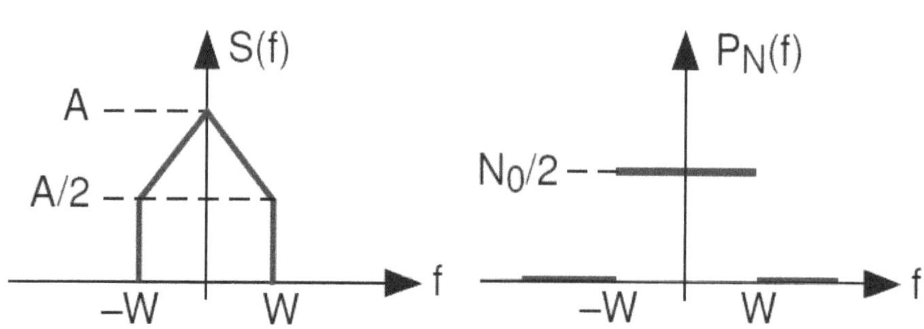


**Figure 6.32**

- What is the message signal's power? What is the signal-to-noise ratio?
- Because the power in the message decreases with frequency, the signal-to-noise ratio is not constant within subbands. What is the signal-to-noise ratio in the upper half of the frequency band?
- A clever 241 student suggests filtering the message before the transmitter modulates it so that the signal spectrum is balanced (constant) across frequency. Realizing that this filtering affects the message signal, the student realizes that the receiver must also compensate for the message to arrive intact. Draw a block diagram of this communication system. How does this system›s signal-to-noise ratio compare with that of the usual AM radio?

### Problem 6.3: Complementary Filters

Complementary filters usually have "opposite" filtering characteristics (like a lowpass and a highpass) and have transfer functions that add to one. Mathematically, H1(f)+H2(f) are complementary if

$$H_1(f) + H_2(f) = 1$$

We can use complementary filters to separate a signal into two parts by passing it through each filter. Each output can then be transmitted separately and the original signal reconstructed at the receiver. Let's assume the message is bandlimited to W Hz and that $H_1(f) = \frac{a}{a+j2\pi f}$.

- What circuits would be used to produce the complementary filters?
- Sketch a block diagram for a communication system (transmitter and receiver) that employs complementary signal transmission to send a message m(t).
- What is the receiver's signal-to-noise ratio? How does it compare to the standard system that sends the signal by simple amplitude modulation?

### Problem 6.4: Phase Modulation

A message signal m(t) phase modulates a carrier if the transmitted signal equals

$$x(t) = A\sin(2\pi f_c t + \phi_d m(t))$$

where φd is known as the phase deviation. In this problem, the phase deviation is small. As with all analog modulation schemes, assume that |m(t)|<1, the message is bandlimited to W Hz, and the carrier frequency fc is much larger than W.

- What is the transmission bandwidth?
- Find a receiver for this modulation scheme.

- What is the signal-to-noise ratio of the received signal?

**Hint:** Use the facts that $\cos(x) \simeq 1$ and $\sin(x) \simeq x$ for small x.

### Problem 6.5: Digital Amplitude Modulation

Two ELEC 241 students disagree about a homework problem. The issue concerns the discrete-time signal s(n) cos (2πf0n), where the signal s(n) has no special characteristics and the modulation frequency f0 is known. Sammy says that he can recover s(n) from its amplitude-modulated version by the same approach used in analog communications. Samantha says that approach won't work.

- What is the spectrum of the modulated signal?
- Who is correct? Why?
- The teaching assistant does not want to take sides. He tells them that if s(n) cos(2πf0n) and s(n) sin(2πf0n) were both available, s(n) can be recovered. What does he have in mind?

### Problem 6.6: Anti-Jamming

One way for someone to keep people from receiving an AM transmission is to transmit noise at the same carrier frequency. Thus, if the carrier frequency is fc so that the transmitted signal is AT (1+m(t)) sin (2πfct) the jammer would transmit AJn (t) sin (2πfct+φ). The noise n(t) has a constant power density spectrum over the bandwidth of the message m(t). The channel adds white noise of spectral height $\frac{N_0}{2}$.

- What would be the output of a traditional AM receiver tuned to the carrier frequency fc?
- RU Electronics proposes to counteract jamming by using a different modulation scheme. The scheme's transmitted signal has the form AT (1+m(t))c(t) where c(t) is a periodic carrier signal (period $\frac{1}{f_c}$) having the waveform as indicated in Fig 6.33. What is the spectrum of the transmitted signal with the proposed scheme? Assume the message bandwidth W is much less than the fundamental carrier frequency fc.
- The jammer, unaware of the change, is transmitting with a carrier frequency of fc, while the receiver tunes a standard AM receiver to a harmonic of the carrier frequency. What is the signal-to-noise ratio of the receiver tuned to the harmonic having the largest power that does not contain the jammer?

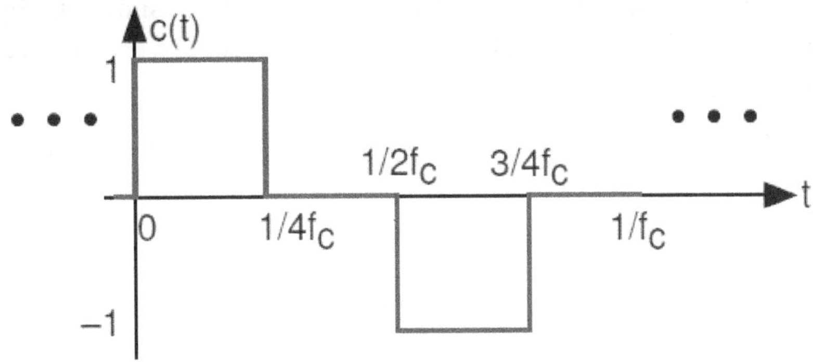


Figure 6.33

## Problem 6.7: Secret Communications

A system for hiding AM transmissions has the transmitter randomly switching between two carrier frequencies f1 and f2. "Random switching" means that one carrier frequency is used for some period of time, switches to the other for some other period of time, back to the first, etc. The receiver knows what the carrier frequencies are but not when carrier frequency switches occur. Consequently, the receiver must be designed to receive the transmissions regardless of which carrier frequency is used. Assume the message signal has bandwidth W. The channel adds white noise of spectral height $\frac{N_0}{2}$.

- How different should the carrier frequencies be so that the message could be received?
- What receiver would you design?
- What signal-to-noise ratio for the demodulated signal does your receiver yield?

## Problem 6.8: AM Stereo

Stereophonic radio transmits two signals simultaneously that correspond to what comes out of the left and right speakers of the receiving radio. While FM stereo is commonplace, AM stereo is not, but is much simpler to understand and analyze. An amazing aspect of AM stereo is that both signals are transmitted within the same bandwidth as used to transmit just one. Assume the left and right signals are bandlimited to W Hz.

$$x(t) = A(1 + m_l(t))\cos(2\pi f_c t) + A m_r(t)\sin(2\pi f_c t)$$

- Find the Fourier transform of x(t). What is the transmission bandwidth and how does it compare with that of standard AM?

- Let us use a coherent demodulator as the receiver, shown in Figure 6.34. Show that this receiver indeed works: It produces the left and right signals separately.
- Assume the channel adds white noise to the transmitted signal. Find the signal-to-noise ratio of each signal.

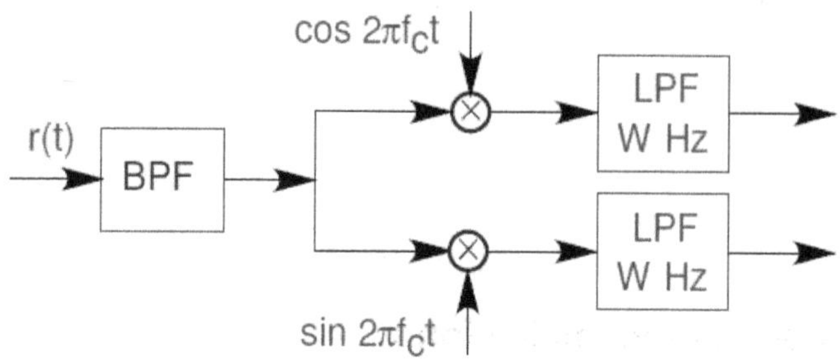


**Figure 6.34**

### Problem 6.9: A Novel Communication System

A clever system designer claims that the transmitter depicted in Figure 6.35 has, despite its complexity, advantages over the usual amplitude modulation system. The message signal m(t) is bandlimited to W Hz, and the carrier frequency fc >> W. The channel attenuates the transmitted signal x(t) and adds white noise of spectral height $\frac{N_0}{2}$.

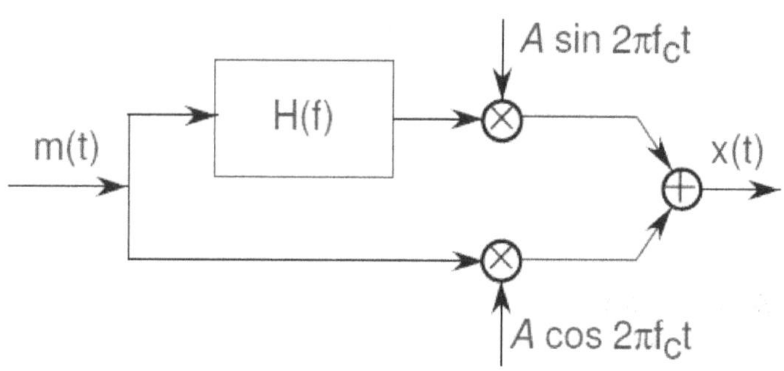


**Figure 6.35**

The transfer function H(f) is given by $H(f)=\begin{cases} j & \text{if } f<0 \\ -j & \text{if } f>0 \end{cases}$

- Find an expression for the spectrum of x(t). Sketch your answer.
- Show that the usual coherent receiver demodulates this signal.
- Find the signal-to-noise ratio that results when this receiver is used.
- Find a superior receiver (one that yields a better signal-to-noise ratio), and analyze its performance.

## Problem 6.10: Multi-Tone Digital Communication

In a so-called multi-tone system, several bits are gathered together and transmitted simultaneously on different carrier frequencies during a T second interval. For example, B bits would be transmitted according to

$$x(t) = A\sum_{k=0}^{B-1} b_k \sin(2\pi(k+1)f_0 t) \quad , \quad 0 \le t < T \tag{6.64}$$

Here, f0 is the frequency offset for each bit and it is harmonically related to the bit interval T. The value of bk is either -1 or 1.

- Find a receiver for this transmission scheme.
- An ELEC 241 almuni likes digital systems so much that he decides to produce a discrete-time version. He samples the received signal (sampling interval $T_s = \frac{T}{N}$). How should N be related to B, the number of simultaneously transmitted bits?
- The alumni wants to find a simple form for the receiver so that his software implementation runs as efficiently as possible. How would you recommend he implement the receiver?

## Problem 6.11: City Radio Channels

In addition to additive white noise, metropolitan cellular radio channels also contain multipath: the attenuated signal and a delayed, further attenuated signal are received superimposed. As shown in Figure 6.36, multipath occurs because the buildings reflect the signal and the reflected path length between transmitter and receiver is longer than the direct path.

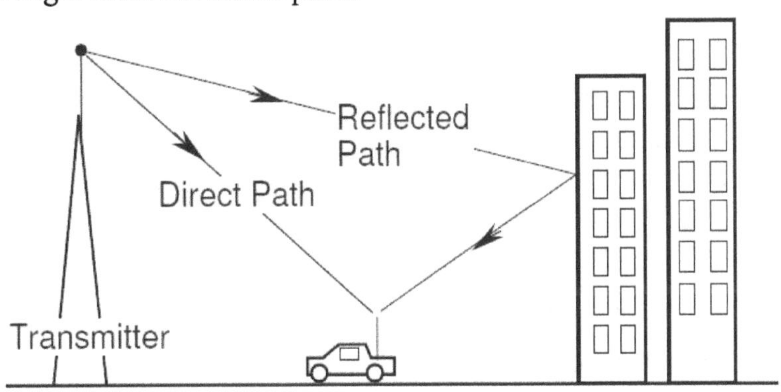


**Figure 6.36**

- Assume that the length of the direct path is d meters and the reflected path is 1.5 times as long. What is the model for the channel, including the multipath and the additive noise?

- Assume d is 1 km. Find and sketch the magnitude of the transfer function for the multipath component of the channel. How would you characterize this transfer function?
- Would the multipath affect AM radio? If not, why not; if so, how so? Would analog cellular telephone, which operates at much higher carrier frequencies (800 MHz vs. 1 MHz for radio), be affected or not? Analog cellular telephone uses amplitude modulation to transmit voice.
- How would the usual AM receiver be modified to minimize multipath effects? Express your modified receiver as a block diagram.

### Problem 6.12: Downlink Signal Sets

Within digital cellular telephone systems, the base station (transmitter) is required to broadcast several speech signals simultaneously to various telephones. Instead of transmitting signals at varying frequencies, a resourceful Rice engineer proposes employing a distinct signal set for each data stream. As an illustration, she proposes using BPSK signal sets for two concurrent data streams, which are represented by the fundamental signal shown in Figure 6.37.

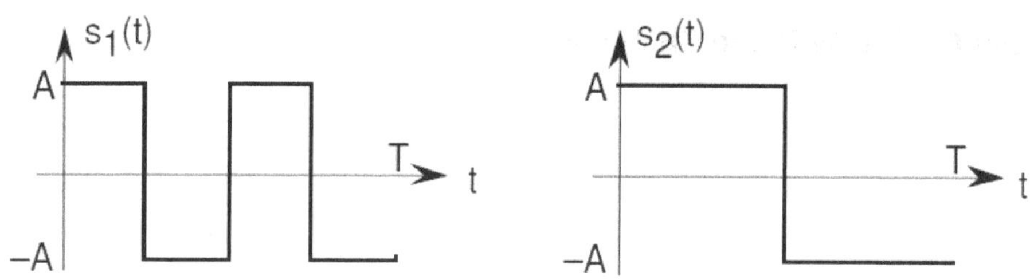


**Figure 6.37**

Thus, bits are represented in data stream 1 by s1(t) and -s1(t) and in data stream 2 by s2(t) and -s2(t), each of which are modulated by 900 MHz carrier. The transmitter sends the two data streams so that their bit intervals align. Each receiver uses a matched filter for its receiver. The requirement is that each receiver not receive the other›s bit stream.

- What is the block diagram describing the proposed system?
- What is the transmission bandwidth required by the proposed system?
- Will the proposal work? Does the fact that the two data streams are transmitted in the same bandwidth at the same time mean that each receiver's performance is affected? Can each bit stream be received without interference from the other?

## Problem 6.13: Mixed Analog and Digital Transmission

A signal m(t) is transmitted using amplitude modulation in the usual way. The signal has bandwidth W Hz, and the carrier frequency is fc. In addition to sending this analog signal, the transmitter also wants to send ASCII text in an auxiliary band that lies slightly above the analog transmission band. Using an 8-bit representation of the characters and a simple baseband BPSK signal set (the constant signal +1 corresponds to a 0, the constant -1 to a 1), the data signal d(t) representing the text is transmitted as the same time as the analog signal m(t). The transmission signal spectrum is as shown in Figure 6.38 and has a total bandwidth B.

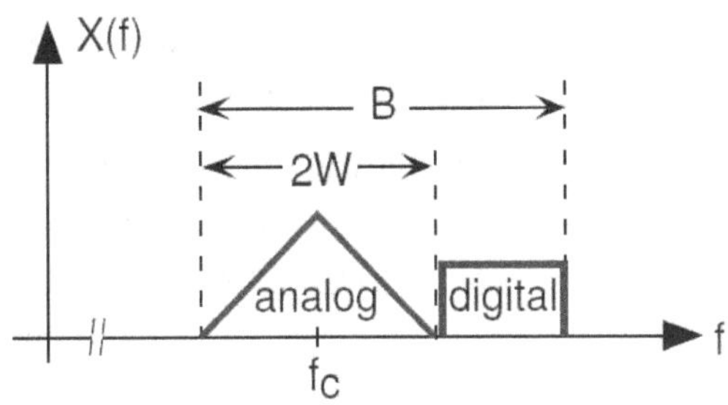


**Figure 6.38**

- Write an expression for the time-domain version of the transmitted signal in terms of m(t) and the digital signal d(t).
- What is the maximum datarate the scheme can provide in terms of the available bandwidth?
- Find a receiver that yields both the analog signal and the bit stream.

## Problem 6.14: Digital Stereo

Just as with analog communication, it should be possible to send two signals simultaneously over a digital channel. Assume you have two CD-quality signals (each sampled at 44.1 kHz with 16 bits/sample). One suggested transmission scheme is to use a quadrature BPSK scheme. If b(1)n and b(2)n each represent a bit stream, the transmitted signal has the form

$$x(t) = A\sum_{n} b^{(1)}(n)\sin(2\pi f_c(t-nT))\,p(t-nT) + b^{(2)}(n)\cos(2\pi f_c(t-nT))\,p(t-nT)$$

where p(t) is a unit-amplitude pulse having duration T and b(1)n and b(2)n equal either +1 or -1 according to the bit being transmitted for each signal. The channel adds white noise and attenuates the transmitted signal.

- What value would you choose for the carrier frequency fc?
- What is the transmission bandwidth?
- What receiver would you design that would yield both bit streams?

### Problem 6.15: Digital and Analog Speech Communication

Suppose we transmit speech signals over comparable digital and analog channels. We want to compare the resulting quality of the received signals. Assume the transmitters use the same power, and the channels introduce the same attenuation and additive white noise. Assume the speech signal has a 4 kHz bandwidth and, in the digital case, is sampled at an 8 kHz rate with eight-bit A/D conversion. Assume simple binary source coding and a modulated BPSK transmission scheme.

- What is the transmission bandwidth of the analog (AM) and digital schemes?
- Assume the speech signal's amplitude has a magnitude less than one. What is maximum amplitude quantization error introduced by the A/D converter?
- In the digital case, each bit in quantized speech sample is received in error with probability pe that depends on signal-to-noise ratio $\frac{E_b}{N_0}$. However, errors in each bit have a different impact on the error in the reconstructed speech sample. Find the mean-squared error between the transmitted and received amplitude.
- In the digital case, the recovered speech signal can be considered to have two noise sources added to each sample's true value: One is the A/D amplitude quantization noise and the second is due to channel errors. Because these are separate, the total noise power equals the sum of these two. What is the signal-to-noise ratio of the received speech signal as a function of pe?
- Compute and plot the received signal's signal-to-noise ratio for the two transmission schemes as a function of channel signal-to-noise ratio.
- Compare and evaluate these systems.

### Problem 6.16: Source Compression

Consider the following 5-letter source.

**Table 6.5**

| Letter | Probability |
|---|---|
| a | 0.5 |
| b | 0.25 |
| c | 0.125 |
| d | 0.0625 |
| e | 0.0625 |

Find this source's entropy.

- Show that the simple binary coding is inefficient.

- Find an unequal-length codebook for this sequence that satisfies the Source Coding Theorem. Does your code achieve the entropy limit?
- How much more efficient is this code than the simple binary code?

## Problem 6.17: Source Compression

Consider the following 5-letter source.

**Table 6.6**

| Letter | Probability |
|---|---|
| a | 0.4 |
| b | 0.2 |
| c | 0.15 |
| d | 0.15 |
| e | 0.1 |

Find this source's entropy.

- Show that the simple binary coding is inefficient.
- Find the Huffman code for this source. What is its average code length?

## Problem 6.18: Speech Compression

When we sample a signal, such as speech, we quantize the signal's amplitude to a set of integers. For a b-bit converter, signal amplitudes are represented by 2b integers. Although these integers could be represented by a binary code for digital transmission, we should consider whether a Huffman coding would be more efficient.

- Load into Matlab the segment of speech contained in y.mat. Its sampled values lie in the interval (-1, 1). To simulate a 3-bit converter, we use Matlab's round function to create quantized amplitudes corresponding to the integers [0 1 2 3 4 5 6 7].
    - y_quant = round(3.5*y + 3.5);

Find the relative frequency of occurrence of quantized amplitude values. The following Matlab program computes the number of times each quantized value occurs.

- for n=0:7; count(n+1) = sum(y_quant == n); end;

Find the entropy of this source.

- Find the Huffman code for this source. How would you characterize this source code in words?

- How many fewer bits would be used in transmitting this speech segment with your Huffman code in comparison to simple binary coding?

## Problem 6.19: Digital Communication

In a digital cellular system, a signal bandlimited to 5 kHz is sampled with a two-bit A/D converter at its Nyquist frequency. The sample values are found to have the shown relative frequencies.

**Table 6.7**

| Sample Value | Probability |
|---|---|
| 0 | 0.15 |
| 1 | 0.35 |
| 2 | 0.3 |
| 3 | 0.2 |

We send the bit stream consisting of Huffman-coded samples using one of the two depicted signal sets (Figure 6.39).

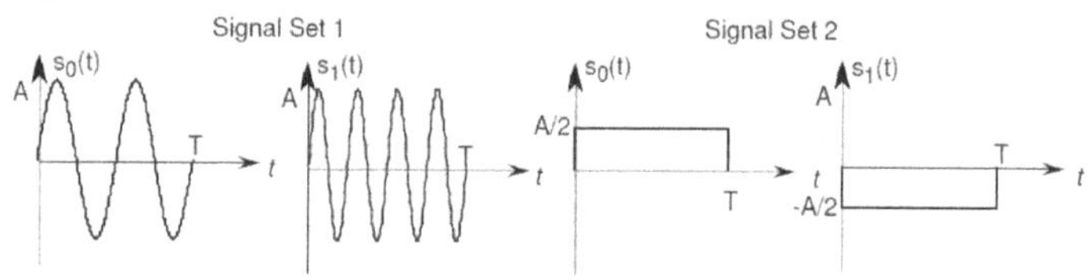


**Figure 6.39**

- What is the datarate of the compressed source?
- Which choice of signal set maximizes the communication system's performance?
- With no error-correcting coding, what signal-to-noise ratio would be needed for your chosen signal set to guarantee that the bit error probability will not exceed 10-3? If the receiver moves twice as far from the transmitter (relative to the distance at which the 10-3 error rate was obtained), how does the performance change?

## Problem 6.20: Signal Compression

Letters drawn from a four-symbol alphabet have the indicated probabilities.

**Table 6.8**

| Letter | Probability |
|---|---|
| a | 1/3 |
| b | 1/3 |
| c | 1/4 |
| d | 1/12 |

What is the average number of bits necessary to represent this alphabet?

- Using a simple binary code for this alphabet, a two-bit block of data bits naturally emerges. Find an error correcting code for two-bit data blocks that corrects all single-bit errors.
- How would you modify your code so that the probability of the letter a being confused with the letter d is minimized? If so, what is your new code; if not, demonstrate that this goal cannot be achieved.

## Problem 6.21: Universal Product Code

The Universal Product Code (UPC), often known as a bar code, labels virtually every sold good. An example (Figure 6.40) of a portion of the code is shown.

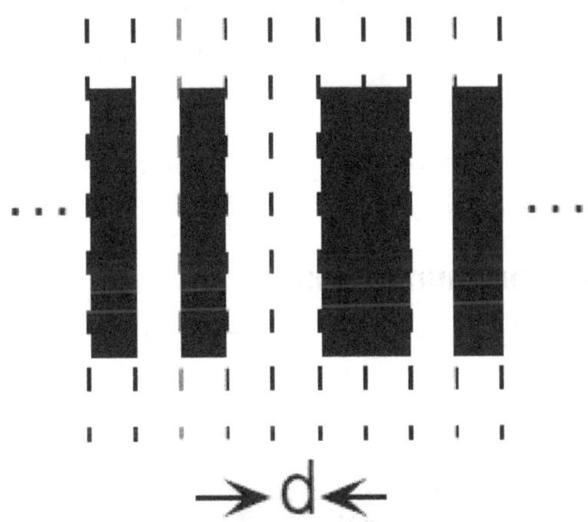


**Figure 6.40**

Here a sequence of black and white bars, each having width d, presents an 11-digit number (consisting of decimal digits) that uniquely identifies the product. In retail stores, laser scanners read this code, and after accessing a database of prices, enter the price into the cash register.

- How many bars must be used to represent a single digit?
- A complication of the laser scanning system is that the bar code must be read either forwards or backwards. Now how many bars are needed to represent each digit?
- What is the probability that the 11-digit code is read correctly if the probability of reading a single bit incorrectly is pe?
- How many error correcting bars would need to be present so that any single bar error occurring in the 11-digit code can be corrected?

## Problem 6.22: Error Correcting Codes

A code maps pairs of information bits into codewords of length 5 as follows.

**Table 6.9**

| Data | Codeword |
|---|---|
| 00 | 00000 |
| 01 | 01101 |
| 10 | 10111 |
| 11 | 11010 |

What is this code's efficiency?

- Find the generator matrix G and parity-check matrix H for this code.
- Give the decoding table for this code. How many patterns of 1, 2, and 3 errors are correctly decoded?
- What is the block error probability (the probability of any number of errors occurring in the decoded codeword)?

## Problem 6.23: Digital Communication

A digital source produces sequences of nine letters with the following probabilities.

**Table 6.10**

| letter | a | b | c | d | e | f | g | h | i |
|---|---|---|---|---|---|---|---|---|---|
| probability | $\frac{1}{4}$ | $\frac{1}{8}$ | $\frac{1}{8}$ | $\frac{1}{8}$ | $\frac{1}{8}$ | $\frac{1}{16}$ | $\frac{1}{16}$ | $\frac{1}{16}$ | $\frac{1}{16}$ |

Find a Huffman code that compresses this source. How does the resulting code compare with the best possible code?

- A clever engineer proposes the following (6,3) code to correct errors after transmission through a digital channel.

**Table 6.11**

| | |
|---|---|
| c1=d1 | c4=d1⊕d2⊕d3 |
| c2=d2 | C5=d2⊕d3 |
| c3=d3 | c6=d1 |

- What is the error correction capability of this code?
- The channel's bit error probability is 1/8. What kind of code should be used to transmit data over this channel?

## Problem 6.24: Overly Designed Error Correction Codes

An Aggie engineer wants not only to have codewords for his data, but also to hide the information from Rice engineers (no fear of the UT engineers). He decides to represent 3-bit data with 6-bit codewords in which none of the data bits appear explicitly.

**Table 6.12**

| | |
|---|---|
| $c1=d1\oplus d2$ | $c4=d1\oplus d2\oplus d3$ |
| $c2=d2\oplus d3$ | $c5=d1\oplus d2$ |
| $c3=d1\oplus d3$ | $c6=d1\oplus d2\oplus d3$ |

Find the generator matrix G and parity-check matrix H for this code.

- Find a 3×6 matrix that recovers the data bits from the codeword.
- What is the error correcting capability of the code?

## Problem 6.25: Error Correction?

It is important to realize that when more transmission errors than can be corrected, error correction algorithms believe that a smaller number of errors have occurred and correct accordingly. For example, consider a (7,4) Hamming code having the generator matrix

$$G = \begin{pmatrix} 1 & 0 & 0 & 0 \\ 0 & 1 & 0 & 0 \\ 0 & 0 & 1 & 0 \\ 0 & 0 & 0 & 1 \\ 1 & 1 & 1 & 0 \\ 0 & 1 & 1 & 1 \\ 1 & 0 & 1 & 1 \end{pmatrix}$$

This code corrects all single-bit error, but if a double bit error occurs, it corrects using a single-bit error correction approach.

- How many double-bit errors can occur in a codeword?
- For each double-bit error pattern, what is the result of channel decoding? Express your result as a binary error sequence for the data bits.

### Problem 6.26: Selective Error Correction

We have found that digital transmission errors occur with a probability that remains constant no matter how "important" the bit may be. For example, in transmitting digitized signals, errors occur as frequently for the most significant bit as they do for the least significant bit. Yet, the former errors have a much larger impact on the overall signal-to-noise ratio than the latter. Rather than applying error correction to each sample value, why not concentrate the error correction on the most important bits? Assume that we sample an 8 kHz signal with an 8-bit A/D converter. We use single-bit error correction on the most significant four bits and none on the least significant four. Bits are transmitted using a modulated BPSK signal set over an additive white noise channel.

- How many error correction bits must be added to provide single-bit error correction on the most significant bits?
- How large must the signal-to-noise ratio of the received signal be to insure reliable communication?
- Assume that once error correction is applied, only the least significant 4 bits can be received in error. How much would the output signal-to-noise ratio improve using this error correction scheme?

### Problem 6.27: Compact Disk

Errors occur in reading audio compact disks. Very few errors are due to noise in the compact disk player; most occur because of dust and scratches on the disk surface. Because scratches span several bits, a single-bit error is rare; several consecutive bits in error are much more common. Assume that scratch and dust-induced errors are four or fewer consecutive bits long. The audio CD standard requires 16-bit, 44.1 kHz analog-to-digital conversion of each channel of the stereo analog signal.

- How many error-correction bits are required to correct scratch-induced errors for each 16-bit sample?
- Rather than use a code that can correct several errors in a codeword, a clever 241 engineer proposes interleaving consecutive coded samples. As the cartoon shows, the bits representing coded samples are interpersed before they are written on the CD. The CD player de-interleaves the coded data, then performs error-correction. Now, evaluate this proposed scheme with respect to the non-interleaved one.

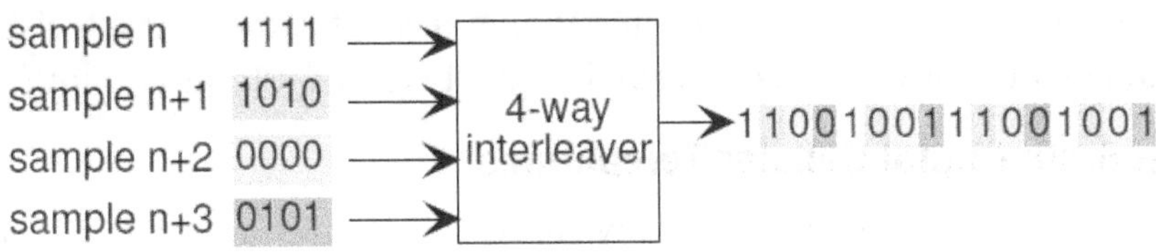


**Figure 6.41**

## Problem 6.28: Communication System Design

RU Communication Systems has been asked to design a communication system that meets the following requirements.

- The baseband message signal has a bandwidth of 10 kHz.
- The RUCS engineers find that the entropy H of the sampled message signal depends on how many bits b are used in the A/D converter (see table below).
- The signal is to be sent through a noisy channel having a bandwidth of 25 kHz channel centered at 2 MHz and a signal-to-noise ration within that band of 10 dB.
- Once received, the message signal must have a signal-to-noise ratio of at least 20 dB.

**Table 6.13**

| b | H |
|---|---|
| 3 | 2.19 |
| 4 | 3.25 |
| 5 | 4.28 |
| 6 | 5.35 |

Can these specifications be met? Justify your answer.

## Problem 6.29: HDTV

As HDTV (high-definition television) was being developed, the FCC restricted this digital system to use in the same bandwidth (6 MHz) as its analog (AM) counterpart. HDTV video is sampled on a 1035×1840 raster at 30 images per second for each of the three colors. The least-acceptable picture received by television sets located at an analog station's broadcast perimeter has a signal-to-noise ratio of about 10 dB.

- Using signal-to-noise ratio as the criterion, how many bits per sample must be used to guarantee that a high-quality picture, which achieves a signal-to-noise ratio of 20 dB, can be received by any HDTV set within the same broadcast region?

- Assuming the digital television channel has the same characteristics as an analog one, how much compression must HDTV systems employ?

### Problem 6.30: Digital Cellular Telephones

In designing a digital version of a wireless telephone, you must first consider certain fundamentals. First of all, the quality of the received signal, as measured by the signal-to-noise ratio, must be at least as good as that provided by wireline telephones (30 dB) and the message bandwidth must be the same as wireline telephone. The signal-to-noise ratio of the allocated wirelss channel, which has a 5 kHz bandwidth, measured 100 meters from the tower is 70 dB. The desired range for a cell is 1 km. Can a digital cellphone system be designed according to these criteria?

### Problem 6.31: Optimal Ethernet Random Access Protocols

Assume a population of N computers want to transmit information on a random access channel. The access algorithm works as follows.

- Before transmitting, flip a coin that has probability p of coming up heads
- If only one of the N computer's coins comes up heads, its transmission occurs successfully, and the others must wait until that transmission is complete and then resume the algorithm.
- If none or more than one head comes up, the N computers will either remain silent (no heads) or a collision will occur (more than one head). This unsuccessful transmission situation will be detected by all computers once the signals have propagated the length of the cable, and the algorithm resumes (return to the beginning):
    - What is the optimal probability to use for flipping the coin? In other words, what should p be to maximize the probability that exactly one computer transmits?
    - What is the probability of one computer transmitting when this optimal value of
    - Using this optimal probability, what is the average number of coin flips that will be necessary to resolve the access so that one computer successfully transmits?
    - Evaluate this algorithm. Is it realistic? Is it efficient?

### Problem 6.32: Repeaters

Because signals attenuate with distance from the transmitter, repeaters are frequently employed for both analog and digital communication. For example, let's assume that the transmitter and receiver are D m apart, and a repeater is positioned halfway between them (Figure 6.42). What the repater does is amplify its received signal to exactly cancel the attenuation encountered along the first leg and to re-transmit the signal to the ultimate receiver. However, the signal the repeater receives contains white noise as well as the transmitted signal. The receiver experiences the same amount of white noise as the repeater.

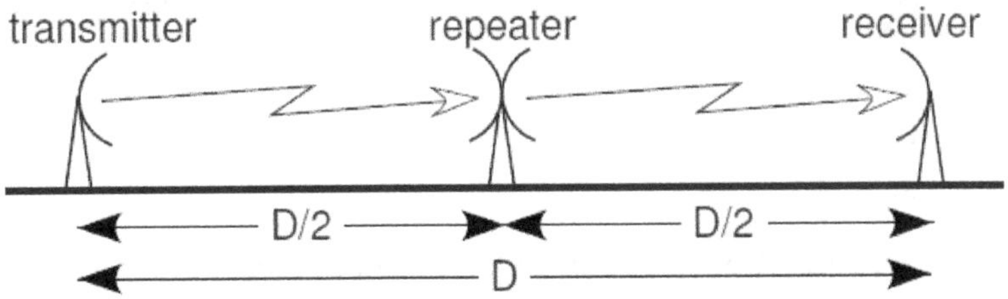


**Figure 6.42**

- What is the block diagram for this system?
- For an amplitude-modulation communication system, what is the signal-to-noise ratio of the demodulated signal at the receiver? Is this better or worse than the signal-to-noise ratio when no repeater is present?
- For digital communication, we must consider the system's capacity. Is the capacity larger with the repeater system than without it? If so, when; if not, why not?

### Problem 6.33: Designing a Speech Communication System

We want to examine both analog and digital communication alternatives for a dedicated speech transmission system. Assume the speech signal has a 5 kHz bandwidth. The wireless link between transmitter and receiver is such that 200 watts of power can be received at a pre-assigned carrier frequency. We have some latitude in choosing the transmission bandwidth, but the noise power added by the channel increases with bandwidth with a proportionality constant of 0.1 watt/kHz.

- Design an analog system for sending speech under this scenario. What is the received signal-to-noise ratio under these design constraints?
- How many bits must be used in the A/D converter to achieve the same signal-to-noise ratio?
- Is the bandwidth required by the digital channel to send the samples without error greater or smaller than the analog bandwidth?

### Problem 6.34: Digital vs. Analog

You are the Chairman/Chairwoman of the FCC. The frequency band 3 MHz to 3.5 MHz has been allocated for a new "high-quality" AM band. Each station licensed for this band will transmit signals having a bandwidth of 10 kHz, twice the message bandwidth of what current stations can send.

- How many stations can be allocated to this band and with what carrier frequencies?
- Looking ahead, conversion to digital transmission is not far in the future. The characteristics of the new digital radio system need to be established and you are the boss! Detail the characteristics of the analog-to-digital converter that must be used to prevent aliasing and ensure a signal-to-noise ratio of 25 dB.
- Without employing compression, how many digital radio stations could be allocated to the band if each station used BPSK modulation? Evaluate this design approach.

## Solutions to Exercises in Chapter 6

### Solution to Exercise 6.3.1 (p. 230)

In both cases, the answer depends less on geometry than on material properties. For coaxial cable, $c=\frac{1}{\sqrt{\mu_d\epsilon_d}}$.

For twisted pair, $c=\frac{1}{\sqrt{\mu\epsilon}}\sqrt{\frac{\text{arccosh}\left(\frac{d}{2r}\right)}{\frac{\delta}{2r}+\text{arccosh}\left(\frac{d}{2r}\right)}}$.

### Solution to Exercise 6.3.2

You can find these frequencies from the spectrum allocation chart (Section 7.3). Light in the middle of the visible band has a wavelength of about 600 nm, which corresponds to a frequency of 5 x 1014Hz. Cable television transmits within the same frequency band as broadcast television (about 200 MHz or 2 x108Hz). Thus, the visible electromagnetic frequencies are over six orders of magnitude higher!

### Solution to Exercise 6.3.3

As frequency increases, $2\pi f\ \tilde{C}\gg\tilde{G}$ and $2\pi f\ \tilde{L}\gg\tilde{R}$. In this high-frequency region,

$$\gamma = j2\pi f\sqrt{\tilde{L}\tilde{C}}\sqrt{\left(1+\frac{\tilde{G}}{j2\pi f\ \tilde{C}}\right)\left(1+\frac{\tilde{R}}{j2\pi f\ \tilde{L}}\right)} \tag{6.65}$$

$$\simeq j2\pi f\sqrt{\tilde{L}\tilde{C}}\times\left(1+\frac{1}{2}\frac{1}{j2\pi f}\left(\frac{\tilde{G}}{\tilde{C}}+\frac{\tilde{R}}{\tilde{L}}\right)\right)$$

$$\simeq j2\pi f\sqrt{\tilde{L}\tilde{C}}+\frac{1}{2}\left(\tilde{G}\sqrt{\frac{\tilde{L}}{\tilde{C}}}+\tilde{R}\sqrt{\frac{\tilde{C}}{\tilde{L}}}\right)$$

Thus, the attenuation (space) constant equals the real part of this expression, and equals $a(f)=\frac{\tilde{G}Z_0+\frac{\tilde{R}}{Z_0}}{2}$.

## Solution to Exercise 6.4.1

As shown previously (6.11), voltages and currents in a wireline channel, which is modeled as a transmission line having resistance, capacitance and inductance, decay exponentially with distance. The inverse-square law governs free-space propagation because such propagation is lossless, with the inverse-square law a consequence of the conservation of power. The exponential decay of wireline channels occurs because they have losses and some filtering.

## Solution to Exercise 6.5.1

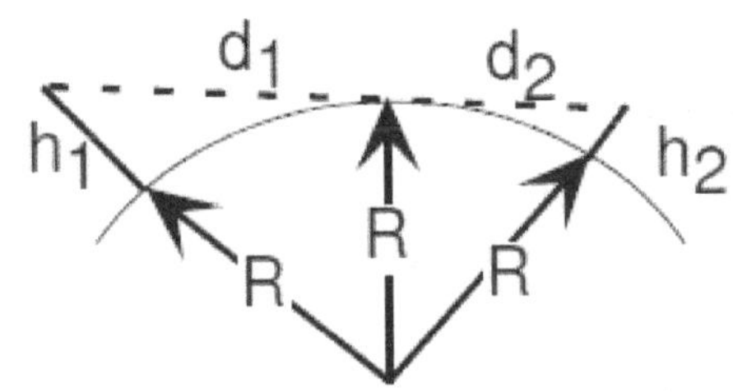


Figure 6.43

Use the Pythagorean Theorem, (h + R)2 = R2+d2, where h is the antenna height, d is the distance from the top of the earth to a tangency point with the earth's surface, and R the earth's radius. The line-of-sight distance between two earth-based antennae equals

$$d_{\text{LOS}} = \sqrt{2h_1R + {h_1}^2} + \sqrt{2h_2R + {h_2}^2} \quad (6.66)$$

As the earth's radius is much larger than the antenna height, we have to a good approximation that $d_{\text{LOS}} = \sqrt{2h_1R} + \sqrt{2h_2R}$ . If one p antenna is at ground elevation, say h2 = 0 , the other antenna's range is $\sqrt{2h_1R}$.

## Solution to Exercise 6.5.2

As frequency decreases, wavelength increases and can approach the distance between the earth's surface and the ionosphere. Assuming a distance between the two of 80 km, the relation $\lambda f = c$ gives a corresponding frequency of 3.75 kHz. Such low carrier frequencies would be limited to low bandwidth analog communication and to low datarate digital communications. The US Navy did use such a communication scheme to reach all of its submarines at once.

## Solution to Exercise 6.7.1

Transmission to the satellite, known as the uplink, encounters inverse-square law power losses. Reflecting off the ionosphere not only encounters the same loss, but twice. Reflection is the same as transmitting exactly what arrives, which means that the total loss is the product of the uplink and downlink losses. The geosynchronous orbit lies at an altitude of 35700km. The ionosphere begins at an altitude of about 50 km. The amplitude loss in the satellite case is proportional to 2:8 x 10–8; for Marconi, it was proportional to 4:4 x 10–10. Marconi was very lucky.

## Solution to Exercise 6.8.1

If the interferer's spectrum does not overlap that of our communications channel – the interferer is out-ofband –we need only use a bandpass filter that selects our transmission band and removes other portions of the spectrum.

## Solution to Exercise 6.9.1

The additive-noise channel is not linear because it does not have the zero-input-zero-output property (even though we might transmit nothing, the receiver's input consists of noise).

## Solution to Exercise 6.11.1

The signal-related portion of the transmitted spectrum is given by $X(f) = \frac{1}{2}M(f - f_c) + \frac{1}{2}M(f + f_c)$. Multiplying at the receiver by the carrier shifts this spectrum to fc and to –fc, and scales the result by half.

$$\begin{aligned} \frac{1}{2}X(f - f_c) + \frac{1}{2}X(f + f_c) &= \frac{1}{4}(M(f - 2f_c) + M(f)) + \frac{1}{4}(M(f + 2f_c) + M(f)) \\ &= \frac{1}{4}M(f - 2f_c) + \frac{1}{2}M(f) + \frac{1}{4}M(f + 2f_c) \end{aligned} \quad (6.67)$$

The signal components centered at twice the carrier frequency are removed by the lowpass filter, while the baseband signal M (f) emerges.

## Solution to Exercise 6.12.1

The key here is that the two spectra $M(f - f_c)$, $M(f + f_c)$ do not overlap because we have assumed that the carrier frequency fc is much greater than the signal's highest frequency. Consequently, the term $M(f - f_c)M(f + f_c)$ normally obtained in computing the magnitude-squared equals zero.

## Solution to Exercise 6.12.2

Separation is 2W. Commercial AM signal bandwidth is 5kHz. Speech is well contained in this bandwidth, much better than in the telephone!

## Solution to Exercise 6.13.1

$$x(t) = \sum_{n=-\infty}^{\infty} s_{b(n)}(t - nT).$$

Solution to Exercise 6.14.1 (p. 243)

K=4.

## Solution to Exercise 6.14.2

$$x(t) = \sum_{nn} (-1)^{b(n)} A p_T (t - nT) \sin\left(\frac{2\pi kt}{T}\right)$$

## Solution to Exercise 6.14.3

The harmonic distortion is 10%.

## Solution to Exercise 6.14.4

Twice the baseband bandwidth because both positive and negative frequencies are shifted to the carrier by the modulation: 3R.

## Solution to Exercise 6.16.1

In BPSK, the signals are negatives of each other: s1 (t) = −s0 (t). Consequently, the output of each multiplier-integrator combination is the negative of the other. Choosing the largest therefore amounts to choosing which one is positive. We only need to calculate one of these. If it is positive, we are done. If it is negative, we choose the other signal.

## Solution to Exercise 6.16.2

The matched filter outputs are $\pm\left(\frac{A^2T}{2}\right)$ because the sinusoid has less power than a pulse having the same amplitude.

**Solution to Exercise 6.17.1** The noise-free integrator outputs differ by aA2T, the factor of two smaller value than in the baseband case arising because the sinusoidal signals have less energy for the same amplitude. Stated in terms of Eb, the difference equals 2aEb just as in the baseband case.

## Solution to Exercise 6.18.1

The noise-free integrator output difference now equals $\alpha A^2 T = \frac{\alpha E_b}{2}$. The noise power remains the same as in the BPSK case, which from the probability of error equation (6.46) yields $p_e = Q\left(\sqrt{\frac{\alpha^2 E_b}{N_0}}\right)$.

## Solution to Exercise 6.20.1

Equallylikelysymbolseachhaveaprobabilityof$\frac{1}{K}$. Thus, $H(A) = -\sum_{kk} \frac{1}{K}\log_2\frac{1}{K} = \log_2 K$. To prove

that this is the maximum-entropy probability assignment, we must explicitly take into account that probabilities sum to one. Focus on a particular symbol, say the first. Pr [a0] appears twice in the entropy formula: the terms $Pr[a_0]\log_2 Pr[a_0]$ and $(1 - Pr[a_0] + \cdots + Pr[a_{K-2}])\log_2(1 - Pr[a_0] + \cdots + Pr[a_{K-2}])$. The derivative with respect to this probability (and all the others) must be zero. The derivative equals $\log_2 Pr[a_0] - \log_2(1 - Pr[a_0] + \cdots + Pr[a_{K-2}])$, and all other derivatives have the same form (just substitute your letter's index). Thus, each probability must equal the others, and we are done. For the minimum entropy answer, one term is 1log21 = 0, and the others are 0log20, which we define to be zero also. The minimum value of entropy is zero.

### Solution to Exercise 6.22.1

The Huffman coding tree for the second set of probabilities is identical to that for the first (Figure 6.18 (Huffman Coding Tree)). The average code length is $\frac{1}{2}1+\frac{1}{4}2+\frac{1}{5}3+\frac{1}{20}3=1.75$ bits. The entropy calculation is straightforward: $H(A)=-\left(\frac{1}{2}\log\frac{1}{2}+\frac{1}{4}\log\frac{1}{4}+\frac{1}{5}\log\frac{1}{5}+\frac{1}{20}\log\frac{1}{20}\right)$, which equals 1:68 bits.

### Solution to Exercise 6.23.1

$$T=\frac{1}{\overline{B(A)}R}.$$

### Solution to Exercise 6.23.2

Because no codeword begins with another's codeword, the first codeword encountered in a bit stream must be the right one. Note that we must start at the beginning of the bit stream; jumping into the middle does not guarantee perfect decoding. The end of one codeword and the beginning of another could be a codeword, and we would get lost.

### Solution to Exercise 6.23.3

Consider the bitstream ......0110111..... taken from the bitstream 0|10|110|110|111|..... We would decode the initial part incorrectly, then would synchronize. If we had a _xed-length code (say 00,01,10,11), the situation is much worse. Jumping into the middle leads to no synchronization at all!

### Solution to Exercise 6.25.1

This question is equivalent to $3p_e\times(1-p_e)+p_e{}^2\leq 1 \text{ or } 2p_e{}^2-3p_e+1\geq 0$. Because this is an upward-going parabola, we need only check where its roots are. Using the quadratic formula, we find that they are located at $\frac{1}{2}$ and 1. Consequently in the range $0\leq p_e\leq\frac{1}{2}$ the error rate produced by coding is smaller.

### Solution to Exercise 6.26.1

With no coding, the average bit-error probability pe is given by the probability of error equation (6.47):

$p_e=Q\left(\sqrt{\frac{2\alpha^2E_b}{N_0}}\right)$. With a threefold repetition code, the bit-error probability is given by $3p_e'^2\times(1-p_e')+p_e'^3$, where $p_e'=Q\left(\sqrt{\frac{2\alpha^2E_b}{3N_0}}\right)$. Plotting this reveals that the increase in bit-error probability out of the channel because of the energy reduction is not compensated by the repetition coding.

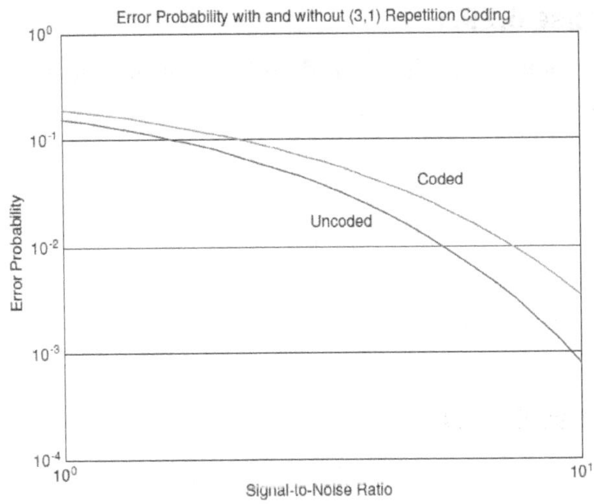


**Figure 6.44**

## Solution to Exercise 6.27.1

In binary arithmetic (see Table 6.2), adding 0 to a binary value results in that binary value while adding 1 results in the opposite binary value.

## Solution to Exercise 6.27.2

$$d_{\min} = 2n + 1$$

## Solution to Exercise 6.28.1

When we multiply the parity-check matrix times any codeword equal to a column of G, the result consists of the sum of an entry from the lower portion of G and itself that, by the laws of binary arithmetic, is always zero.

Because the code is linear – sum of any two codewords is a codeword – we can generate all codewords as sums of columns of G. Since multiplying by H is also linear, Hc = 0.

## Solution to Exercise 6.28.2

In binary arithmetic see this table, adding 0 to a binary value results in that binary value while adding 1 results in the opposite binary value.

## Solution to Exercise 6.28.3

The probability of a single-bit error in a length-N block is $Np_e(1-p_e)^{N-1}$ and a triple-bit error has probability $\binom{N}{3} p_e^3(1-p_e)^{N-3}$. For the first to be greater than the second, we must have

$$p_e < \frac{1}{\sqrt{\frac{(N-1)(N-2)}{6}} + 1}$$

For N = 7, pe < 0:31.

**Solution to Exercise 6.29.1**

In a length-N block, N single-bit and $\frac{N(N-1)}{2}$ double-bit errors can occur. The number of non-zero vectors resulting from $H\hat{c}$ must equal or exceed the sum of these two numbers.

$$2^{N-K} - 1 \geq N + \frac{N(N-1)}{2} \quad \text{or} \quad 2^{N-K} \geq \frac{N^2 + N + 2}{2} \tag{6.68}$$

The first two solutions that attain equality are (5,1) and (90,78) codes. However, no perfect code exists other than the single-bit error correcting Hamming code. (Perfect codes satisfy relations like (6.68) with equality.)

**Solution to Exercise 6.31.1**

To convert to bits/second, we divide the capacity stated in bits/transmission by the bit interval duration T.

**Solution to Exercise 6.33.1**

The network entry point is the telephone handset, which connects you to the nearest station. Dialing the telephone number informs the network of who will be the message recipient. The telephone system forms an electrical circuit between your handset and your friend's handset. Your friend receives the message via the same device – the handset – that served as the network entry point.

**Solution to Exercise 6.36.1**

The transmitting op-amp sees a load or $R_{\text{out}} + Z_0 \parallel \frac{R_{\text{out}}}{N}$, where N is the number of transceivers other than this one attached to the coaxial cable. The transfer function to some other transceiver's receiver circuit is Rout divided by this load.

**Solution to Exercise 6.36.2**

The worst-case situation occurs when one computer begins to transmit just before the other's packet arrives. Transmitters must sense a collision before packet transmission ends. The time taken for one computer's packet to travel the Ethernet's length and for the other computer's transmission to arrive equals the roundtrip, not one-way, propagation time.

**Solution to Exercise 6.36.3**

The cable must be a factor of ten shorter: It cannot exceed 100 m. Different minimum packet sizes means different packet formats, making connecting old and new systems together more complex than need be.

### Solution to Exercise 6.37.1

When you pick up the telephone, you initiate a dialog with your network interface by dialing the number. The network looks up where the destination corresponding to that number is located, and routes the call accordingly. The route remains fixed as long as the call persists. What you say amounts to high-level protocol while establishing the connection and maintaining it corresponds to low-level protocol.

# APPENDIX

## 7.1 DECIBELS

The decibel scale expresses amplitudes and power values logarithmically. The definitions for these differ, but are consistent with each other.

$$\text{power}(s, \text{in decibels}) = 10\log\frac{\text{power}(s)}{\text{power}(s_0)} \tag{7.1}$$

$$\text{amplitude}(s, \text{in decibels}) = 20\log\frac{\text{amplitude}(s)}{\text{amplitude}(s_0)}$$

Here power (s0) and amplitude (s0) represent a reference power and amplitude, respectively. Quantifying power or amplitude in decibels essentially means that we are comparing quantities to a standard or that we want to express how they changed. You will hear statements like «The signal went down by 3 dB» and «The filter›s gain in the stopband is -60" (Decibels is abbreviated dB.).

### Exercise 7.1.1

The prefix "deci" implies a tenth; a decibel is a tenth of a Bel. Who is this measure named for?

The consistency of these two definitions arises because power is proportional to the square of amplitude:

$$\text{power}(s) \propto \text{amplitude}^2(s) \tag{7.2}$$

Plugging this expression into the definition for decibels, we find that

$$\begin{aligned} 10\log\frac{\text{power}(s)}{\text{power}(s_0)} &= 10\log\frac{\text{amplitude}^2(s)}{\text{amplitude}^2(s_0)} \\ &= 20\log\frac{\text{amplitude}(s)}{\text{amplitude}(s_0)} \end{aligned} \tag{7.3}$$

Because of this consistency, stating relative change in terms of decibels is unambiguous. A factor of 10 increase in amplitude corresponds to a 20 dB increase in both amplitude and power!

**Decibel table**

| Power Ratio | dB |
|---|---|
| 1 | 0 |
| $\sqrt{2}$ | 1.5 |
| 2 | 3 |
| $\sqrt{10}$ | 5 |
| 4 | 6 |
| 5 | 7 |
| 8 | 9 |
| 10 | 10 |
| 0.1 | -10 |

**Figure 7.1: Common values for the decibel. The decibel values for all but the powers of ten are approximate, but are accurate to a decimal place.**

The accompanying table provides "nice" decibel values. Converting decibel values back and forth is fun, and tests your ability to think of decibel values as sums and/or differences of the well-known values and of ratios as products and/or quotients. This conversion rests on the logarithmic nature of the decibel scale. For example, to find the decibel value for $\sqrt{2}$, we halve the decibel value for 2; 26 dB equals 10+10+6 dB that corresponds to a ratio of 10×10×4=400. Decibel quantities add; ratio values multiply.

One reason decibels are used so much is the frequency-domain input-output relation for linear systems: $Y(f) = X(f)\,H(f)$. Because the transfer function multiplies the input signal's spectrum, to find the output amplitude at a given frequency we simply add the filter's gain in decibels (relative to a reference of one) to the input amplitude at that frequency. This calculation is one reason that we plot transfer function magnitude on a logarithmic vertical scale expressed in decibels.

## 7.2 PERMUTATIONS AND COMBINATIONS

### 7.2.1 Permutations and Combinations

The lottery "game" consists of picking k numbers from a pool of n. For example, you select 6 numbers out of 60. To win, the order in which you pick the numbers doesn't matter; you only have to choose the right set of 6 numbers. The chances of winning equal the number of different length-k sequences that can be chosen. A related, but different, problem is selecting the batting lineup for a baseball team. Now the order matters, and many more choices are possible than when order does not matter.

Answering such questions occurs in many applications beyond games. In digital communications, for example, you might ask how many possible double-bit errors can occur in a codeword. Numbering the bit positions from 1 to N, the answer is the same as the lottery problem with k=6. Solving these kind of problems amounts to understanding permutations - the number of ways of choosing things when order matters as in baseball lineups - and combinations - the number of ways of choosing things when order does not matter as in lotteries and bit errors.

Calculating permutations is the easiest. If we are to pick k numbers from a pool of n, we have n choices for the first one. For the second choice, we have n-1. The number of length-two ordered sequences is therefore be n(n-1). Continuing to choose until we make k choices means the number of permut $\frac{n!}{(n-k)!}$, 's n (n-1) (n-2)...(n-k+1). This result can be written in terms of factorials as with n! = n(n-1) (n-2)...1. For mathematical convenience, we define 0!=1.

When order does not matter, the number of combinations equals the number of permutations divided by the number of orderings. The number of ways a pool of k things can be ordered equals k!. Thus, once we choose the n for our baseball game, we have 9!=362, 880 differe $\binom{n}{k}$ ieups! The $\frac{n!}{(n-k)!k!}$. the combination of k things drawn from a pool of n is and equals

## Exercise 7.2.1

What are the chances of winning the lottery? Assume you pick 6 numbers from the numbers 1-60.

Combinatorials occur in interesting places. For example, Newton derived that the n-th power of a sum obeyed the formula $(x+y)^n = \binom{n}{0}x^n + \binom{n}{1}x^{n-1}y + \binom{n}{2}x^{n-2}y^2 + \cdots + \binom{n}{n}y^n.$

## Exercise 7.2.2

What does the sum of binomial coefficients equal? In other words, what is

$$\sum_{k=0}^{n} \binom{n}{k}$$

A related problem is calculating the probability that any two bits are in error in a length-n codeword when p is the probability of any bit being in error. The probability of any particular two-bit error sequence is p2(1-p)n-2.The probability of a two-bit error occurring anywhere equals this probability times the number of combinations: $\binom{n}{2}p^2(1-p)^{n-2}$. Note that the probability that zero or one or two, etc. errors occurring must be one; in other words, something must happen to the codeword! That means that we must have $\binom{n}{0}(1-p)^n + \binom{n}{1}p(1-p)^{n-1} + \binom{n}{2}p^2(1-p)^{n-2} + \cdots + \binom{n}{n}p^n = 1.$ Can you prove this?

## 7.3 FREQUENCY ALLOCATIONS

To prevent radio stations from transmitting signals "on top of each other," the United States and other national governments in the 1930s began regulating the carrier frequencies and power outputs stations could use. With increased use of the radio spectrum for both public and private use, this regulation has become increasingly important. This is the so-called Frequency Allocation Chart, which shows what kinds of broadcasting can occur in which frequency bands. Detailed radio carrier frequency assignments are much too detailed to present here.

### Frequency Allocation Chart

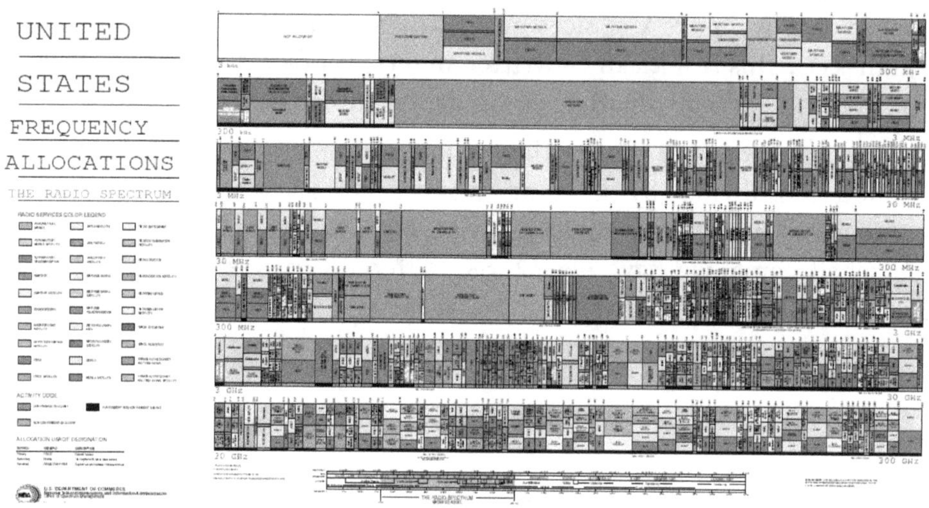


Figure 7.2

### Solutions to Exercises in Chapter 7

### Solution to Exercise 7.1.1

Alexander Graham Bell. He developed it because we seem to perceive physical quantities like loudness and brightness logarithmically. In other words, percentage, not absolute differences, matter to us. We use decibels today because common values are small integers. If we used Bels, they would be decimal fractions, which aren't as elegant.

### Solution to Exercise 7.2.1

$$\binom{60}{6} = \frac{60!}{54!6!} = 50,063,860.$$

### Solution to Exercise 7.2.2

Because of Newton's binomial theorem, the sum equals $(1+1)n = 2n$.

# INDEX

www.ingramcontent.com/pod-product-compliance
Lightning Source LLC
LaVergne TN
LVHW062113200926
848484LV00007B/1659

* 9 7 8 1 7 8 7 1 5 3 7 1 4 *